Recent Advances in Applied Microbiology

Pratyoosh Shukla
Editor

Recent Advances in Applied Microbiology

Springer

Editor
Pratyoosh Shukla
Department of Microbiology
Maharshi Dayanand University
Rohtak, Haryana
India

ISBN 978-981-10-5274-3   ISBN 978-981-10-5275-0 (eBook)
DOI 10.1007/978-981-10-5275-0

Library of Congress Control Number: 2017954324

Printed on acid-free paper

This Springer imprint is published by Springer Nature
The registered company is Springer Nature Singapore Pte Ltd.
The registered company address is: 152 Beach Road, #21-01/04 Gateway East, Singapore 189721, Singapore

Foreword

This important new book describes key recent advances in applied microbiology. Microbes provide abundant benefits to mankind. Our knowledge of microbial applications must continue to advance through research such as that described in this book in order for us to expand our use of microbes for the benefit of the world's growing population. Many microbial processes are sustainable and have the potential to reduce carbon emissions compared with existing chemical processes. The rapid advances in genomics and the increased use of engineering principles, computer science, and bioinformatics in applied microbiology make this a particularly exciting time for the field.

The book comprises 14 chapters that are divided into 4 sections. "Microbial Biotechnology" includes chapters discussing microbial enzymes and useful products such as polyhydroxyalkanoates that can be used in the production of biodegradable plastics and microbial surfactants with application in bioprocessing. "Microbes in Health" tackles the topics of multidrug-resistant bacteria, probiotics that can be used to improve human health, and processes of microbial pathogenesis. The section "Microbial Interactions" comprises chapters discussing the application of microbes in improving plant growth and crop yields, microbes that can tolerate metal contamination and may have application in bioremediation, plant pathogen interactions, and the use of bacteria in the transformation of isoflavones. The final section "Computational Approaches in Microbiology" includes chapters highlighting advanced studies on proteins, enzymes, and peptides.

Dr. Pratyoosh Shukla, the editor of this volume, is to be congratulated on doing a fine job of bringing together a diverse range of topics to effectively highlight some of the important contributions in applied microbiology that will lead to new microbial biotechnology ventures. The practical discoveries described in this book help to provide new products and solutions to some of the challenges facing us in the twenty-first century.

Russell T. Hill

Director and Professor
Institute of Marine and Environmental Technology
University of Maryland Center for Environmental Science
Columbus Center, Baltimore, MD, USA

Contents

Contributors

Aneesh Balakrishna Pillai Environmental Biology Laboratory, Rajiv Gandhi Centre for Biotechnology, Poojappura, Thiruvananthapuram, Kerala, India

Mehak Baweja Enzyme Technology and Protein Bioinformatics Laboratory, Department of Microbiology, Maharshi Dayanand University, Rohtak, Haryana, India

Trinad Chakraborty Institute of Medical Microbiology, Justus-Liebig University, Giessen, Germany

Dhritiman Chanda Microbiology Laboratory, Assam University, Silchar, India

Tushar Kanti Dangar Microbiology Laboratory, Crop Production Division, National (Formerly Central), Rice Research Institute, Cuttack, India

Nilima Dash Microbiology Laboratory, Crop Production Division, National (Formerly Central), Rice Research Institute, Cuttack, Odisha, India

A.N. Ganeshamurthy ICAR- Indian Institute of Horticultural Research, Bengaluru, India

Anchal Goel Enzyme and Microbial Biochemistry Laboratory, Department of Chemistry, Indian Institute of Technology, Hauz Khas, New Delhi, India

Mohamed Hijri Department of Biological Sciences, Institut de recherche en biologie végétale (IRBV), University De Montreal, Montreal, Canada

D.K. Jha Department of Botany, Gauhati University, Guwahati, Assam, India

Chandandeep Kaur ICAR- Indian Institute of Horticultural Research, Bengaluru, India

Sunil K. Khare Enzyme and Microbial Biochemistry Laboratory, Department of Chemistry, Indian Institute of Technology, Hauz Khas, New Delhi, Kerala, India

Hari Krishnan Kumarapillai Environmental Biology Laboratory, Rajiv Gandhi Centre for Biotechnology, Poojappura, Thiruvananthapuram, Kerala, India

Dhananjay Kumar Department of Bio-Engineering, Birla Institute of Technology, Mesra, Ranchi, Jharkhand, India

Vishal Kumar Enzyme Technology and Protein Bioinformatics Laboratory, Department of Microbiology, Maharshi Dayanand University, Rohtak, Haryana, India

Hao Liu State Key Laboratory of Pulp and Paper Engineering, South China University of Technology, Guangzhou, China

Kunal Mukhopadhyay Department of Bio-Engineering, Birla Institute of Technology, Mesra, Ranchi, Jharkhand, India

Kasinathan Narayanan Department of Pharmaceutical Biotechnology, Manipal College of Pharmaceutical Sciences, Manipal University, Manipal, India

Harinder Singh Oberoi Division of Post Harvest Technology and Agricultural Engineering, ICAR- Indian Institute of Horticultural Research, Bengaluru, India

Avishek Pahari Microbiology Laboratory, Crop Production Division, National (Formerly Central), Rice Research Institute, Cuttack, Odisha India

Helena Pillich Institute of Medical Microbiology, Justus-Liebig University, Giessen, Germany

B.V. Pradeep Department of Microbiology, Karpagam University (Karpagam Academy of Higher Education), Coimbatore, Tamil Nadu, India

Madhu Puri Institute of Medical Microbiology, Justus-Liebig University, Giessen, Germany

Lopamudra Satapathy Department of Bio-Engineering, Birla Institute of Technology, Mesra, Ranchi, Jharkhand, India

G. Selvakumar ICAR- Indian Institute of Horticultural Research, Bengaluru, India

G.D. Sharma Department of Environmental Science, Dr YS Parmar University of Horticulture and Forestry, Solan, India

Reetika Sharma Department of Environmental Science, Dr YS Parmar University of Horticulture and Forestry, Solan, Himachal Pradesh, India

Pratyoosh Shukla Enzyme Technology and Protein Bioinformatics Laboratory, Department of Microbiology, Maharshi Dayanand University, Rohtak, Haryana, India

Rajeshwari Sinha Food Safety and Toxins Programme, Centre for Science and Environment, New Delhi, India

Nagarathenam Sivagurunathan Department of Pharmaceutical Biotechnology, Manipal College of Pharmaceutical Sciences, Manipal University, Manipal, India

Sinosh Skariyachan Department of Biotechnology Engineering, Dayananda Sagar Institutions, Bangalore, India

Visvesvaraya Technological University, Belagavi, Karnataka, India

Volety Mallikarjuna Subrahmanyam Department of Pharmaceutical Biotechnology, Manipal College of Pharmaceutical Sciences, Manipal University, Manipal, Karnataka, India

Nagendra Suryanarayana Microbiology Division, Defence Research and Development Establishment, Gwalior, Madhya Pradesh, India

Kulanthaivel Thavachelvam Microbiology Division, Defence Research and Development Establishment, Gwalior, Madhya Pradesh, India

Urmil Tuteja Microbiology Division, Defence Research and Development Establishment, Gwalior, India

V. Usha Rani Department of Microbiology, Karpagam University (Karpagam Academy of Higher Education), Coimbatore, Tamil Nadu, India

Vanlalhmuaka Entomology Division, Defence Research Laboratory, Tezpur, Assam, India

Josyula Venkata Rao Department of Pharmaceutical Biotechnology, Manipal College of Pharmaceutical Sciences, Manipal University, Manipal, Karnataka, India

Ruby Yadav Enzyme Technology and Protein Bioinformatics Laboratory, Department of Microbiology, Maharshi Dayanand University, Rohtak, Haryana, India

About the Editor

Prof. Pratyoosh Shukla, PhD, is a professor and the head of the Department of Microbiology at Maharshi Dayanand University, Rohtak, India. His primary research interests are in enzyme technology, microbial biotechnology, and protein bioinformatics. He completed his PhD in the field of microbiology and fungal biotechnology at APS University, Rewa, India. Following his PhD, he pursued his postdoctoral studies at the Department of Biotechnology and Food Technology, Durban University of Technology, South Africa. He has 15 years of research and 17 years of teaching experience. He has produced 73 scientific publications and has authored or edited 6 books. He also has filed a patent on novel β-1, 4-endoxylanase from *Thermomyces lanuginosus* SS-8 and the mode of action thereof.

He is a life member of a number of academic bodies, including the Indian Science Congress Association (ISCA), India Society for Technical Education (ISTE), Mycological Society of India (MSI), Asian Federation of Biotechnology (AFOB), American Society for Microbiology (ASM), European Federation of Biotechnology (EFB), etc. He also holds the roles of associate editor, *BMC Microbiology*; editor, *Indian Journal of Microbiology* (Springer); editor in chief, *Journal of Microbiology*, Internet Scientific Publications, USA (2007–2009); reviewer and member of the editorial board for the *Journal of Applied Sciences in Environmental Sanitation*, ITS, Indonesia; etc. He is currently the general secretary of the Association of Microbiologists of India (AMI) (since 2014). He has also been presented with a number of academic awards, such as the ASM-IUSSTF Indo-US Professorship Award in Microbiology by the American Society for Microbiology (2014); AMI Alembic Award in industrial microbiology; and the Fast Track Young Scientist by DST, Govt. of India (2012). He was also selected as a scientist/project investigator and participated in the Southern Ocean Antarctica Expedition (Ministry of Earth Sciences, Govt. of India) (January to March, 2011).

Part I

Microbial Biotechnology

Immobilization of *A. oryzae* β-galactosidase on Silica Nanoparticles: Development of an Effective Biosensor for Determination of Lactose in Milk Whey

1

Anchal Goel, Rajeshwari Sinha, and Sunil K. Khare

Abstract

The present study demonstrates the covalent immobilization of β-galactosidase on functionalized silica nanoparticles for its application in lactose and whey hydrolysis. Under optimal conditions of 1% (w/v) glutaraldehyde, protein to carrier ratio of 66.6 mg/g and pH 7.0, a very high immobilization efficiency of 94% was obtained. The pH and temperature optimum of the immobilized β-gal was 4.5 and 50 °C with ONPG as substrate. Compared to the soluble enzyme, covalently bonded nanosilica-β-gal conjugate exhibited greater stability against inhibition by galactose and a higher thermal stability at 40 °C with a $t_{1/2}$ of 15.8 h. A lower K_m and increased catalytic efficiency indicated higher substrate affinity and reactivity upon enzyme attachment to nanoparticle surface. Reusability of the immobilized preparation extended up to 14 cycles. The immobilized preparation effectively hydrolyzed whey and lactose to soluble simple sugars with 50% of hydrolysis occurring in 6 h. The rate of lactose and whey hydrolysis by immobilized β-gal was 1.5 and 2.5 times higher than that for the free enzyme, respectively. Immobilized β-gal preparation may be advantageously and commercially explored for effective bioremediation of dairy waste, devising biosensors or analytical tools for food and environmental technology or conversion of whey into value-added products.

A. Goel • S.K. Khare (✉)
Enzyme and Microbial Biochemistry Laboratory, Department of Chemistry, Indian Institute of Technology, Hauz Khas, New Delhi 110016, India
e-mail: anchal.gl90@gmail.com; skhare@rocketmail.com

R. Sinha
Enzyme and Microbial Biochemistry Laboratory, Department of Chemistry, Indian Institute of Technology, Hauz Khas, New Delhi 110016, India
e-mail: rajeshwari.sinha@yahoo.com

P. Shukla (ed.), *Recent Advances in Applied Microbiology*,
DOI 10.1007/978-981-10-5275-0_1

Keywords
β-galactosidase • Immobilization • Silica nanoparticles • Lactose • Whey hydrolysis

1.1 Introduction

The worldwide production of milk whey, a by-product of the dairy industry, is estimated to be about 180–190 × 10^6 ton/year (Baldasso et al. 2011). Whey primarily comprises of 94–95% water, 5–6% dry matter, 3.8–4.3% lactose, 0.8–1.0% total protein, 0.6–0.65% whey protein, and 0.5–0.7% minerals (Tsakali et al. 2010). Release of this whey into the environment leads to deterioration of soil structure thus impacting crop yields detrimentally and also depletes the dissolved oxygen from water bodies thereby disrupting the aquatic life (Shukla and Wierzbicki 1975; Becerra and Gonzalez Siso 1996). Given the serious environmental threat associated with the disposal of whey, generated as dairy waste, it becomes essential to devise newer methods of utilizing or pretreating the whey before being disposed.

An important approach which could be employed for treatment of whey involves the enzymatic hydrolysis of lactose present in whey. β-galactosidase (β-gal) is one such enzyme, widely used in food technology as enzyme supplements for people suffering from lactose intolerance (Heyman 2006). Treatment of lactose in milk or milk-based products with β-gal has also been used in addressing problems of insolubility and low sweetening ability of lactose (Husain 2010; Oliveira et al. 2011), generation of sweet syrups for manufacture of soft drinks and pastries (Mustafa et al. 2014), and transglycosylation of lactose to galactooligosaccharides (GOS) (Colinas et al. 2014; Maischberger et al. 2008; Rosenberg 2006).

Immobilized β-gal preparations present an efficient and commonly employed approach for hydrolysis of lactose. Their importance stems from the improved stability, higher activity, and reusability, resistance to catalyst poisoning, reduced microbial contamination, easy recovery, and separation properties offered by such immobilized preparations. So far, β-gal has been immobilized on a wide range of supports like chitosan, cotton cloth, epoxy support, cellulose beads, cross-linked enzyme aggregates, as well as glutaraldehyde-agarose (Klein et al. 2013; Albayrak and Yang 2002; Marín-Navarro et al. 2014; Roy and Gupta 2003; Klein et al. 2012; Gaur et al. 2006; Li et al. 2015; Cardelle-Cobas et al. 2016). Nanomaterials, owing to their high surface to volume ratio, provide immensely attractive surfaces for enzyme immobilization and development of robust nano-biocatalytic preparations with myriad of applications (Ansari and Husain 2012). β-galactosidase, primarily sourced from *Aspergillus oryzae* and *Kluyveromyces lactis*, has been immobilized on a wide range of nanoparticles (NPs) including Fe_3O_4-chitosan, silver, ZnO, chitosan-hydroxyapatite, polystyrene nanofibers, concanavalin A layered Al_2O_3, and silica (Pan et al. 2009; Ansari et al. 2012; Husain et al. 2011; Cabuk et al. 2014; Ansari and Husain 2011; Verma et al. 2012; Misson et al. 2016).

An important strategy for improving aqueous dispersibility and preventing NP aggregation is their surface modification through functionalization (Subbiah et al. 2010). Additionally, functionalization also provides scope for high enzyme loading capacity, uniform distribution of enzyme on the NP surface, and stronger enzyme links with the NPs. Silica NPs, known for their thermal, mechanical, and chemical stability, low toxicity, biocompatibility, and resistance to microbiological attacks, provide sufficient functional groups for surface modifications that allow for control of surface chemistry and efficient enzyme attachment (Hartmann and Kostrov 2013). The use of functionalized silica nanoparticles as a viable scaffold for β-gal immobilization is being recently explored (Verma et al. 2012; Singh et al. 2011).

In consideration of the tremendous relevance of β-gal in the food industry, the development of an active, stable, reusable biocatalyst that cost-effectively addresses the treatment of whey before disposal is relevant. The present study describes the development of an efficient nano-biocatalytic biosensor system using functionalized silica nanoparticles as immobilization support for β-galactosidase, for application in hydrolysis of lactose present in whey. With improved enzymatic properties, the immobilized preparation demonstrates immense potential to facilitate effective bioremediation of dairy waste water, aid in devising biosensors and analytical tools for environmental and food technology, and also enable generation of value-added products from glucose-galactose syrup obtained upon hydrolysis of lactose present in dairy waste.

1.2 Results

1.2.1 β-galactosidase Assay and Protein Estimation

The activity of free and immobilized β-galactosidase toward ONPG (*O*-Nitrophenyl-β-D-galactosidase) was determined following the method of Craven et al. (1965) with slight modifications. One unit of β-gal activity is defined as number of micromoles of o-nitrophenol released by hydrolysis of substrate per minute per ml of enzyme.

Protein concentration was determined by Bradford assay using bovine serum albumin as standard (Bradford 1976).

1.2.2 Activation of Silica Nanoparticles

A covalent coupling-based method was used to immobilize the enzyme on functionalized silica nanoparticles (Zhang et al. 2011). Five hundred microliters (15 mg) of commercial silica nanoparticles were washed thoroughly in sodium phosphate buffer (0.2 M, pH 7.0), suspended in 1.0 ml of the same buffer, and incubated by addition of 1% (v/v) glutaraldehyde for 2.5 h at 30 °C. The activated preparation was then centrifuged, washed at least five times with sodium phosphate buffer (0.2 M, pH 7.0) to remove any unbound glutaraldehyde, and finally resuspended in the same buffer solution.

1.2.3 Immobilization of β-galactosidase on Activated Silica Nanoparticles

To 1.0 ml of activated support, β-gal solution containing 1.0 mg protein and 200.0 U of activity was added and left for overnight incubation at 30 °C under constant shaking at 200 rpm. The enzyme nanoparticle suspension was then centrifuged at 5000×*g* for 15 min and supernatant collected. Any non-covalently adsorbed protein was further removed by repeated washing with same buffer. The immobilized matrix was resuspended in 1.0 ml of sodium phosphate buffer (0.2 M, pH 7.0) and directly used for the determination of activity and stability. Immobilization efficiency was calculated as per Sinha and Khare (2015):

$$\text{Immobilization efficiency} = \frac{\text{Total activity of the immobilized } \beta\text{-gal}}{\text{Total activity of the free } \beta\text{-gal}} \times 100$$

1.2.4 Physical Characterization of Immobilized Silica Nanoparticles

Morphological changes on the silica nanoparticles before and after immobilization were observed using scanning electron microscopy (SEM; Carl Zeiss, EVO50, UK) and transmission electron microscopy (TEM; FEI Tecnai Transmission Electron Microscope, Europe). The samples were diluted and sonicated for 1 min before being micrographed. The size and morphology of the silica nanoparticles are shown in Fig. 1.1. Uniform spherical particles of size 15 nm were observed in TEM, which had increased to 35 nm upon formation of the silica NP- β-gal conjugate. The shape of the silica nanoparticles however did not change upon immobilization.

1.2.5 Optimization of Immobilization Conditions

Immobilization parameters were optimized for maximum efficiency. Activation of the functionalized silica NP was performed by varying the glutaraldehyde concentrations in the range of 0.5–4% (v/v). The pH of the immobilizing medium was varied between pH 3.0 and 10.0. The protein loading on 15.0 mg nanoparticles was also varied from 0.1 to 3.0 mg/ml.

The finally optimized conditions are summarized below in Table 1.1. The functional–NH_2 groups on 3-aminopropyl functionalized silica cross-link with the aldehyde groups of glutaraldehyde forming three-dimensional cross-linked aggregates, which prevent desorption and keep the enzyme strongly bound to NP surface. The

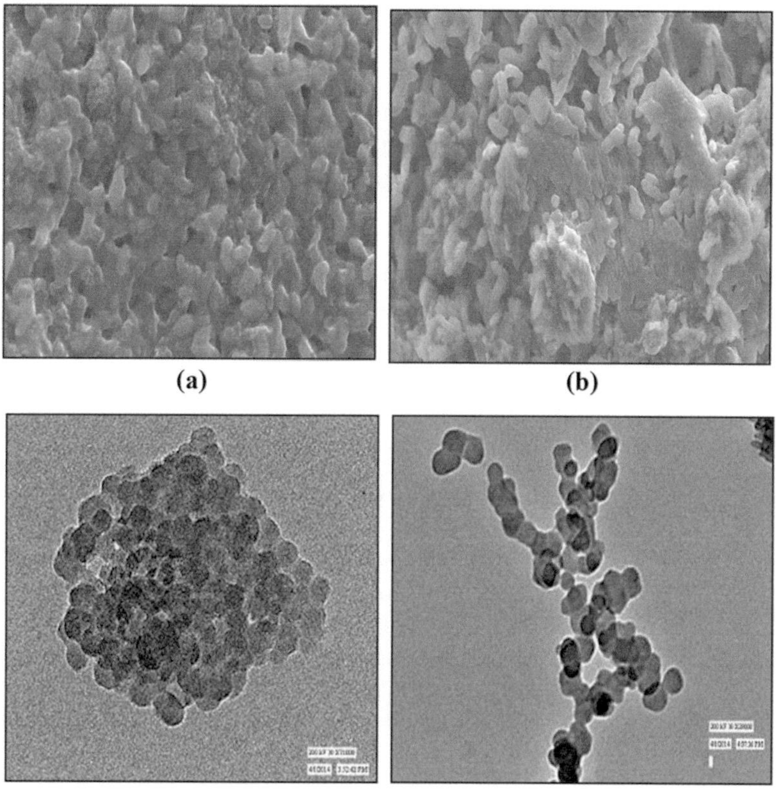

Fig. 1.1 Scanning and transmission electron micrographs of silica nanoparticles. (1) SEM with 20,000× magnification and (bar 200 nm) **(a)** washed silica nanoparticles; **(b)** silica nanoparticle-β-gal bioconjugate (2) TEM (bar 50 nm) with 110,000× magnification; **(c)** washed silica nanoparticles; **(d)** silica nanoparticle-β-gal bioconjugate

Table 1.1 Optimized conditions for β-galactosidase immobilization on silica nanoparticles

Parameters	Optimized conditions
Cross-linker for functionalization	Glutaraldehyde (1%, v/v)
pH of immobilizing medium	Sodium phosphate buffer (0.2 M, pH 7.0)
Amount of functionalized nanoparticles	15.0 mg
Protease loading	66.66 mg/g of silica nanoparticles

highest immobilization efficiency was obtained with 1.0% (v/v) glutaraldehyde. The decrease in enzyme activity with increasing glutaraldehyde concentration may be attributed to the aldol condensation of glutaraldehyde (when in excess), which not only affects the nanoparticle surface but may also alter the enzyme conformation (Zhang et al. 2010). Bioaffinity supports con A-cellulose, activated with 1.0% (v/v) glutaraldehyde that led to retention of 90% of initial activity of β-gal after 44 h at pH 7.0 (Ansari and Husain 2010). The optimal pH for immobilization was achieved with sodium phosphate buffer (0.2 M) at pH 7.0. Maximum immobilization efficiency was obtained at a protein carrier ratio of 66.0 mg protein/g of silica nanoparticles. Increasing enzyme concentration tends to saturate the binding sites on nanoparticles which reduce additional binding of the enzyme to nanoparticle surface. Higher protein/carrier ratio of 100 mg/g was observed in the case of *A. oryzae* β-gal immobilized on silica cross-linked micellar nanoparticles (Wu et al. 2013).

Under optimized conditions, the immobilization efficiency increased by 1.42 times, from 66 to 94%. While a range of immobilization efficiencies ranging from 19 to 66% have been reported for β-gal attachment on different supports (Verma et al. 2012; Facin et al. 2015; Wentworth et al. 2004), this study reports one of the highest immobilization efficiencies observed so far.

1.2.6 Enzymatic Characterization

The effect of pH, temperature on free and immobilized β-gal was determined by varying the assay pH in the range 3.0–7.0 (at 50 °C) and the assay temperature in the range 30–70 °C (at pH 4.0) (Fig. 1.2a, b). Other experimental conditions were kept constant. The maximum activity was considered as 100%. While overall β-gal activity was observed to be comparatively higher for immobilized enzyme than free enzyme at all pH, the optimum pH shifted from 4.0 to 4.5 upon immobilization. Similar increase in pH optimum from 4.5 to 5.5 was observed in the case of β-gal immobilized onto κ-Carrageenan gel beads (Elnashar et al. 2014). The temperature optimum for both forms of β-gal remained at 50 °C, with bound β-gal retaining 70% of its activity at 60 °C. While similar optimum operational temperature of 50 °C has been reported earlier (Chen and Duan 2015; Ansari et al. 2015), broader shifts in temperature optima from 40 °C to 60 °C and 50 °C to 60 °C have also been documented (Ansari and Husain 2011; Ansari and Husain 2010).

The thermal and pH stability was compared by incubating the free and the immobilized β-gal at different temperatures (30–60 °C) and different pH (3.0–10.0, at 30 °C) up to 12 h in the absence of any substrate. Both soluble and nanosilica immobilized β-gal exhibited similar pH stability in the pH range of 4.0–5.5, which gradually declined with increasing alkalinity of the medium. At 50 °C, β-gal was thermally unstable having lost more than 80% of its initial activity within 2 h, both in free and bound forms. The silica NP- β-gal bioconjugate however exhibited improved thermal stability at 40 °C with a 2.3-fold increase in $t_{1/2}$ (15.8 h) over the free counterpart (Fig. 1.2c). Enhancement in thermal stability of β-gal induced upon immobilization has been previously observed (Pan et al. 2009; Verma et al. 2012; Crescimbeni et al. 2010). The higher stability is possibly due to the multipoint covalent binding of β-gal to the support which restricts the conformational mobility of the enzyme and prevents unfolding or denaturation. It may also be that immobilization leads to favorable conformational transitions in the protein structure, thus orienting the active site flexibly toward its substrate leading to enhanced activity.

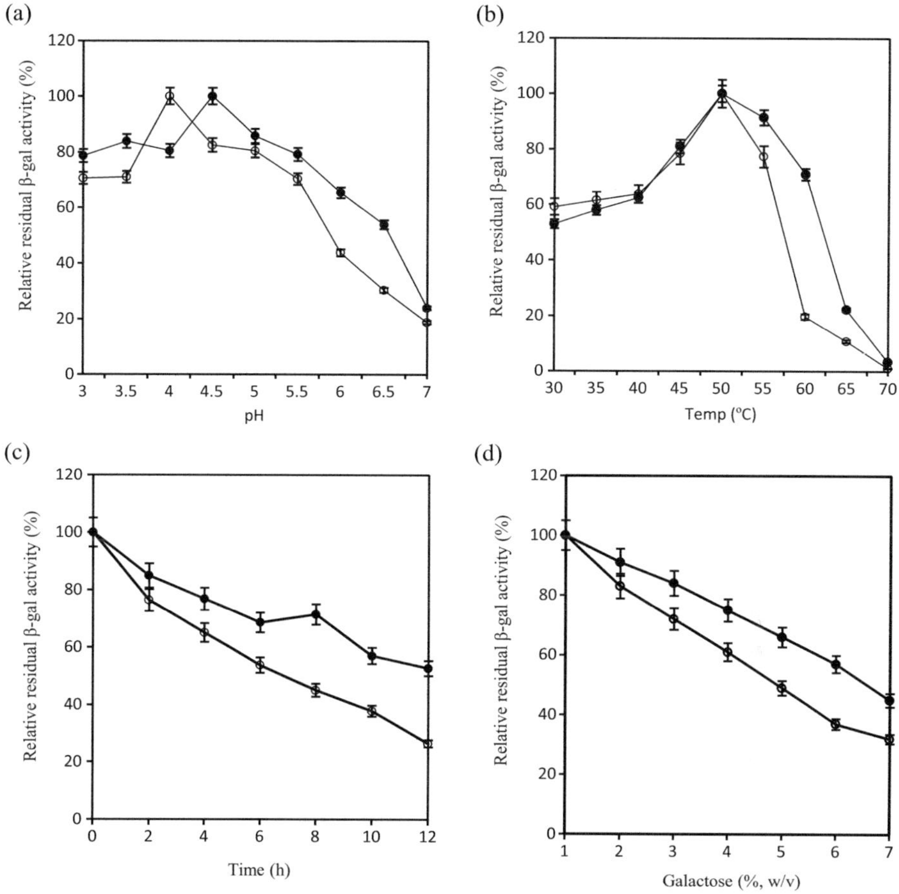

Fig. 1.2 Effect of pH, temperature, and galactose on enzymatic activities of free and immobilized β-gal **(a)** pH optimum; **(b)** temperature optimum; **(c)** thermal stability at 40 °C; **(d)** effect of galactose concentrations. (-○-) free β-gal; (-●-) immobilized β-gal

The effect of various concentrations of galactose (1.0–5.0%, w/v) on the activity of free and immobilized β-gal was measured under standard assay conditions (Fig. 1.2d). The activity of the enzyme in the absence of galactose was taken as control (100%). Although galactose caused inhibition in enzymatic activity, the immobilized enzyme exhibited increased stability over the free enzyme, retaining almost 50% of its original activity after 6 h.

Table 1.2 Kinetic parameters of the free and silica immobilized β-galactosidase

Kinetic parameters	Free enzyme	Immobilized enzyme
K_m (mM)	3.53	2.80
V_{max} ($mmolmL^{-1}min^{-1}$)	1219.51	1190.47
k_{cat} (s^{-1})	6122.95	5977.15
k_{cat}/K_m ($mM^{-1}s^{-1}$)	1734.54	2134.69

1.2.7 Determination of Kinetic Parameters

The catalytic parameters K_m, V_{max}, k_{cat}, and k_{cat}/K_m were determined by assaying both forms of the β-gal using varying substrate concentrations (ONPG) from 0.5 to 8.0 mM at 50 °C. Kinetic parameters, determined using the Lineweaver–Burk plot, are summarized in Table 1.2. The K_m of the β-gal decreased from 3.53 mM, in its native form, to 2.80 mM upon immobilization on silica NPs. K_m is a measure of the affinity of the enzymes to substrates, and the decrease in the value of K_m for immobilized β-gal indicates a higher affinity of the enzyme for the substrate. A higher catalytic efficiency was observed for the immobilized β-gal indicating enhanced reactivity upon immobilization onto a solid support. Contrary to our observations, many studies have reported a higher K_m for immobilized β-gal over that of the free enzyme suggesting reduced substrate affinity (Ansari and Husain 2011; Verma et al. 2012; Selvarajan et al. 2015). V_{max} of the immobilized enzyme however remained almost similar to that for the native enzyme.

1.2.8 Reusability of the Immobilized β-galactosidase

Reusability of the immobilized enzyme is an important factor that contributes considerably to reducing the processing cost involved and providing economical benefits. The reusability of the immobilized enzyme was determined at 40 °C by carrying out hydrolysis of ONPG under standard assay conditions. After each cycle, the immobilized enzyme was removed (by centrifuging at 5000×*g* for 15 min), washed with Milli-Q water followed by sodium acetate buffer (0.2 mM, pH 4.5), and resuspended in the same buffer to be added to a fresh batch of substrate (ONPG). The activity of the enzyme after first cycle was taken as control (100%). Each cycle here is defined as the hydrolysis of the substrate present in the reaction mixture, achieved in a reaction time of 5 min. The cycles were repeated till the immobilized preparation lost more than 40% of its initial activity. The silica NP-β-gal conjugate was reusable almost up to 14 cycles. The immobilized β-gal exhibited 100% activity in the first three cycles while having retained more than 50% of its initial activity up to nine cycles (Fig. 1.3). Decline in reusability may be due to the dropping efficiency of the conjugate and a simultaneous loss of enzyme during subsequent washings. In a recent study, residual activities of *Agaricus bisporus* lactase immobilized on polyaniline nanofiber, magnetically separable polyaniline nanofiber, and magnetically

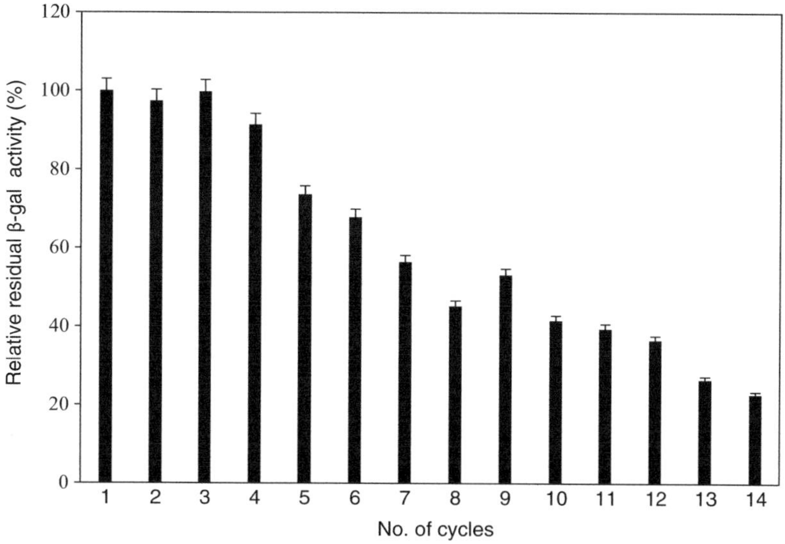

Fig. 1.3 Operational stability of *A. oryzae* β-gal immobilized on functionalized silica nanoparticles. Activity obtained in the first cycle was taken as 100% activity

separable DEAE cellulose fiber after ten times of recycling were 98%, 96%, and 97%, respectively (Jin et al. 2015).

1.2.9 Application of Immobilized β-galactosidase in Lactose and Whey Hydrolysis

The applicability of immobilized β-gal in the hydrolysis of lactose solution and milk whey was explored and compared with that of the soluble enzyme. Given that β-gal was thermally unstable at 50 °C, the experiment was performed at 40 °C. Free and immobilized β-gal preparation (containing 500.0 U) was added to 25.0 ml of lactose solution (4.5%, w/v) made in sodium acetate buffer (0.2 M, pH 4.5). Three controls were set up wherein enzyme was replaced with equal amounts of sodium acetate buffer (0.2 M, pH 4.5), denatured free β-gal, and denatured immobilized β-gal. The reaction mixture was monitored over a period of 6 h by incubating at 40 °C with continuous shaking.

Milk whey, obtained from the nearest dairy, was prefiltered for removal of any insoluble impurities and pH measured. A 25.0 ml of this acidic whey from milk (pH 4.5–5.0) was similarly hydrolyzed by free and immobilized β-gal at 40 °C up to 6 h.

Sample aliquots (1.0 ml each) were withdrawn at regular time intervals, and the reaction in these aliquots was stopped by heating at 100 °C for 10 min. Glucose concentration in these samples was measured using a glucose oxidase (GO) assay kit (Sigma Aldrich chemicals, USA) following standard protocol as provided by the manufacturer. From the glucose concentration, relative residual level for the lactose concentrations was derived based on the reaction stoichiometry. The rate of lactose hydrolysis was then calculated as follows:

$$\text{Rate of lactose hydrolysis}\,(\%) = \frac{\text{Amount of lactose hydrolyzed}}{\text{Initial amount of lactose present}} \times 100$$

The hydrolysis of lactose and whey (a natural substrate) at 40 °C up to 6 h is shown in Fig. 1.4a. In both cases, immobilized β-gal showed improved lactose/whey hydrolysis as compared to free β-gal. Maximum hydrolysis rate was achieved in 6 h in the case of immobilized enzyme, beyond which a steady state was achieved. The rate of lactose and whey hydrolysis was 1.5 and 2.5 times higher than that for the free enzyme, respectively. After 6 h, 50% of both the substrates were hydrolyzed by the immobilized enzyme, while only 35% of lactose and 20% of whey were hydrolyzed by soluble form of β-gal. The overall increase in rates of lactose hydrolysis was observed to be comparatively higher in the case of whey as compared to lactose. The degree of whey lactose hydrolysis with galactosidase, immobilized onto a modified polypropylene membrane, was 91% after 10 h (Vasileva et al. 2016). In another study, immobilized β-gal from *Pseudoalteromonas* sp. 22b recombinantly expressed in *E. coli* hydrolyzed 90% of lactose in continuous and batch systems (Makowski et al. 2007). Optimization of hydrolysis parameters in present case will eventually lead to similar high hydrolysis rates.

The qualitative estimation of the products formed on lactose hydrolysis was done using thin-layer chromatography (TLC) (Fig. 1.4b). One microliter sample of each sugar solution (standard stock of 1% (w/v) lactose, glucose, and galactose made in sodium acetate buffer [0.2 M, pH 4.5]) was spotted onto the silica TLC plate. In the case of the hydrolyzed samples, aliquots were withdrawn at regular time intervals, and 1.0 μl of these samples was then similarly spotted. The plates were then placed in the chamber containing the developing solvent (*n*-butanol-isopropanol-water [3:12:4 v/v/v]). The spots were developed by spraying the air-dried plate with 2% orcinol in H_2SO_4: water (1:9 v/v) and keeping the plate in hot air oven for 15 min. The fractions obtained upon hydrolysis matched those of the standard samples indicating efficient lactose hydrolysis to yield mono- and oligosaccharides. The improved hydrolytic activity observed in the case of immobilized enzyme could be attributed to its better thermal stability at 40 °C as well as increased activity.

A graphical representation of hydrolysis of whey lactose by nanosilica immobilized β-gal is shown in Fig. 1.5.

1.3 Conclusion

The study describes the fabrication of an active, stable, and reusable nanobiocatalytic preparation for applicability in pretreatment of dairy waste and whey before its disposal as industry effluent. The key outcomes of this study are summarized below:

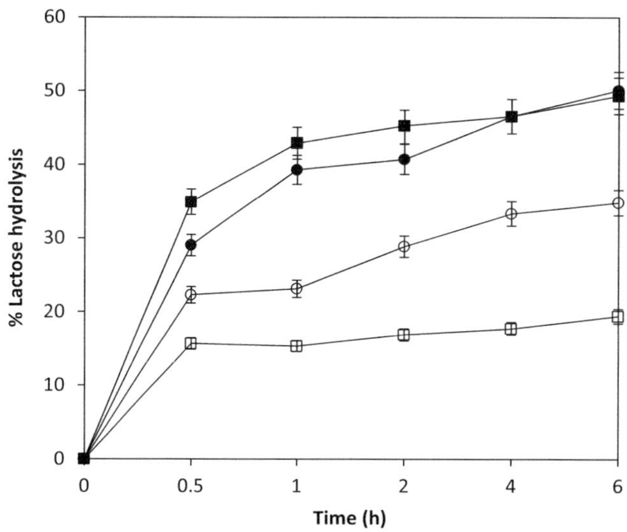

(a)

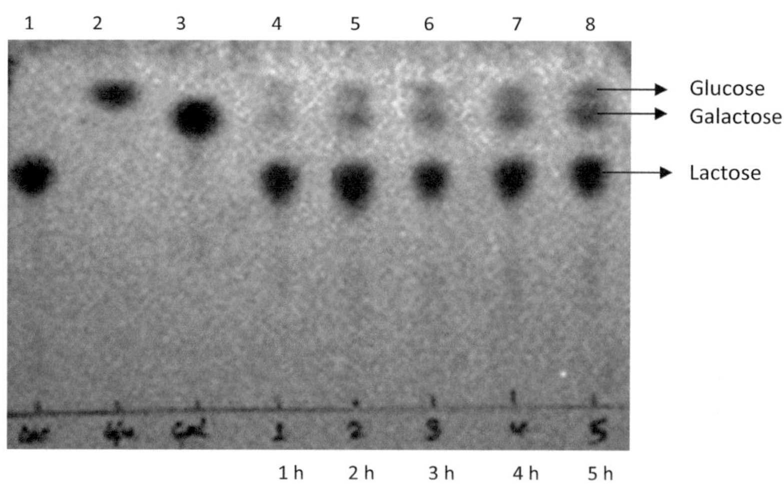

(b)

Fig. 1.4 (**a**) Effect of β-gal on hydrolysis of milk whey. Free and immobilized β-gal preparation (375 U) was added to 25.0 ml of lactose solution (4.5%, w/v) and whey solution, and the hydrolysis was monitored over a period of 6 h by incubating at 40 °C. The amount of glucose generated upon hydrolysis was quantified as described in the Methods section. (-○-) Free β-gal + lactose, (-●-) immobilized β-gal + lactose, (□) free β-gal + whey, (■) immobilized β-gal + whey. (**b**) Thin-layer chromatogram of products from enzymatic hydrolysis of lactose by immobilized β-gal. Lane1, standard solution of 1% (w/v) lactose; Lane 2, standard solution of 1% (w/v) glucose; Lane 3, standard solution of 1% (w/v) galactose; Lane 4, lactose hydrolysis for 1 h; Lane 5, lactose hydrolysis for 2 h; Lane 6, lactose hydrolysis for 3 h; Lane 7, lactose hydrolysis for 4 h; Lane 8, lactose hydrolysis for 5 h

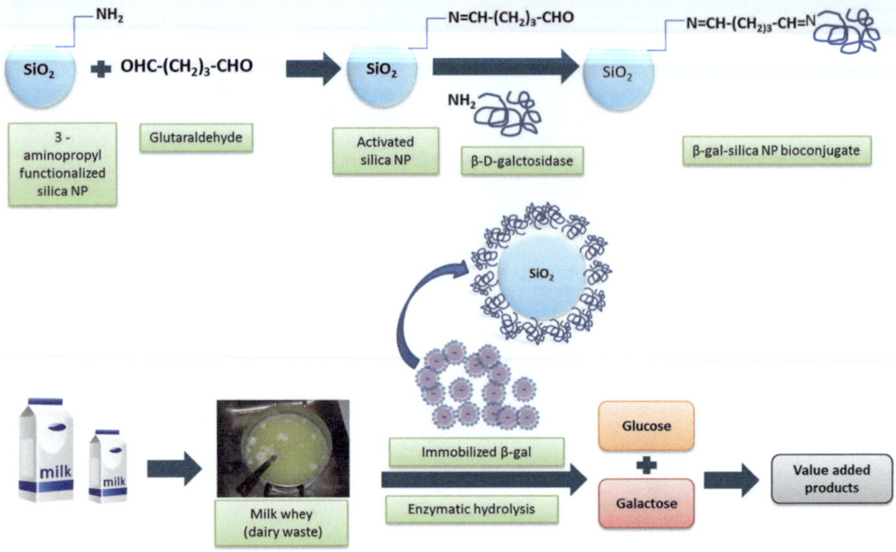

Fig. 1.5 Graphical representation of hydrolysis of whey lactose by nanosilica immobilized β-gal

- β-D-galactosidase (β-gal) was covalently immobilized on functionalized silica nanoparticles with a very 94% immobilization efficiency.
- The immobilized β-gal exhibited improved thermal stability, substrate affinity, and reusability as compared to that of soluble enzyme. $t_{1/2}$ for bound β-gal showed 2.3-fold enhancement over free β-gal.
- Nanosilica-β-gal conjugate was reusable up to 14 cycles.
- Higher enzyme affinity and increased catalytic efficiency of immobilized β-gal were observed.
- The immobilized β-gal effectively hydrolyzed lactose and milk whey with better rates of hydrolysis.

The immobilized preparation presents an important approach for improvement of lactose hydrolysis technology and provides tremendous scope for being advantageously exploited for its potential in obtaining dairy products suitable for people suffering from lactose intolerance and converting whey into value-added products.

Acknowledgments The financial support provided by the Department of Biotechnology (Government of India) is gratefully acknowledged. Author Dr. Rajeshwari Sinha is grateful to the Council of Scientific and Industrial Research (CSIR) for research fellowship.

Conflict of Interest The authors declare that there is no conflict of interest.

Author Contributions Conceived and designed the experiments: AG, RS, and SKK. Performed the experiments: AG. Analyzed the data: AG, RS, and SKK. Contributed to the writing of the manuscript: AG, RS, and SKK. All authors have approved of the article.

References

Albayrak N, Yang ST (2002) Production of galacto-oligosaccharides from lactose by *Aspergillus oryzae* β-galactosidase immobilized on cotton cloth. Biotechnol Bioeng 77:8–19

Ansari SA, Husain Q (2010) Lactose hydrolysis by β-galactosidase immobilized on concanavalin a-cellulose in batch and continuous mode. J Mol Catal B Enzym 6:68–74

Ansari SA, Husain Q (2011) Immobilization of *Kluyveromyces lactis* β galactosidase on concanavalin A layered aluminium oxide nanoparticles—its future aspects in biosensor applications. J Mol Catal B Enzym 70:119–126

Ansari SA, Husain Q (2012) Potential applications of enzymes immobilized on/in nano materials: a review. Biotechnol Adv 30:512–523

Ansari SA, Satar R, Alam F, Alqahtani MH, Chaudhary AG, Naseer MI, Karim S, Sheikh IA (2012) Cost effective surface functionalization of silver nanoparticles for high yield immobilization of *Aspergillus oryzae* β-galactosidase and its application in lactose hydrolysis. Process Biochem 47:2427–2433

Ansari SA, Satar R, Zaidi SK, Naseer MI, Karim S, Alqahtani MH, Rasool M (2015) Nanodiamonds as an effective and novel matrix for immobilizing β galactosidase. Food Bioprod Process 95:298–303

Baldasso C, Barros TC, Tessaro IC (2011) Concentration and purification of whey proteins by ultrafiltration. Desalination 278:381–386

Becerra M, Gonzalez Siso MI (1996) Yeast β-galactosidase in solid-state fermentation. Enzyme Microb Technol 19:39–44

Bradford MM (1976) A rapid and sensitive method for the quantization of microgram quantities of protein utilizing the principle of protein-dye binding. Anal Biochem 72:248–254

Cabuk B, Tari C, Harsa ST (2014) β-Galactosidase immobilization on chitosan-hydroxyapatite complex: effects of immobilization conditions. J Nutr Health Food Eng 1:00004

Cardelle-Cobas A, Olano A, Irazoqui G, Giacomini C, Batista-Viera F, Corzo N, Corzo-Martínez M (2016) Synthesis of oligosaccharides derived from lactulose (OsLu) using soluble and immobilized Aspergillus oryzae β-Galactosidase. Front Bioeng Biotechnol 4:21. doi:10.3389/fbioe.2016.00021

Chen SC, Duan KJ (2015) Production of galactooligosaccharides using β-galactosidase immobilized on chitosan-coated magnetic nanoparticles with Tris (hydroxymethyl) phosphine as an optional coupling agent. Int J Mol Sci 16:12499–12512

Colinas BR, Arrojo LF, Ballesteros AO, Plou FJ (2014) Galactooligosaccharides formation during enzymatic hydrolysis of lactose: towards a prebiotic-enriched milk. Food Chem 145:388–394

Craven GR, Steers E Jr, Anfinsen CB (1965) Purification, composition and molecular weight of the β-galactosidase of *Escherichia coli* K12. J Biol Chem 240:2469–2477

Crescimbeni MC, Nolan V, Clop PD, Marin GN, Perillo A (2010) Activity modulation and reusability of β-D-galactosidase confined in sol-gel derived porous silicate glass. Coll Surf B 76:387–396

Elnashar MM, Awad GE, Hassan ME, Eldin MSM, Haroun BM, El-Diwany AI (2014) Optimal immobilization of β-galactosidase onto κ-carrageenan gel beads using response surface methodology and its applications. Sci World J. doi:10.1155/2014/571682

Facin BR, Moret B, Baretta D, Belfiore LA, Paulino AT (2015) Immobilization and controlled release of β-galactosidase from chitosan-grafted hydrogels. Food Chem 179:44–51

Gaur R, Pant H, Jain R, Khare SK (2006) Galacto-oligosaccharide synthesis by immobilized *Aspergillus oryzae* β-galactosidase. Food Chem 97:426–430

Hartmann M, Kostrov X (2013) Immobilization of enzymes on porous silicas–benefits and challenges. Chem Soc Rev 42:6277–6289

Heyman MB (2006) Lactose intolerance in infants, children, and adolescents. Pediatrics 118:1279–1286

Husain Q (2010) β-Galactosidases and their potential applications: a review. Crit Rev Biotechnol 30:41–62

Husain Q, Ansari SA, Alam F, Azam A (2011) Immobilization of *Aspergillus oryzae* β-galactosidase on zinc oxide nanoparticles via simple adsorption mechanism. J Biol Macromol 49:37–43
Jin L, Li Y, Ren XH, Lee JH (2015) Immobilization of lactase onto various polymer nanofibers for enzyme stabilization and recycling. J Microbiol Biotechnol 25:1291–1298
Klein MP, Nunes MR, Rodrigues RC, Benvenutti EV, Costa TM, Hertz PF, Ninow JL (2012) Effect of the support size on the properties of β-galactosidase immobilized on chitosan: advantages and disadvantages of macro and nanoparticles. Biomacromolecules 13:2456–2464
Klein MP, Fallavena LP, Schöffer JDN, Ayub MA, Rodrigues RC, Ninow JL, Hertz PF (2013) High stability of immobilized β-D-galactosidase for lactose hydrolysis and galactooligosaccharides synthesis. Carbohydr Polym 95:465–470
Li L, Li G, Cao LC, Ren GH, Kong W, Wang SD, Guo GS, Liu Y (2015) Characterization of the cross-linked enzyme aggregates of a novel β-galactosidase, a potential catalyst for the synthesis of galacto-oligosaccharides. J Agric Food Chem 63:894–901
Maischberger T, Nguyen TH, Sukyai P, Kittl R, Riva S, Ludwig R, Haltrich D (2008) Production of lactose-free galacto-oligosaccharide mixtures: comparison of two cellobiose dehydrogenases for the selective oxidation of lactose to lactobionic acid. Carbohydr Res 343:2140–2147
Makowski K, Białkowska A, Szczęsna-Antczak M, Kalinowska H, Kur J, Cieśliński H, Turkiewicz M (2007) Immobilized preparation of cold-adapted and halotolerant Antarctic β-galactosidase as a highly stable catalyst in lactose hydrolysis. FEMS Microbiol Ecol 59:535–542
Marín-Navarro J, Talens-Perales D, Oude-Vrielink A, Polaina J (2014) Immobilization of thermostable β-galactosidase on epoxy support and its use for lactose hydrolysis and galactooligosaccharides biosynthesis. World J Microbiol Biotechnol 30:989–998
Misson M, Dai S, Jin B, Chen BH, Zhang H (2016) Manipulation of nanofiber-based β-galactosidase nanoenvironment for enhancement of galacto-oligosaccharide production. J Biotechnol 222:56–64
Mustafa L, Alsaed AK, A-Domi H (2014) Drying of sweet whey using drum dryer technique and utilization of the produced powder in French type bread and butter cookies. Pak J Biol Sci 17:812–820
Oliveira C, Guimaraes PM, Domingues L (2011) Recombinant microbial systems for improved β-galactosidase production and biotechnological applications. Biotechnol Adv 29:600–609
Pan C, Hu B, Li W, Sun YI, Ye H, Zeng X (2009) Novel and efficient method for immobilization and stabilization of β-D-galactosidase by covalent attachment onto magnetic Fe_3O_4–chitosan nanoparticles. J Mol Catal B Enzym 61:208–215
Rosenberg ZMM (2006) Current trends of β-galactosidase application in food technology. J Food Nutr Res 45:47–54
Roy I, Gupta MN (2003) Lactose hydrolysis by Lactozym™ immobilized on cellulose beads in batch and fluidized bed modes. Process Biochem 39:325–332
Selvarajan E, Mohanasrinivasan V, Devi CS, Doss CG (2015) Immobilization of β-galactosidase from *Lactobacillus plantarum* HF571129 on ZnO nanoparticles: characterization and lactose hydrolysis. Bioprocess Biosyst Eng 38:1655–1669
Shukla TP, Wierzbicki LE (1975) Beta-galactosidase technology: a solution to the lactose problem. Crit Rev Food Sci Nutr 5:325–356
Singh RK, Zhang YW, Jeya M, Lee JK (2011) Covalent immobilization of β-1, 4-glucosidase from *Agaricus arvensis* onto functionalized silicon oxide nanoparticles. Appl Microbiol Biotechnol 89:337–344
Sinha R, Khare SK (2015) Immobilization of halophilic *Bacillus* sp. EMB9 protease on functionalized silica nanoparticles and application in whey protein hydrolysis. Bioprocess Biosyst Eng 38:739–748
Subbiah R, Veerapandian M, Yun KS (2010) Nanoparticles: functionalization and multifunctional applications in biomedical sciences. Curr Med Chem 17:4559–4577
Tsakali E, Petrotos K, D'Allessandro A, Goulas P (2010) A review on whey composition and the methods used for its utilization for food and pharmaceutical products. In: 6th International conference on simulation and modelling in the food and Bio-Industry, Braganca, Portugal

Vasileva N, Ivanov Y, Damyanova S, Kostova I, Godjevargova T (2016) Hydrolysis of whey lactose by immobilized β-galactosidase in a bioreactor with a spirally wound membrane. Int J Biol Macromol 82:339–346

Verma ML, Barrow CJ, Kennedy JF, Puri M (2012) Immobilization of β-D-galactosidase from *Kluyveromyces lactis* on functionalized silicon dioxide nanoparticles: characterization and lactose hydrolysis. Int J Biol Macromol 50:432–437

Wentworth DS, Skonberg D, Donahue DW, Ghanem A (2004) Application of chitosan-entrapped β-galactosidase in a packed-bed reactor system. J Appl Polym Sci 91:1294–1299

Wu Z, Wang Z, Guan B, Wang X, Zhang Y, Xiao Y, Zhi B, Liu Y, Li Z, Huo Q (2013) Improving the properties of β-galactosidase from *Aspergillus oryzae* via encapsulation in aggregated silica nanoparticles. New J Chem 37:3793–3797

Zhang S, Gao S, Gao G (2010) Immobilization of β-galactosidase onto magnetic beads. Appl Biochem Biotechnol 160:1386–1393

Zhang YW, Tiwari MK, Jeya M, Lee JK (2011) Covalent immobilization of recombinant *Rhizobium etli* CFN42 xylitol dehydrogenase onto modified silica nanoparticles. Appl Microbial Biotechnol 90:499–507

Anchal Goel completed her B.Sc. (Chemistry) from Hans Raj College, Delhi University, and her M.Sc. (Chemistry) from the Indian Institute of Technology, Delhi. During her M.Sc, she worked in the area of enzyme immobilization on nanoparticles under the supervision of Prof. Sunil Khare.

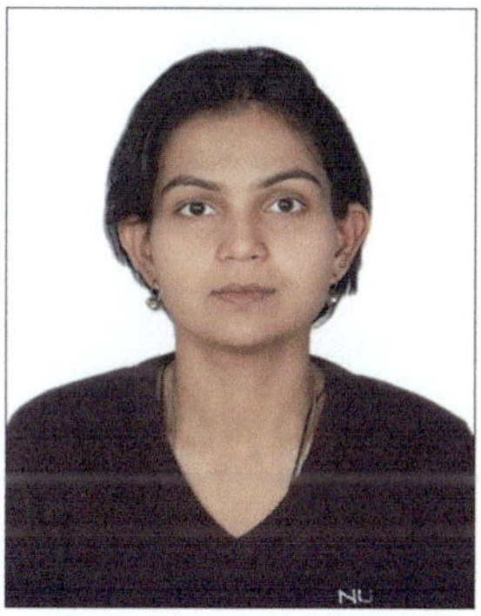

Rajeshwari Sinha received her PhD from the Indian Institute of Technology Delhi. Her primary research interests include deciphering the interaction of halophilic and non-halophilic microorganisms with nanoparticles through proteomic, structural, functional, and enzymatic approaches. Presently she is working as a Program Officer at the Centre for Science and Environment, New Delhi, where she contributes through science-based policy research in the area of antimicrobial resistance, emanating particularly due to antibiotic use in farm animals.

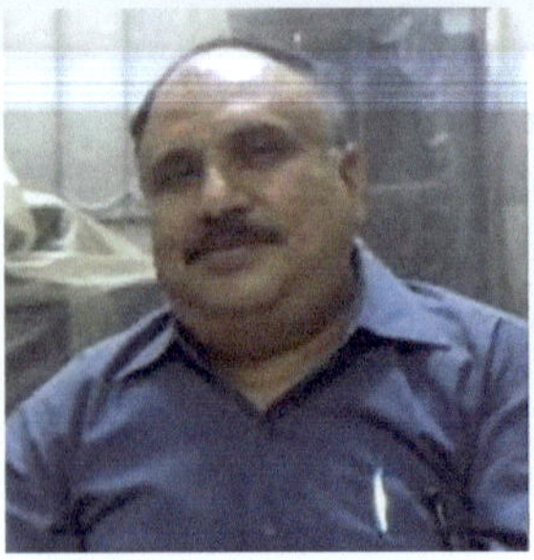

Sunil K. Khare received his PhD from IIT Delhi, carried out Postdoctoral Research at National Food Research Institute, Tsukuba, Japan, and has been DBT visiting scientist at Northern Regional Research Laboratory, Illinois USA. He is presently working as Professor of Biochemistry at the Department of Chemistry, IIT Delhi. He has been working in the area of solvent tolerant and halophilic class of extremophiles and their enzymes. His current noteworthy contributions have been in differential proteomics of solvent tolerant and halophilic class of extremophiles and deciphering nanotoxicity mechanisms in microbial systems.

Bacterial Polyhydroxyalkanoates: Recent Trends in Production and Applications

2

Aneesh Balakrishna Pillai and Hari Krishnan Kumarapillai

Abstract

Polyhydroxyalkanoates (PHAs) are biodegradable and biocompatible polymers accumulated in microorganisms as intracellular carbon and energy reserve, which are utilized when the external carbon supply is limited. PHAs have gained popularity as 'green polymers' which can be a substitute for petroleum-derived plastics due to their plastic-like properties, possibility to produce from renewable resources, and complete biodegradability in environment. The high production cost is the main hindrance to the wide spread use of these materials. Research is progressing with an aim to produce PHAs from cheap and easily available carbon sources and from waste materials and thereby make them economically competitive with conventional plastics. This review is focused on recent advances in the field of bacterial production of polyhydroxyalkanoates and their applications.

Keywords

Biopolymers • Polyhydroxyalkanoates • Poly-3-hydroxybutyrate • PHA biosynthesis • PHA copolymers • Bacterial fermentation

2.1 Introduction

Polyhydroxyalkanoates (PHAs) are polyoxoesters accumulated in eubacteria and Archaea as carbon and energy storage components in response to nutrient stress conditions (Rehm 2003). They are synthesized when a carbon source is available in

A. Balakrishna Pillai • H.K. Kumarapillai (✉)
Environmental Biology Laboratory, Rajiv Gandhi Centre for Biotechnology,
Poojappura, Thiruvananthapuram 695014, Kerala, India
e-mail: harikrishnan@rgcb.res.in

P. Shukla (ed.), *Recent Advances in Applied Microbiology*,
DOI 10.1007/978-981-10-5275-0_2

excess, and other essential nutrients such as nitrogen and phosphorus are growth limiting (Anderson and Dawes 1990). They are stored in their cytoplasm as water insoluble PHA granules (0.2–0.5 μm diameter) with a phospholipid monolayer (Rehm 2003; Urtuvia et al. 2014), and this carbon sink is mobilized by intracellular PHA depolymerases when carbon starvation occurs (Gao et al. 2001). They have unique combination of biodegradability, biocompatibility and controllable thermal-mechanical properties (Chen and Wu 2005). PHAs can be synthesized from renewable resources and are entirely degraded to CO_2 and H_2O under environmental conditions (Jendrossek and Handrick 2002). Hence they have attracted extensive interest as recyclable, eco-friendly, biodegradable alternatives to petroleum-based plastics (Wang et al. 2012b). In spite of the intensive research carried out on bacterial PHAs world over, their production cost is still far higher than the price of petroleum-derived plastics (Castilho et al. 2009).

PHAs are thermoplastic or elastomeric polyesters having molecular weights on the order of 2×10^5 to 3×10^6 and are made up of enantiomerically pure R-hydroxyalkanoic acid monomers (Philip et al. 2007). Generally, a PHA molecule consists of 600–35,000 (R)-hydroxy-fatty acid monomer units (Khanna and Srivastava 2005), and the number and types of monomers may vary depending on the conditions. This creates huge diversity in PHAs and was termed as ‘PHAome’ (Chen and Hajnal 2015). Till now, more than 150 structurally different monomers have been found in PHAs from different bacterial strains offering them a vast expanse of material characteristics for a range of applications (Steinbüchel and Valentin 1995; Steinbüchel and Lütke-Eversloh 2003). After the discovery of PHAs, plenty of similar polymers and copolymers have been reported so far, but a few of them are commercially available for application developments (Chen et al. 2015a). This review covers the recent advances in the different areas of biopolymer research and their applications.

2.2 Classification

The composition of PHAs may differ depending on the microorganism, culture conditions and carbon source (Urtuvia et al. 2014). These monomer composition variations influence the properties of PHAs, and depending on the number of carbon atoms in the monomer units, PHAs can be categorized into short-chain-length (SCL), medium-chain-length (MCL) and SCL-MCL PHA copolymers (containing SCL and MCL monomers) (Phithakrotchanakoon et al. 2013).

2.2.1 Short-Chain-Length (SCL) PHAs

This category of PHAs consists of 3–5 carbon atoms in the monomer units. Generally they can be poly-3-hydroxypropionate (P-3-HP, C_3), poly-3-hydroxybutyrate (P-3-HB, C_4) and poly-3-hydroxyvalerate (P-3-HV, C_5). Based on the position of the hydroxyl group in the monomer, the polymer can also be poly-4-hydroxybutyrate (P-4-HB), poly-5-hydroxyvalerate (P-5-HV), etc. (Steinbüchel and Valentin 1995). PHAs of this category are stiff, brittle, show high degree of crystallinity and have a limited range of applications (Urtuvia et al. 2014). Some bacterial species accumulate PHAs with more than one type of monomer units and are known as copolymers. Poly-3-hydroxybutyrate-co-3-hydroxyvalerate [P(3HB-co-3HV)] is an example for copolymer with 3-hydroxybutyrate and 3-hydroxyvalerate as monomers. These kinds of copolymers have improved ductility and flexibility (Wang et al. 2013b).

2.2.2 Medium-Chain-Length (MCL) PHAs

MCL PHAs have 6–14 carbon atoms in the monomer units. Poly-3-hydroxyhexanoate (P3HHx, C_6), poly-3-hydroxyheptanoate (P3HHp, C_7), poly-3-hydroxy-octanoate (P3HO, C_8), etc. are examples for MCL PHAs (Wang et al. 2009). Copolymers like poly-3-hydroxyhexanoate-co-3-hydroxyoctanoate [P(3HHx-co-3HO)] with 3-hydroxyhexanoate and 3-hydroxyoctanoate monomers also come under this category. Polymers belonging to this group are flexible with low tensile strength and crystallinity and show high melting point (Urtuvia et al. 2014).

2.2.3 Short-Chain-Length Medium-Chain-Length (SCL-MCL) PHAs

SCL-MCL PHAs are copolymers with SCL and MCL monomers. For example, poly-3-hydroxybutyrate-co-3-hydroxyhexanoate [P(3HB-co-3HHx)] consists of 3-hydroxybutyrate and 3-hydroxyhexanoate monomers (Akaraonye et al. 2010). The properties of SCL-MCL PHAs may vary with varying monomer composition, and depending on the molar fraction of these monomer constituents, the properties may range from high crystallinity to elasticity (Phithakrotchanakoon et al. 2013; Nomura et al. 2004b). For example, a copolymer made up of high content of 3-hydroxybutyrate and a low level of 3-hydroxyhexanoate has low density polyethylene-like properties and overcomes the brittleness of P(3HB) homopolymer (Matsusaki et al. 2000). There is a report of *Pseudomonas aeruginosa* accumulating PHAs with long-chain-length 3-hydroxyhexadecanoate (C_{16}) and 3-hydroxyoctadecanoate (C_{18}) units with SCL 3-hydroxybutyrate and 3-hydroxyvalerate as constituents P(3HB-co-3HV-co-3HHD-co-3HOD) (Singh and Mallick 2008).

More than 150 different monomer constituents of PHAs have been reported so far which could be homopolymers, copolymers and combination thereof (Khanna and Srivastava 2005; Ojumu et al. 2004; Steinbüchel and Lütke-Eversloh 2003). This indicates the diversity of PHAs with varying properties and applicability.

2.3 General Properties

PHAs are biocompatible, biodegradable, enantiomerically pure, non-toxic thermoplastics with piezoelectricity and are inert, indefinitely stable in air and water insoluble (Rai et al. 2011a; Reddy et al. 2012). Their properties may vary depending on their chemical composition. Generally, molecular weight (Mw) of PHAs ranges between 200 and 3000 kDa and is dependent on microorganism, media ingredients, inoculum, fermentation conditions, mode of fermentation and downstream processing techniques (Rai et al. 2011a). The side chain length and its functional group in the monomers determine the PHA properties, like glass transition temperature, melting temperature and crystallinity (Akaraonye et al. 2010). Structurally diverse PHA varieties show high variability in their thermal properties. They have melting temperature (T_m) value between 60 and 177 °C, and their polydispersity ranges between 1.2 and 6.0 (Steinbüchel 1991). Thermal and mechanical features of selected PHAs are given in Table 2.1.

Poly-3-hydroxybutyrate (P3HB) is the first discovered and the most studied PHA (Madison and Huisman 1999). PHB molecule is a compact right-handed helix with a twofold screw axis and a fibre repeat of 0.596 nm (Okamura and Marchessault 1967; Cobntbekt and Mabchessault 1972; Anderson and Dawes 1990). It is optically active, with the chiral centre of the monomer unit always in the R configuration which is essential for biodegradability and biocompatibility (Mauclaire et al. 2010). Comparing to polypropylene, PHB possesses better natural resistance to UV weathering but far inferior solvent resistance. Physically, it is more brittle and stiffer than polypropylene (Holmes 1988). When 3HB monomers combine with 3HV units, copolymers are formed and the crystallinity and melting point (T_m) decreases. These changes bring about variations in the mechanical properties of the material, such as reduction in stiffness and increased toughness, which make them better suited for commercial applications. Hence, by adjusting the monomer composition of PHAs, it is possible to make tailor-made polymer materials for variable applications (Anderson et al. 1990).

PHAs are water insoluble but soluble in chlorinated hydrocarbons like chloroform and dichloromethane. Chemical modification of PHAs with polar functional

Table 2.1 Physical properties of PHAs

PHA	T_m (°C)	T_g (°C)	Young's modulus (GPa)	Tensile strength (MPa)	References
P-3-HP	77	−20	0.3	27	Shimamura et al. (1994b) and Zhou et al. (2011)
P-3-HB	175	0–5	3.5	40	Ha and Cho (2002)
P-4-HB	69	−43	ND	104	Choi et al. (1999) and Zhu et al. (2013)
P-3-HV	119	−15	ND	31	Zhu et al. (2013)
PHB-co-HV (20 mol% HV)	145	−1	1.2	32	Holmes (1988)
PHB-co-HHx (17 mol% HHx)	130	−2	ND	ND	Shimamura et al. (1994a)

groups and the block copolymerization of PHAs with hydrophilic groups in various polymeric architectures resulted in development of PHA-based water soluble polymers. They have high impact on polymer engineering and may lead to the development of smart biomaterials in emerging areas (Li and Loh 2015).

2.4 PHA-Producing Bacteria

In 1926, French scientist Maurice Lemoigne of Pasteur Institute discovered intracellular accumulation of 3-hydroxybutyric acid polymers in *Bacillus megaterium*, and this was the first report of PHB accumulation in bacteria (Lemoigne 1926). From then, a large number of Gram-positive and Gram-negative bacteria were reported to be accumulating PHAs intracellularly including species from genera *Cupriavidus*, *Ralstonia*, *Pseudomonas*, *Aeromonas*, *Bacillus*, *Alcaligenes*, *Burkholderia*, *Enterobacter* and *Rhodobacter* (Leong et al. 2014; Philip et al. 2007; Gumel et al. 2013) as well as some cyanobacteria such as *Synechococcus*, *Nostoc* and *Spirulina* (Nishioka et al. 2001; Panda et al. 2005) and halophiles such as *Halobacterium* (Kirk and Ginzburg 1972), *Haloferax* (Lillo and Rodriguez-Valera 1990) and *Halomonas* (Quillaguaman et al. 2005). Even though bacterial PHA accumulation is related to stress conditions such as nutrient limitation with excess level of carbon source (Ojumu et al. 2004; Lee 1996; Dawes and Senior 1972), some bacteria such as *Alcaligenes eutrophus*, *Azotobacter vinelandii* UWD, *Alcaligenes latus* and recombinant *Escherichia coli* are able to synthesize PHAs irrespective of stress conditions (Keshavarz and Roy 2010). Generally *A. latus*, *Cupriavidus necator*, *Bacillus cereus*, *B. megaterium*, *Burkholderia cepacia*, *Caulobacter crescentus*, *Rhizobium meliloti*, etc. accumulate SCL PHAs (Tsuge et al. 2015; Łabużek and Radecka 2001; Keenan et al. 2004; Qi and Rehm 2001; Mercan and Beyatli 2005), whereas bacteria such as *Aeromonas hydrophila*, *Aeromonas caviae*, species of *Pseudomonas*, etc. accumulate MCL PHAs.

2.5 Mechanism of PHA Biosynthesis

Carbon sources supplied for PHA synthesis have a major role in the structure of accumulated PHAs. Thereby the properties of the polymer are determined by PHA biosynthetic pathways through which they were synthesized. PHA biosynthetic mechanisms mainly comprise of three pathways (Chen et al. 2015a) (Fig. 2.1).

2.5.1 Pathway I (The Acetyl-CoA to 3-hydroxybutyryl-CoA Pathway)

This is the classical SCL PHA synthesis mechanism with acetyl-CoA produced from sugar, fatty acids or amino acids as the precursors. The enzyme β-ketoacyl-CoA thiolase (acetyl-CoA/acetyl-CoA acetyltransferase; EC 2.3.1.9) catalyzes the

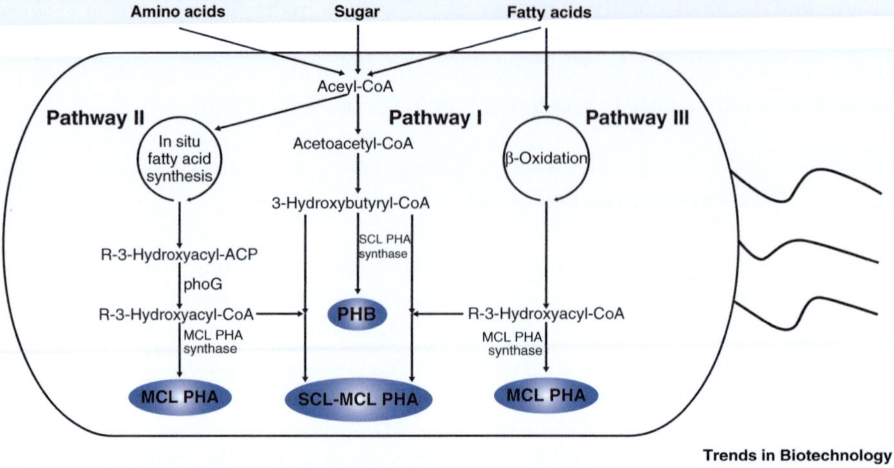

Fig. 2.1 PHA biosynthetic pathways in bacteria, Chen et al. 2015a; Reproduced with permission, Copyright (2015), Elsevier

first step in P(3HB) formation in which a pair of acetyl-CoA molecules is condensed into acetoacetyl-CoA. Acetoacetyl-CoA reductase (EC 1.1.1.36) catalyzes the second step by reducing acetoacetyl-CoA into 3-hydroxybutyryl-CoA. At the final stage, 3-hydroxybutyryl-CoA monomers are polymerized by P(3HB) polymerase (PHB synthase) to yield PHB (Doi et al. 1988; Chen et al. 2015a; Wang et al. 2012b). *Cupriavidus necator* is a typical representative of this pathway with a SCL PHA synthase ($PhaC_{SCL}$) which specifically targets C_3–C_5 substrates (Haywood et al. 1989).

2.5.2 Pathway II (The β-Oxidation Pathway)

This pathway of MCL PHA biosynthesis makes use of intermediate compounds derived from the β-oxidation of fatty acids (Sudesh et al. 2000). In β-oxidation pathway, fatty acids are first converted to enoyl-CoA. Subsequently, R-3-hydroxyacyl-CoA hydratase converts enoyl-CoA to R-3-hydroxyacyl-CoA, which is the precursor for MCL PHA polymerization. The final step of polymerization of R-3-hydroxyacyl-CoA is catalyzed by MCL PHA synthase ($PhaC_{MCL}$). *Pseudomonas putida*, *P. oleovorans* and *P. aeruginosa* are the representatives of this pathway (Sudesh et al. 2000; Tsuge et al. 2003).

2.5.3 Pathway III (The In Situ Fatty Acid Synthesis Pathway)

This pathway involves the in situ fatty acid synthesis cycle which delivers R-3-hydroxyacyl-ACP for PHA synthesis. The key enzyme for this pathway is 3-hydroxyacyl-acyl carrier protein-CoA transferase (PhaG), which converts

3-hydroxyacyl-ACP (acyl carrier protein) to 3-hydroxyacyl-CoA (Chen et al. 2015a; Wang et al. 2009; Rehm and Steinbüchel 1999). Through metabolic engineering of MCL PHA biosynthesis employing PhaG, several tailor-made bio-polyesters have been developed (Hoffmann et al. 2002; Zheng et al. 2004; Matsumoto et al. 2001; Hoffmann et al. 2000; Fiedler et al. 2000). In bacteria, β-oxidation cycles and the fatty acid in situ synthesis can occur concurrently to supply precursors for PHA synthesis (Huijberts et al. 1994). In addition to these three, new pathways can be established for biosynthesis of PHAs through metabolic engineering (Chen et al. 2015a).

2.6 Major Enzymes Involved in PHA Biosynthesis

2.6.1 β-Ketoacyl-CoA Thiolase (PhaA)

The gene *phaA* codes for the enzyme β-ketoacyl-CoA thiolase (acetyl-CoA/acetyl-CoA acetyltransferase; EC 2.3.1.9) which catalyzes the first step in P(3HB) formation and also a key enzyme in many other important biosynthetic pathways (Modis and Wierenga 1999; Thompson et al. 1989). They are found in almost all the living cells ranging from higher eukaryotes to yeasts and prokaryotes (Madison and Huisman 1999). Based on their substrate specificity, β-ketoacyl-CoA thiolases can be of two types, type I degradative (EC 2.3.1.16) and type II biosynthetic (EC 2.3.1.9) thiolases (Kim et al. 2014a). Type I thiolases mainly involved in the catabolism of fatty acids with broad substrate specificity for β-ketoacyl-CoA with carbon atoms ranging from four to 16. The second type of β-ketoacyl-CoA thiolase is specialized for a number of reactions, such as biosynthesis of P(3HB), ketone bodies, steroids, isoprenoids, etc. Their chain length specificity generally ranges from three to five carbon atoms (Masamune et al. 1989b; Madison and Huisman 1999).

R. eutropha genome harbours genes coding for two β-ketoacyl-CoA thiolases, enzyme A and enzyme B. β-ketoacyl-CoA thiolase A has the substrate specificity to acetoacetyl-CoA and 3-ketopentanoyl-CoA. On the other hand, enzyme B has vast substrate specificity and acts on acetoacetyl-CoA as well as 3-ketoalkanoyl-CoAs of C_5-C_8 and C_{10} (Madison and Huisman 1999). The enzyme catalysis of PhaA is a two-stage process. In the first half reaction, a cysteine residue at the active site acts on an acetyl-S-CoA molecule which results in the formation of an acetyl-*S*-enzyme intermediate. In the second half reaction, another cysteine acts as the catalytic base and oxidizes another acetyl-CoA, resulting in an activated acetyl-CoA intermediate which is capable to act on the acetyl-*S*-enzyme intermediate to form acetoacetyl-CoA (Masamune et al. 1989a; Palmer et al. 1991; Madison and Huisman 1999).

The crystal structure of β-keto thiolase B (ReBktB) from *R. eutropha* H16 revealed that catalytic site of ReBktB contains three conserved residues, Cys90, His350 and Cys380, which may function as a covalent nucleophile, a general base and a second nucleophile, respectively (Kim et al. 2014a). These findings were in agreement with the previous work on *Zoogloea ramigera* biosynthetic thiolases (Modis and Wierenga 2000).

2.6.2 Acetoacetyl-CoA Reductase

Acetoacetyl-CoA reductase (EC 1.1.1.36) catalyzes the second step in the PHB biosynthetic pathway by stereo selective reduction of the 3-ketone group of acetoacetyl-CoA, converting acetoacetyl-CoA into 3-hydroxybutyryl-CoA (Madison and Huisman 1999; Keshavarz and Roy 2010; Matsumoto et al. 2013). The acetoacetyl-CoA reductases involved in PHB synthetic pathway are known to be specific for NADPH (Chohan and Copeland 1998; Anderson and Dawes 1990; Fukui et al. 1987; Mansfield et al. 1995; Saito et al. 1977; Shuto et al. 1981), but in some PHA producers, the enzyme has been reported to have activity with NADH also (Amos and McInerney 1993; Manchak and Page 1994; Ritchie et al. 1971; Yabutani et al. 1995).

In addition to a gene coding for acetoacetyl-CoA reductase (*phaB*), the genome of *R. eutropha* H16 harbours isologs *phaB2*, *phaB3* and 15 other potential reductase genes. The gene sequence analyses revealed them as paralogs originated from gene duplication events (Budde et al. 2010). The crystallographic data of (R)-3-hydroxybutyryl-CoA dehydrogenase from *R. eutropha* H16 (PhaB) (Kim et al. 2014b) showed a fan-shaped tetrameric structure with each of the two active sites back to back with twofold symmetries. The $NADP^+$ and acetoacetyl-CoA bind at a Rossmann fold and a clamp domain, respectively. The clamp domain is involved in placing the substrate in the cleft and stabilizing the substrate conformation (Kim et al. 2014b).

2.6.3 PHA Synthase

PHA polymerases [PHA synthases (PhaCs)] are the enzymes that catalyze the polymerization of 3-R-hydroxyalkyl CoA thioesters to PHAs with collateral release of CoA (Zhang et al. 2015b). It was initially identified in 1964 by Merrick and Doudoroff in their studies with *B. megaterium* and *Rhodospirillum rubrum*, and it was characterized by Griebel et al. in 1968 in their studies on the production of PHB in *B. megaterium* (Merrick and Doudoroff 1964; Griebel et al. 1968). PHA synthases are categorized into four types based on their amino acid sequence, substrate specificity and subunit constitution (Zhang et al. 2015b; Rehm 2003). Schematic representation of the genetic organization of different classes of bacterial PHA synthases is given in Fig. 2.2.

Class I PHA synthases (e.g. *R. eutropha*) are made up of a single type of subunit (PhaC) having molecular weight ranging from 61 to 73 kDa. They preferentially polymerize CoA thioesters of different (R)-3-hydroxy-fatty acids made up of 3–5 carbon atoms and produce SCL PHAs (Rehm 2003). Class II PHA synthases also comprise of a single type of subunit (PhaC) and are found in fluorescent pseudomonads (e.g. *P. aeruginosa*) (Timm and Steinbüchel 1990). They specifically utilize CoA thioester of (*R*)-3-hydroxy-fatty acids with 6–14 carbon atoms to produce MCL PHAs (Amara and Bernd 2003). Recently two PHA synthases belonging to class I and class II, exhibiting only class I PHA synthase substrate specificity to

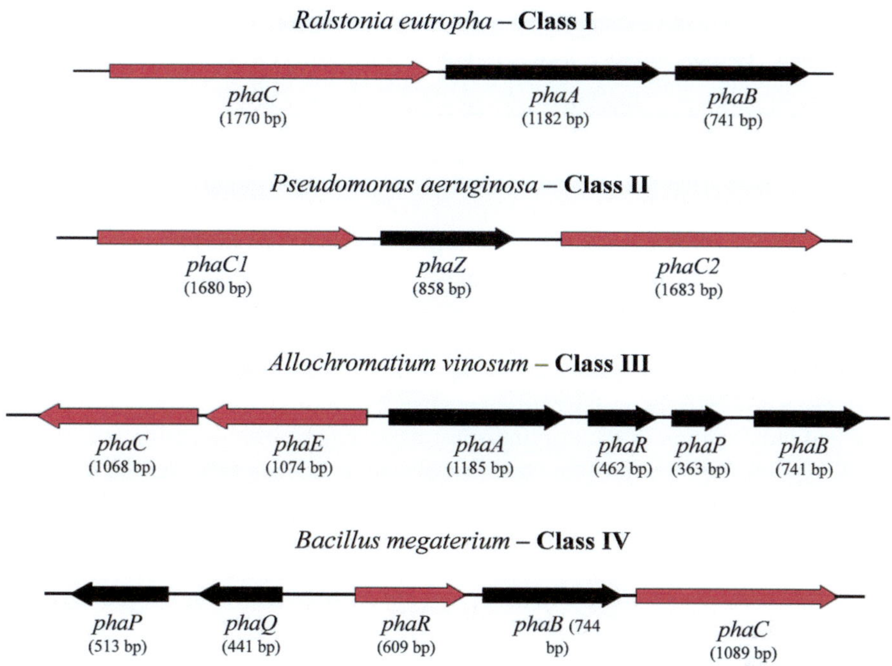

Fig. 2.2 Genetic background of PHA synthase genes in bacteria coding four different classes of enzymes

polymerize SCL PHA from 3-hydroxybutyryl-CoA, were identified and purified (Jiang et al. 2015). Class III PHA synthases (e.g. *Allochromatium vinosum*) are composed of two subunits: one catalytic subunit PhaC (40–53 kDa) with primary structure similar to class I and II PHA synthases and a PhaE subunit (approximately 20 or 40 kDa) which form a functional PhaEC complex (Liebergesell and Steinbüchel 1992; Tsuge et al. 2015). Though PhaE subunit is essential for the PhaC to be functional, its exact function in PHA polymerization is still unclear (Tsuge et al. 2015). Class IV PHA synthases (e.g. *B. megaterium*) also comprise two subunits as in class III PHA synthases, with a 41.5 kDa catalytic subunit, PhaC, and a different second subunit, PhaR (22 kDa) instead of PhaE (McCool and Cannon 2001).

The exact protein structure and enzyme catalytic mechanism of PHA synthases are still unknown since its crystallographic structure data are unavailable. However, Rehm et al. developed a threading model of PHA synthase from *R. eutropha*, based on the homology to *Burkholderia glumae* lipase (Rehm et al. 2002). Primary structure analysis of PHA synthases shows six conserved regions and eight conserved amino acids (Rehm 2003). The C-terminal regions of these synthases are highly conserved, and N-terminal regions are less conserved (Rehm 2015). The first 100 amino acid residues of class I PHA synthases are not involved in the enzyme activity (Schubert et al. 1991; Rehm et al. 2002). Later, studies with single amino acid substitution at the N-terminus of the PhaC from *R. eutropha* demonstrated elevated copy numbers of PHA synthase and better polymer production, which indicates the functional importance of the variable N-terminus (Normi et al. 2005). The C-terminal region of the class I and class II PHA synthases is rich in hydrophobic amino acid residues, and this region plays an important role in enzyme activity by anchoring the PhaC to the hydrophobic core of the PHA granule (Rehm et al. 2002).

The catalytic triad amino acid residues, cysteine, aspartate and histidine (C319, D480 and H508) in PhaC of *R. eutropha* are directly involved in the reaction mechanism (Rehm 2015). Upon substrate binding, class I and II synthases undergo dimerization which indicate the participation of two activated thiol groups each contributed by one subunit (Wodzinska et al. 1996; Rehm et al. 2001). PHA polymerization is initiated by activation of active-site cysteine by histidine residue. This triggers the nucleophilic action on the thioester bond of (R)-3-hydroxyacyl-CoA substrate and subsequent thioester bond formation between the thiol group and 3-hydroxy-fatty acid and dimerization of the enzyme-substrate complex. The initiation is followed by elongation step in which the conserved aspartate residue activates the hydroxyl group of the bound 3-hydroxy-fatty acid that in turn acts on the thioester bond between the cysteine and the hydroxyl-fatty acid of the second subunit (Rehm 2015). This results in the ester bond formation between two 3-hydroxy-fatty acid molecules. Now the nascent PHA chain is bounded to the active dimer, and the other subunit seeks another 3-hydroxy-fatty acid monomer (Rehm 2015). Thus the growing PHA chain moves from one active site to the other while being extended by one building block as in the mechanism of the fatty acid synthesis (Witkowski et al. 1997). The schematic representation of polymerization of PHA monomers as described by (Rehm 2015) is given in Fig. 2.3.

In addition to these three major enzymes, a large number of other PHA biosynthetic enzymes have also been reported from different bacterial species. List of such enzymes is given in Table 2.2.

2.7 Large-Scale PHA Production Strategies

Mainly three different cultivation strategies are employed for the large-scale production of PHAs using native bacterial isolates or recombinant strains.

2.7.1 Batch Fermentation

Batch fermentation is the primary and simplest method to assess a bioprocess due to its adaptability and lower operational costs (Kaur et al. 2012; Amache et al. 2013). In this closed system of cultivation, substrates are initially added in the fermenter at the beginning of the fermentation process, and products are recovered at the end. The system is kept undisturbed throughout the process, without the addition of any substrates or removal of biomass or media at any point of the fermentation. In recent years, this is the most widely used fermentation strategy to study the influence of various bioprocess parameters, microbial conversion of new carbon sources

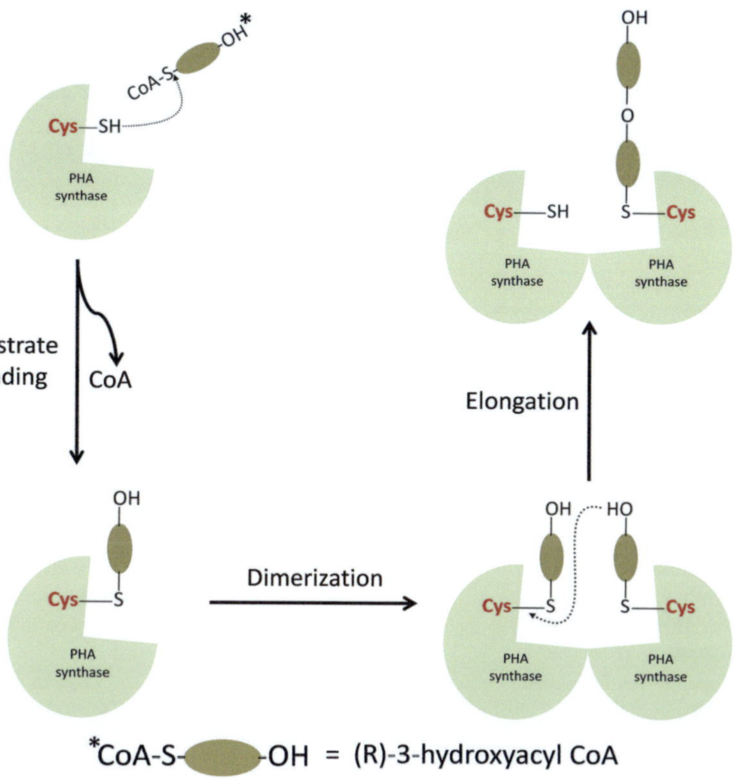

Fig. 2.3 Enzymatic mechanism of class I PHA synthase

for the production of different types of PHAs (Kaur and Roy 2015). In order to produce PHAs cost-effectively through batch process, recently, the choice of carbon sources has been shifted from refined simple sugars (Narayanan and Ramana 2012; Liu et al. 2014; Gahlawat and Srivastava 2012; García et al. 2014) to unrefined, cheap carbon sources such as agro-industrial wastes including cane molasses, whey, sunflower stalk hydrolysate, sugar beet juice, rice straw hydrolysate, grass biomass hydrolysate, plant oils, olive mill wastewater, etc. (Obruca et al. 2011; Akaraonye et al. 2012; Sindhu et al. 2013; Chen et al. 2013; Tripathi et al. 2013; Davis et al. 2013; Sathiyanarayanan et al. 2013; Zhang et al. 2013; Wang et al. 2013a; Kim et al. 2016; Alsafadi and Al-Mashaqbeh 2016). The mesophilic bacteria such as *Alcaligenes* and *Bacillus* sp. were the most commonly used organisms for the studies with cheap carbon sources (Kaur and Roy 2015).

Recently, several studies have been reported on high level PHA accumulating batch fermentation strategies. A thermophilic bacteria *Bacillus shackletoni* accumulated a high biopolymer content of 72.6% dry cell weight (DCW) in a batch process (Liu et al. 2014). Similarly another thermophile *Chelatococcus* strain was grown up to a biomass concentration of 4.8 gL^{-1} and a P(3HB) content of 73% DCW (Ibrahim and Steinbüchel 2010). In *Chelatococcus daeguensis* TAD1, a growth-associated

Table 2.2 Enzymes involved in biosynthesis of PHAs

No.	Enzyme	Abbreviation	Species	Reference
1	Glyceraldehyde-3-phosphate dehydrogenase	–	*Cupriavidus necator*	Raberg et al. (2011)
2	Pyruvate dehydrogenase complex	–	*Cupriavidus necator* and *Burkholderia cepacia*	Raberg et al. (2011)
3	3-Ketothiolase	PhaA	*Cupriavidus necator*	Peoples and Sinskey (1989)
4	NADPH-dependent acetoacetyl-CoA reductase	PhaB	*Cupriavidus necator*	Peoples and Sinskey (1989)
5	PHA synthase	PhaC	*Cupriavidus necator* and various	Kadouri et al. (2005) and Peoples and Sinskey (1989)
6	Acetyl-CoA carboxylase	ACC	*Escherichia coli* K-12 MG1655	Lee et al. (2011)
7	Malonyl-CoA/ACP transacylase	FabD	*Escherichia coli* K-12 MG1655	Lee et al. (2011)
8	3-Ketoacyl carrier protein synthase	FabH	*Escherichia coli* K-12 MG1655	Lee et al. (2011) and Nomura et al. (2004a)
9	NADPH-dependent 3-ketoacyl reductase	FabG	*Pseudomonas aeruginosa*	Ren et al. (2000)
10	Succinic semialdehyde dehydrogenase	SucD	*Clostridium kluyveri*	Valentin and Dennis (1997)
11	4-Hydroxybutyrate dehydrogenase	4HbD	*Clostridium kluyveri*	Valentin and Dennis (1997)
12	4-Hydroxybutyrate-CoA/CoA transferase	OrfZ	*Clostridium kluyveri*	Valentin and Dennis (1997)
13	Alcohol dehydrogenase, putative	–	*Aeromonas hydrophila* 4AK4	Xie and Chen (2008)
14	Hydroxyacyl-CoA synthase, putative	–	Mutants and recombinants of *Cupriavidus necator*	Valentin and Steinbüchel (1995)
15	Methylmalonyl-CoA mutase	Sbm	*Escherichia coli* W3110	Aldor et al. (2002)
16	Methylmalonyl-CoA racemase	–	*Nocardia corallina*	Valentin and Dennis (1996)
17	Methylmalonyl-CoA decarboxylase	YgfG	*Escherichia coli* W3110	Aldor et al. (2002)
18	Ketothiolase, putative	–	–	Satoh et al. (1999)
19	3-Ketothiolase	BktB	*Cupriavidus necator*	Slater et al. (1998)
20	Ketothiolase, putative	–	–	Satoh et al. (1999)
21	NADPH-dependent acetoacetyl-CoA reductase	–	*Rhizobium* (Cicer) sp. CC 1192	Chohan and Copeland (1998)
22	Acyl-CoA synthetase	FadD	*Pseudomonas putida* CA-3 and *Escherichia coli* MG1655	Hume et al. (2009) and Yuan et al. (2008)

(continued)

Table 2.2 (continued)

No.	Enzyme	Abbreviation	Species	Reference
23	Acyl-CoA oxidase, putative	–	–	Mittendorf et al. (1998)
24	Enoyl-CoA hydratase I, putative	–	–	Mittendorf et al. (1998)
25	(R)-Enoyl-CoA hydratase	PhaJ	*Pseudomonas putida* KT2440	Sato et al. (2011)
26	Epidermase	–	–	Mittendorf et al. (1998)
27	3-Ketoacyl-CoA thiolase	FadA	*Pseudomonas putida* KT2442	Ouyang et al. (2007)
28	3-Hydroxyacyl-ACP/CoA transacylase	PhaG	*Pseudomonas mendocina*	Zheng et al. (2005)
29	Cyclohexanol dehydrogenase	ChnA	*Acinetobacter* sp. SE19 and *Brevibacterium epidermidis* HCU	Brzostowicz et al. (2002)
30	Cyclohexanone monooxygenases	ChnB	*Acinetobacter* sp. SE19 and *Brevibacterium epidermidis* HCU	Brzostowicz et al. (2002)
31	Caprolactone hydrolase	ChnC	*Acinetobacter* sp. SE19 and *Brevibacterium epidermidis* HCU	Brzostowicz et al. (2002)
32	6-Hydroxyhexanoate dehydrogenase	ChnD	*Acinetobacter* sp. SE19 and *Brevibacterium epidermidis* HCU	Brzostowicz et al. (2002)
33	6-Oxohexanoate dehydrogenase	ChnE	*Acinetobacter* sp. SE19 and *Brevibacterium epidermidis* HCU	Brzostowicz et al. (2002)
34	Semialdehyde dehydrogenase, putative	–	–	Chen (2010)
35	6-Hydroxyhexanoate dehydrogenase, putative	–	–	Chen (2010)
36	Hydroxyacyl-CoA synthase, putative	–	–	Chen (2010)
37	Lactonase, putative	–	Mutants and recombinants of *Cupriavidus necator*	Valentin and Steinbüchel (1995)
38	Beta-hydroxyacyl-ACP dehydratase	FabA	*Pseudomonas putida* LS46	Fu et al. (2015)
39	Beta-hydroxyacyl-ACP dehydratase	FabZ	*Pseudomonas putida* LS46	Fu et al. (2015)

Source: Modified from Tan et al. (2014)

and non-nitrogen limited P(3HB) production pattern was reported (Xu et al. 2014). This property will help in reduction of fermentation period and improve the overall productivity (Liu et al. 2014) and make the process more economical. Besides P(3HB), other homopolymers such as P(3HO) and poly(-3-hydroxy-4-methylvalerate) P(3H4MV) and several copolymers such as P(3HB-co-3HV),

P(3HB-co-4HB), P(3HB-co-3HHx), P(3HP-co-4HB) and P(3HB-co-3HV-co-4HB) have been produced by batch fermentation (Liu et al. 2014; Vigneswari et al. 2010; Rao et al. 2010; Lau et al. 2010; Kulkarni et al. 2010; Rai et al. 2011b; Ng et al. 2011; Kshirsagar et al. 2013; Meng et al. 2012; Bhattacharyya et al. 2012; Aziz et al. 2012; Horng et al. 2013).

However, batch fermentation process is associated with low PHA productivity due to the degradation of the accumulated PHA by the cells after utilization of the carbon source resulting in reduced PHA content (Zinn et al. 2001).

2.7.2 Fed-Batch Fermentation

Fed-batch fermentation involves initiation of the process as a batch, following the supplementation of limiting media components into the bioreactor during microbial growth. The products are recovered from the system only at the end of fermentation. Thus there is an inflow without an outflow and a corresponding increase in the volume with respect to time (Mejía et al. 2010; Peña et al. 2014). This ensures regulation of appropriate nutrient feed into the bioreactor allowing continuous microbial growth and product formation (Kaur et al. 2012). This fermentation strategy has been widely used for the production of P(3HB) and other PHAs (Ruan et al. 2003; Rocha et al. 2008). *A. latus* ATCC 29714 produced a biomass concentration of 35.4 gL^{-1} with a P(3HB) production of 18.2 gL^{-1} in the presence of sucrose in exponentially fed-batch cultures (Grothe and Chisti 2000). Fed-batch culture of *Methylobacterium* sp. ZP24 with a limiting condition of dissolved oxygen accumulated a total concentration of 4.5 gL^{-1} of P(3HB) when processed cheese whey was supplemented with ammonium sulphate (Nath et al. 2008). Maximum biomass production of 90.7 gL^{-1} with 45.84% (w/w) of P(3HB) content and a productivity of 1.73 gL^{-1} h^{-1} was observed when an increase of the C/N ratio at 12.5 and intermittent feeding of the sugarcane molasses in cultures of *B. megaterium* (Kanjanachumpol et al. 2013). In a three-stage fed-batch cultivation study with canola oil as substrate, *Wautersia eutropha* produced 18.27 gL^{-1} of copolymer comprising four monomeric units [P(3HB)-co-P(3HV)-co-P(3HO)-co-P(3HDD)] (López-Cuellar et al. 2011). In a two-stage fed-batch fermentation process with glycerol as the carbon source, *C. daeguensis* TAD1 accumulated a PHB content of 17.4 gL^{-1} with a productivity rate of 0.434 gL^{-1} h^{-1}, which is the highest productivity rate reported for PHB to date (Cui et al. 2015).

In a recent study, using mixed volatile fatty acids (VFAs) as substrate, a three-stage PHA production process with a novel continuous feeding mode was established with mixed microbial cultures (MMCs) (Chen et al. 2015b). This system achieved a maximum intracellular PHA content of 70.4% and PHA yield of 0.81 cmol PHA/cmol VFA in the continuous feeding reactor making it an economical strategy for PHA production in MMCs.

2.7.3 Continuous Fermentation

Continuous culture (chemostat) is the third operation strategy for PHA production. In this method the sterile medium continuously replaces the culture broth. PHA-producing chemostats continuously feed the carbon source in excess, keeping one or more nutrients such as phosphorous or nitrogen in limitation. This type of fermentation system is highly controllable as the specific growth rate can be maintained by adjusting the dilution rate. Continuous fermentation has the potential to give highest PHA productivity levels under appropriate growth conditions if maintained at high dilution rates without a chance for wash-out (Amache et al. 2013; Kaur and Roy 2015). The main drawback of chemostat is the higher chance of contamination (Zinn et al. 2001).

A recent study with *P. putida* KT2440 showed that using elevated pressure, oxygen transfer rate and MCL PHA productivity can be efficiently enhanced. The process involved three-stage continuous cultivation under elevated pressure with a batch cultivation on octanoic acid, fed-batch cultivation on octanoic acid/10-undecenoic acid and finally a continuous cultivation on octanoic acid/10-undecenoic acid at a dilution rate of 0.15 h^{-1}. The polymer yield increased at higher pressure with a volumetric productivity of 11 gL^{-1} h^{-1} (Follonier et al. 2012). Available reports on continuous cultivations suggest that chemostats offer much higher productivities than those attained in batch and even some fed-batch fermentations (Kaur and Roy 2015). Hence by employing continuous fermentation strategy, PHAs can be economically produced in large scale using non-sterile processes and inexpensive substrates (Kaur and Roy 2015).

2.8 Recombinant PHA Production

E. coli is considered as an ideal host for PHA production with a vast knowledge in its rich genetic background and proper experience to cultivate it (Ren et al. 2005; Horng et al. 2010). Wild-type *E. coli* strains cannot synthesize or utilize any type of PHAs. Through genetic engineering, PHAs can be produced in recombinant *E. coli* strains with the newly introduced biosynthetic pathways (Le Meur et al. 2013). *E. coli* cells are unable to utilize accumulated PHA as it does not have any PHA depolymerase enzymes (Li et al. 2007). They have faster growth rate and have the ability to utilize several cheap and easily available carbon sources. They can grow to high cell densities and are capable of accumulating PHA polymers to 80–90% of cell dry weight which enable large-scale production of PHA. The availability of simple downstream processing methods for *E. coli* makes the polymer recovery process easy with inexpensive chemicals (Ren et al. 2005; Suriyamongkol et al. 2007; Kang et al. 2008). These factors make the process of recombinant PHA production more advantageous (Fidler and Dennis 1992).

In 1988, Slater et al. and Schübert et al. independently cloned *A. eutrophus* PHA genes into *E. coli*, and PHB granules were formed in recombinant *E. coli* cells for the first time (Slater et al. 1988; Schubert et al. 1988). Since then its PHB synthesis genes (*phaCAB*$_{Re}$) were most often to be used in recombinant *E. coli* for production of the biopolymer (Horng et al. 2010; Chien et al. 2010; Kim et al. 1992; Hahn et al. 1995).

As *E. coli* is not a natural producer of PHA, there are some unusual phenomena found in recombinant strains during PHA accumulation. One of the observations is filament formation, as the presence of PHA granules deactivates *ftsZ* gene coding for an important cell division protein, FtsZ (filamenting temperature-sensitive mutant Z). In high cell density cultures, the change in cell morphology hinders the PHA accumulation. This problem has been solved by *ftsZ* gene over expression and thereby successfully synthesized high amount of PHA polymers up to 157 gL^{-1} without cell elongation (Wang and Lee 1997). Another phenomenon is high yield limitation due to the inconstancy of the introduced genes and subsequent loss of plasmid from the recombinant cell (Suriyamongkol et al. 2007). By using the strategies such as chromosomal insertion and plasmid addiction system (Andreeßen et al. 2010), this condition can be avoided. Carbon catabolite repression (CCR) is a phenomenon closely related to sugar phosphotransferase system (PTS) where in a mixture of carbon sources, *E. coli* will initially consume substrate that can provide the highest growth rate. This will lead to the sole consumption of glucose even if the system is provided with other carbon sources. To overcome this, *E. coli* phosphotransferase (*ptsG*) mutant was constructed which simultaneously utilize glucose and xylose to produce SCL PHAs with cell density up to 2.3 gL^{-1} and polymer content up to 11.5 wt% (Li et al. 2007). A semi-rational approach by combining genomic library design, development of DNA constructs and their proper screening for overproduction of PHAs in *E. coli* was reported recently as an efficient way to optimize the process (Li et al. 2016).

In recent times, novel SCL PHAs with enhanced mechanical and thermal properties have been developed for a wide range of applications (Leong et al. 2014). Recombinant *E. coli* has been proven as the most promising host microorganism for SCL PHAs biosynthesis (Park et al. 2005; Horng et al. 2010). They can accumulate P(3HB) of much higher molecular weight than those that produced by natural PHA-synthesizing microorganisms (Kang et al. 2008). A list of recently developed SCL PHA-producing recombinant *E. coli* strains with details of carbon sources, cultivation mode, biomass and PHA content is given in Table 2.3.

The enhanced mechanical and thermal properties of the copolymer of 3-hydroxybutyrate (3HB) and 3-hydroxyvalerate (3HV) [poly(3-hydroxybutyrate-co-3-hydroxyvalerate)] [P(3HB-co-3HV)] attracted much commercial interest (Chen et al. 2011). But the wide applications of this copolymer have been limited by the high production costs (Nikel et al. 2006). The two commercial brands of P(3HB-co-3HV) available in the market are Biopol™ and Biocycle™ (Mumtaz et al. 2010). Strategy to produce P(3HB-co-3HV) through threonine biosynthetic pathway using inexpensive, unrelated carbon sources has been developed in recombinant *E. coli* (Chen et al. 2011).

Table 2.3 Recent reports in SCL PHA production with recombinant *E. coli* strains

E. coli strains	Carbon sources	Cultivation mode	DCW (gL^{-1})	PHA content (wt %)	Type of PHA	References
K24KL	Glycerol	Fed-batch	41.9	63	P3HB	Nikel et al. (2010)
K24KP	Glycerol	Batch	9.62 ± 0.02	16.9 ± 0.3	P3HB	de Almeida et al. (2010)
K24KP	Glucose	Batch	9.43 ± 0.01	37.2 ± 0.5	P3HB	de Almeida et al. (2010)
W	Glucose	Batch	3.45	7.94 ± 0.56	P3HB	Arifin et al. (2011)
W	Glucose	Fed-batch	89.8	40.1 ± 0.4	P3HB	Arifin et al. (2011)
W	Sucrose	Fed-batch	87.6	36.2 ± 1.2	P3HB	Arifin et al. (2011)
WΔcscR	Sucrose	Fed-batch	104.1	45.8 ± 3.6	P3HB	Arifin et al. (2011)
JM109	Glucose	Batch	9.3 ± 0.1	80 ± 1	P3HB	Tomizawa et al. (2011)
S17–1	1,3-propanediol, 1,4-butanediol	Batch	9.92 ± 0.30	62.70 ± 5.21	P(3HP-co-4HB)	Meng et al. (2012)
JM109	Glucose	Batch	2.04 ± 0.60	19 ± 6.4	P4HB	Le Meur et al. (2013)
JM109	Xylose	Batch	2.16 ± 0.37	32 ± 3.7	P4HB	Le Meur et al. (2013)
S17–1	1,3-propanediol, 1,4-butanediol	Batch	7.16 ± 1.16	42.26 ± 1.24	Mixture of P3HP and P4HB	Tripathi et al. (2013)
CT2	Glucose, propionic acid	Batch	6.80	58.71	P(3HB-co-3HV)	Horng et al. (2013)
JM 109	Glycolate	Batch	0.47 ± 0.01	0.9 ± 0.2	P(3HB-co-3HV-co-2,3-DHBA)	Insomphun et al. (2016)

Low crystallinity and high elasticity of MCL PHAs make them suitable for novel applications in tissue engineering, medical devices, cosmetic items and paint formulations (Park and Lee 2004). Recombinant *E. coli* system producing MCL PHA has been successfully developed by transforming the MCL PHA synthase gene, co-expressing with enoyl-CoA hydratase or 3-ketoacyl-ACP reductase (Park and Lee 2004). Through genetic manipulation of reversed fatty acid β-oxidation cycle, MCL PHAs can be synthesized directly from glucose in *E. coli* (Zhuang et al. 2014). Recent reports of MCL PHAs biosynthesis by various *E. coli* strains are given in Table 2.4.

Park and Lee developed a recombinant *E. coli* synthesizing SCL-MCL PHA copolymers with composition of 3HHx up to 63 mol% using decanoate as carbon source and proved that monomer composition in SCL-MCL PHAs can be adjusted by controlling the carbon sources (Park and Lee 2004). Using the same carbon

Table 2.4 Recent reports in MCL PHA production with recombinant *E. coli* strains

E. coli strains	Carbon sources	DCW (gL^{-1})	PHA content (wt %)	Composition (mol %)					References
				3HHx	3HO	3HD	3HDD	3HTD	
KNSP1	Glucose, Decanoate	3.54 ± 0.11	4.24 ± 0.5	–	64.2	35.8	–	–	Kang et al. (2011)
KNSP1	Glycerol, Decanoate	4.45 ± 0.21	4.49 ± 0.1	–	69.8	30.2	–	–	Kang et al. (2011)
DH5α	Glucose, Decanoate	4.31 ± 0.17	5.98 ± 0.04	–	39.2	60.8	–	–	Li et al. (2011)
DH5α	Decanoate	3.08 ± 0.13	9.34 ± 0.21	–	43.6	56.4	–	–	Li et al. (2011)
LS5218	Dodecanoate	1.1	19.1	8.2	32.3	32.2	27.3	–	Agnew et al. (2012)
LS5218	Glucose	3.4 ± 0.2	11.6 ± 1.3	1.3	39.2	56.9	2.6	–	Wang et al. (2012b)
LS5218	Glucose	6.6	6.6	1.2	45.6	24.4	22.0	6.7	Zhuang et al. (2014)

source, similar studies have been done on *E. coli* DH5α and observed same results (Li et al. 2011). P(3HB-co-3HHx) production in recombinant *E. coli* using lauric acid as carbon source can be raised by over expression of acyl-CoA dehydrogenase gene (*yafH*) (Lu et al. 2003; Lu et al. 2004). There have been reports for the role of 3-ketoacyl-acyl carrier protein (ACP) synthase III gene (*fabH*) and 3-ketoacyl-acyl carrier protein reductase gene (*fabG*) in enhancement of SCL-MCL PHA copolymer production from unrelated carbon sources, and *fabG* decides the monomer ratio of SCL-MCL PHA copolymers synthesized (Nomura et al. 2004a; Nomura et al. 2004b; Nomura et al. 2005). Recombinant *E. coli* with a low-substrate-specificity PhaC from *Pseudomonas stutzeri* 1317 synthesized 12.10 wt% of cell dry weight SCL-MCL PHA copolymers, with 21.18 mol% 3-hydroxybutyrate and 78.82 mol% MCL monomers (Zhuang et al. 2014). These protein-engineering techniques can be employed for producing tailor-made SCL-MCL PHA copolymers and for increasing flexibility in carbon source utilization (Leong et al. 2014). Recently developed SCL-MCL producing recombinant *E. coli* strains is listed in Table 2.5.

2.9 Applications of PHAs

Earlier applications of PHAs were mainly in the areas of consumer packaging items (e.g. cosmetic containers, shampoo bottles, etc.) (Hocking and Marchessault 1994) and manufacturing disposable items such as cosmetic containers, razors, utensils, cups, upholstery, carpet, compostable bags, diapers, feminine hygiene products, as well as medical surgical garments (Clarinval and Halleux 2005; Miková and Chodák 2006). The economic feasibility of large-scale production of PHAs for such applications is still under investigation, and hence, recently, the applications of PHAs are mainly focused on biomedical field.

Low molecular weight PHB, complexed with other macromolecules, can be found in cytoplasm and intracellular fluids as well as in membranes of living cells (Reusch 1995). Hence PHB and its monomer are considered as non-toxic to the cells (Chen and Wu 2005). They are ideal candidates for tissue engineering because of its high immunotolerance, low toxicity and biodegradability (Lomas et al. 2013). PHAs such as PHB, PHBV, P3HB4HB, PHBHHx and PHBVHHx have good biocompatibility and support cell proliferation without inducing tumour (Peng et al. 2011). Biocompatibility of PHBHHx can be enhanced by genetic improvement through the fusion of evolutionarily preserved cell-binding motifs such as Arg-Gly-Asp (RGD) with PHA-binding protein, phasin (PhaP), on the surface of the scaffold material (Xie et al. 2013).

In biomedical field, PHAs and its composites are used to develop devices such as meniscus regeneration devices; atrial septal defect repair devices; bulking and filling agents; vein valves; pericardial patches; bone marrow scaffolds and ligament; ocular cell implants; spinal fusion cages; dural substitutes; bone graft substitutes; bone dowels; wound dressings (Chen and Wu 2005); skin tissue regeneration (Ying et al. 2008); tendon repair (Webb et al. 2013); cartilage tissue engineering (You et al. 2011); nerve conduits to repair in vivo peripheral nerve damage (Bian et al.

Table 2.5 Recent reports in SCL-MCL PHA production with recombinant *E. coli* strains

E. coli strains	Carbon sources	DCW (gL^{-1})	PHA content (wt %)	Composition (mol %)						References
				3HB	3HHx	3HO	3HD	3HDD	3HTD	
DH5α	Decanoate	2.57 ± 0.12	1.87 ± 0.03	43.2	12.8	10.3	33.6	NG	NG	Li et al. (2011)
DH5α	Decanoate and glucose	4.90 ± 0.21	7.30 ± 0.23	83.4	4.0	5.6	7.0	NG	NG	Li et al. (2011)
LS5218	Dodecanoate	3.59 ± 0.11	3.93 ± 0.11	39.59 ± 0.23	5.80 ± 0.12	6.91 ± 0.12	45.46 ± 0.27	2.24 ± 0.07	NG	Gao et al. (2012)
LS5218	Gluconate and dodecanoate	3.73 ± 0.02	4.18 ± 0.16	52.39 ± 3.40	5.16 ± 0.54	20.24 ± 1.11	14.89 ± 1.75	7.32 ± 0.70	NG	Gao et al. (2012)
$ABC2_{Pp}J1_{Pp}$	Glycerol and dodecanoate	NG	0.1 ± 0.02	52 ± 4.9	7 ± 3.2	30 ± 1.5	11 ± 2.1	NG	NG	Phithakrotchanakoon et al. (2013)
$ABC2_{Pp}J4_{Pp}$	Glycerol and dodecanoate	NG	0.1 ± 0.01	65 ± 7.5	6 ± 1.9	16 ± 5.8	13 ± 3.3	NG	NG	Phithakrotchanakoon et al. (2013)
$ABC2_{Pp}J_{Ah}$	Glycerol and dodecanoate	NG	0.1 ± 0.01	36 ± 4.6	25 ± 1.6	30 ± 1.0	9 ± 3.9	NG	NG	Phithakrotchanakoon et al. (2013)
LS5218	Glucose	6.5	12.1	21.2	6.1	45.8	11.0	9.2	6.8	Zhuang et al. (2014)

2009; Wang et al. 2010); liver tissue engineering (Zhu et al. 2007); spinal cord repair (Novikov et al. 2002; Xu et al. 2010); periodontal tissue engineering (Wang et al. 2012a); manufacturing of sutures, suture fasteners, surgical mesh, orthopaedic pins, adhesion barriers, repair patches, screws, staples and bone plating systems; slings, cardiovascular patches (Dai et al. 2009); treatment of infected bone defects (Wang et al. 2016); guided tissue regeneration devices, articular cartilage repair devices, stents, etc. (Wang et al. 2008; Chen 2010).

In recent years, several studies have been conducted to improve the biocompatibility of PHAs and its composites. PHBHHx, when modified with maleic anhydride, showed improved biocompatibility, reasonable mechanical properties, as well as accelerated biodegradation as a biomaterial for biomedical applications when compared with PHBHHx (Li et al. 2008). In another study, PHB, PHBV and ozone-treated poly-3-hydroxy-10-undecenoate were made into scaffolds loaded with murine fibroblast cells for ligament and tendon tissue engineering (Rathbone et al. 2010). PHBHHx/collagen scaffolds with human embryonic stem cells (hESCs) have been proven as a biocompatible alternative for damaged tissues (Lomas et al. 2013). Studies with polyhydroxy propionate-co-polyhydroxy octadecanoate [P(HP-co-HOD)]–polyethylene glycol (PEG) scaffolds showed that they support cell growth better than the conventional plastic surface (Shabna et al. 2014). When PHBV microspheres were used as scaffolds for neural tissue engineering, greater axon–dendrite segregation was observed (Chen and Tong 2012). Bio-functionalized and aligned PHB/PHBV nanofibrous scaffolds can elicit essential signals for Schwann cells activity and can be used as a potential scaffold for nerve regeneration (Masaeli et al. 2014). For application in bone tissue engineering, electrospun poly-3-hydroxybutyrate-co-4-hydroxybutyrate [P(3HB-co-4HB)] fibres supporting survival, proliferation and differentiation of mouse adipose-derived stem cells (mASCs) were developed (Fu et al. 2014). Attempts to bio-valorize waste glycerol for the fabrication of electrospun scaffolds with P(3HB-4HB-3HV) were successful and observed their ability to support human mesenchymal stem cell (hMSC) growth at acceptable proliferation levels (Canadas et al. 2014). In a recent study, it was observed that PHBVHHx scaffolds loaded with umbilical cord-derived mesenchymal stem cells (UC-MSCs) significantly promote recovery of injured liver in mice (Su et al. 2014). PHBHHx scaffolds loaded with PHA-binding protein fused with arginyl-glycyl-aspartic acid (PhaP-RGD) better facilitate the proliferation and chondrogenic differentiation of human umbilical cord-derived mesenchymal stem cells (hUC-MSCs) and were found useful for cartilage tissue engineering (Li et al. 2015).

Since the PHA-based materials are biocompatible, non-toxic and biodegradable, they are used for microsphere-, microparticle- and nanoparticle-mediated targeted drug delivery (Francis et al. 2011; Lu et al. 2010; Mendes et al. 2012). For sustained release studies, MCL PHAs are found more suitable due to their low crystallinity rather than SCL PHAs (Masood et al. 2015). Xiong et al. proved for the first time that PHB and PHBHHx can be effectively used for intracellular controlled drug releases by targeted releasing of a lipid-soluble colourant, rhodamine B isothiocyanate (RBITC) (Xiong et al. 2010). By chemical coupling of hydrophobic PHAs to hydrophilic substances such as monomethoxy polyethylene glycol (mPEG), amphiphilic biodegradable nanoparticles with an outer hydrophilic shell and an interior hydrophobic core can be developed for safe and controlled release of variety of hydrophobic drugs (Shah et al. 2010). Such an amphiphilic P3HV4HB-mPEG nanoparticle system was used for encapsulation of a chemotherapeutic agent, cisplatin, and the in vitro studies proved the effect of alleviating the toxicity associated with the multiple dosing of the drug (Shah et al. 2012). In targeted drug delivery using P(HB-co-HV) nanoparticles, it is possible to control the bioavailability of drug by varying the copolymer composition in nanoparticles (Masood et al. 2013). Studies showed that PHB can be used for the preparation of magnetic nanoparticles to use in cancer therapy (Erdal et al. 2012). In other two recent studies with the PHA nanoparticle-mediated delivery of anticancer drugs, doxorubicin and etoposide, good drug uptake by tumour cells and better antitumour activity were demonstrated (Kılıçay et al. 2011; Zhang et al. 2010). In a more recent study, a sustained drug release system with folate-decorated P(HB-HO) nanoparticles was successfully developed. This system showed effective inhibition of tumour cells, precise targeting and reduced side effects and toxicity to normal tissues (Zhang et al. 2015a). Dong et al. designed a growth factor-loaded PHBHHx nanoparticles containing 3D collagen matrix for stem cell tissue-engineering application which showed sustained release of growth factors for long-term stimulation of human mesenchymal stem cell (hMSCs) proliferation/differentiation (Dong et al. 2015).

2.10 Conclusion and Future Perspectives

The review discussed about various aspects of the bacterial polyhydroxyalkanoates, such as diversity of PHA producers, mechanisms behind their biosynthesis, large-scale cost-effective production strategies, recombinant production of the polymer and their applications with special emphasis on recent advances in the field. During the last two decades, several innovative ideas have been developed to improve the cost-effective production of the biopolymer. In biomedical applications, PHAs and their copolymers have become the material of choice especially in the area of tissue engineering and targeted drug delivery due to their higher biocompatibility, biodegradability and better processibility. Apart from this, PHAs have a bright future ahead in the manufacture of various consumer products also.

The progress in the field of process engineering and metabolic engineering has revolutionized the microbial production of polyhydroxyalkanoates. The diversity of PHAs has increased to a great extend with the advances in the field of metabolic engineering. New combinations of copolymers and polymer blends with improved properties and wider applicability are being reported every year. Even though research in this field has advanced to a greater extend in recent years, challenges such as lower efficiency of bacterial fermentation processes and the higher cost of production when compared to their petrochemical counterparts are still remain unsolved. Therefore, interest is being focused on developing a single bacterial strain capable of accumulating a wide range of PHAs in a cost-effective manner, with low molecular polydispersity and predictable monomer composition. This will open up new opportunities in the field of low cost bacterial polymer production with added values. Environmental impact due to our prodigious consumption of petroleum-based plastics and their improper disposal methods in its current form has become a major concern, and hence a novel trend is being emerged globally for the utilization of environmental friendly biodegradable materials as an alternative. Therefore bacterial PHAs seem to be a sustainable eco-friendly biomaterial for the future.

Acknowledgements The authors are grateful to Prof. M. Radhakrishna Pillai, Director, RGCB, for the facilities provided. We acknowledge the financial support from Department of Biotechnology, Govt. of India, and Council of Scientific and Industrial Research (CSIR), India (SRF; 09/716(0149)/2012-EMR-I).

References

Agnew DE, Stevermer AK, Youngquist JT et al (2012) Engineering *Escherichia coli* for production of C 12–C 14 polyhydroxyalkanoate from glucose. Metab Eng 14(6):705–713

Akaraonye E, Keshavarz T, Roy I (2010) Production of polyhydroxyalkanoates: the future green materials of choice. J Chem Technol Biotechnol 85(6):732–743

Akaraonye E, Moreno C, Knowles JC et al (2012) Poly (3-hydroxybutyrate) production by *Bacillus cereus* SPV using sugarcane molasses as the main carbon source. Biotechnol J 7(2):293–303

Aldor IS, Kim S-W, Prather KLJ et al (2002) Metabolic engineering of a novel propionate-independent pathway for the production of poly (3-hydroxybutyrate-co-3-hydroxyvalerate) in recombinant *Salmonella enterica* serovar typhimurium. Appl Environ Microbiol 68(8):3848–3854

Alsafadi D, Al-Mashaqbeh O (2016) A one-stage cultivation process for the production of poly-3-(hydroxybutyrate-co-hydroxyvalerate) from olive mill wastewater by *Haloferax mediterranei*. New Biotechnol. doi:http://dx.doi.org/10.1016/j.nbt.2016.05.003

Amache R, Sukan A, Safari M et al (2013) Advances in PHAs production. Chem Eng 32. doi:10.3303/CET1332156

Amara AA, Bernd H (2003) Replacement of the catalytic nucleophile cysteine-296 by serine in class II polyhydroxyalkanoate synthase from *Pseudomonas aeruginosa*-mediated synthesis of a new polyester: identification of catalytic residues. Biochem J 374(2):413–421

Amos DA, McInerney MJ (1993) Formation of D-3-hydroxybutyryl-coenzyme A by an acetoacetyl-coenzyme A reductase in *Syntrophomonas wolfei* subsp. *wolfei*. Arch Microbiol 159(1):16–20

Anderson AJ, Dawes EA (1990) Occurrence, metabolism, metabolic role, and industrial uses of bacterial polyhydroxyalkanoates. Microbiol Rev 54(4):450–472

Anderson AJ, Haywood GW, Dawes EA (1990) Biosynthesis and composition of bacterial poly (hydroxyalkanoates). Int J Biol Macromol 12(2):102–105

Andreeßen B, Lange AB, Robenek H et al (2010) Conversion of glycerol to poly (3-hydroxypropionate) in recombinant *Escherichia coli*. Appl Environ Microbiol 76(2):622–626

Arifin Y, Sabri S, Sugiarto H et al (2011) Deletion of cscR in *Escherichia coli* W improves growth and poly-3-hydroxybutyrate (PHB) production from sucrose in fed batch culture. J Biotechnol 156(4):275–278

Aziz NA, Sipaut CS, Abdullah AAA (2012) Improvement of the production of poly (3-hydroxyb utyrate-co-3-hydroxyvalerate-co-4-hydroxybutyrate) terpolyester by manipulating the culture condition. J Chem Technol Biotechnol 87(11):1607–1614

Bhattacharyya A, Pramanik A, Maji SK et al (2012) Utilization of vinasse for production of poly-3-(hydroxybutyrate-co-hydroxyvalerate) by *Haloferax mediterranei*. AMB Express 2(34):189. doi:10.1186/2191-0855-2-34

Bian Y-Z, Wang Y, Aibaidoula G et al (2009) Evaluation of poly (3-hydroxybutyrate-co-3-hydroxyhexanoate) conduits for peripheral nerve regeneration. Biomaterials 30(2):217–225

Brzostowicz P, Blasko M, Rouvière P (2002) Identification of two gene clusters involved in cyclohexanone oxidation in *Brevibacterium epidermidis* strain HCU. Appl Microbiol Biotechnol 58(6):781–789

Budde CF, Mahan AE, Lu J et al (2010) Roles of multiple acetoacetyl coenzyme A reductases in polyhydroxybutyrate biosynthesis in *Ralstonia eutropha* H16. J Bacteriol 192(20):5319–5328

Canadas RF, Cavalheiro JM, Guerreiro JD et al (2014) Polyhydroxyalkanoates: waste glycerol upgrade into electrospun fibrous scaffolds for stem cells culture. Int J Biol Macromol 71:131–140

Castilho LR, Mitchell DA, Freire DM (2009) Production of polyhydroxyalkanoates (PHAs) from waste materials and by-products by submerged and solid-state fermentation. Bioresour Technol 100(23):5996–6009

Chen G-Q (2010) Plastics completely synthesized by bacteria: polyhydroxyalkanoates. In: Chen G-Q (ed) Plastics from bacteria natural functions and applications. Springer, Berlin Heidelberg, pp 17–37

Chen G-Q, Hajnal I (2015) The 'PHAome'. Trends Biotechnol 33(10):559–564

Chen W, Tong YW (2012) PHBV microspheres as neural tissue engineering scaffold support neuronal cell growth and axon–dendrite polarization. Acta Biomater 8(2):540–548

Chen G-Q, Wu Q (2005) The application of polyhydroxyalkanoates as tissue engineering materials. Biomaterials 26(33):6565–6578

Chen Q, Wang Q, Wei G et al (2011) Production in *Escherichia coli* of poly (3-hydroxybutyrate-co-3-hydroxyvalerate) with differing monomer compositions from unrelated carbon sources. Appl Environ Microbiol 77(14):4886–4893

Chen B-Y, Hung J-Y, Shiau T-J et al (2013) Exploring two-stage fermentation strategy of polyhydroxyalkanoate production using *Aeromonas hydrophila*. Biochem Eng J 78:80–84

Chen G-Q, Hajnal I, Wu H et al (2015a) Engineering biosynthesis mechanisms for diversifying polyhydroxyalkanoates. Trends Biotechnol 33(10):565–574

Chen Z, Huang L, Wen Q et al (2015b) Efficient polyhydroxyalkanoate (PHA) accumulation by a new continuous feeding mode in three-stage mixed microbial culture (MMC) PHA production process. J Biotechnol 209:68–75

Chien C-C, Hong C-C, Soo P-C et al (2010) Functional expression of phaCAB genes from *Cupriavidus taiwanensis* strain 184 in *Escherichia coli* for polyhydroxybutyrate production. Appl Biochem Biotechnol 162(8):2355–2364

Chohan SN, Copeland L (1998) Acetoacetyl coenzyme A reductase and polyhydroxybutyrate synthesis in *Rhizobium* (Cicer) sp. strain CC 1192. Appl Environ Microbiol 64(8):2859–2863

Choi MH, Yoon SC, Lenz RW (1999) Production of poly (3-hydroxybutyric acid-co-4-hydroxybutyric acid) and poly (4-hydroxybutyric acid) without subsequent degradation by *Hydrogenophaga pseudoflava*. Appl Environ Microbiol 65(4):1570–1577

Clarinval AM, Halleux J (2005) Classification of biodegradable polymers. In: Smith R (ed) Biodegradable polymers for industrial applications. CRC Press, Boca Raton

Cobntbekt J, Mabchessault R (1972) Physical properties of poly-β-hydroxybutyrate: IV. Conformational analysis and crystalline structure. J Mol Biol 71(3):735–756

Cui B, Huang S, Xu F et al (2015) Improved productivity of poly (3-hydroxybutyrate)(PHB) in thermophilic *Chelatococcus daeguensis* TAD1 using glycerol as the growth substrate in a fed-batch culture. Appl Microbiol Biotechnol 99(14):6009–6019

Dai Z-W, Zou X-H, Chen G-Q (2009) Poly (3-hydroxybutyrate-co-3-hydroxyhexanoate) as an injectable implant system for prevention of post-surgical tissue adhesion. Biomaterials 30(17):3075–3083

Davis R, Kataria R, Cerrone F et al (2013) Conversion of grass biomass into fermentable sugars and its utilization for medium chain length polyhydroxyalkanoate (mcl-PHA) production by *Pseudomonas* strains. Bioresour Technol 150:202–209

Dawes EA, Senior PJ (1972) The role and regulation of energy reserve polymers in microorganisms. Adv Microb Physiol 10:135–266 de Almeida A, Giordano AM, Nikel PI et al (2010) Effects of aeration on the synthesis of poly (3-hydroxybutyrate) from glycerol and glucose in recombinant *Escherichia coli*. Appl Environ Microbiol 76(6):2036–2040

Doi Y, Tamaki A, Kunioka M et al (1988) Production of copolyesters of 3-hydroxybutyrate and 3-hydroxyvalerate by *Alcaligenes eutrophus* from butyric and pentanoic acids. Appl Microbiol Biotechnol 28(4–5):330–334

Dong CL, Webb WR, Peng Q et al (2015) Sustained PDGF-BB release from PHBHHx loaded nanoparticles in 3D hydrogel/stem cell model. J Biomed Mater Res A 103(1):282–288

Erdal E, Kavaz D, Şam M et al (2012) Preparation and characterization of magnetically responsive bacterial polyester based nanospheres for cancer therapy. J Biomed Nanotechnol 8(5):800–808

Fidler S, Dennis D (1992) Polyhydroxyalkanoate production in recombinant *Escherichia coli*. FEMS Microbiol Rev 9(2–4):231–235

Fiedler S, Steinbüchel A, Rehm BH (2000) PhaG-mediated synthesis of poly (3-hydroxyalkanoates) consisting of medium-chain-length constituents from nonrelated carbon sources in recombinant *Pseudomonas fragi*. Appl Environ Microbiol 66(5):2117–2124

Follonier S, Henes B, Panke S et al (2012) Putting cells under pressure: a simple and efficient way to enhance the productivity of medium-chain-length polyhydroxyalkanoate in processes with *Pseudomonas putida* KT2440. Biotechnol Bioeng 109(2):451–461

Francis L, Meng D, Knowles J et al (2011) Controlled delivery of gentamicin using poly (3-hydroxybutyrate) microspheres. Int J Mol Sci 12(7):4294–4314

Fu N, Deng S, Fu Y et al (2014) Electrospun P34HB fibres: a scaffold for tissue engineering. Cell Prolif 47(5):465–475

Fu J, Sharma P, Spicer V et al (2015) Quantitative 'Omics analyses of medium chain length polyhydroxyalkanaote metabolism in *Pseudomonas putida* LS46 cultured with waste glycerol and waste fatty acids. PLoS One 10 (11):e0142322. doi:http://dx.doi.org/10.1371/journal.pone.0142322

Fukui T, Ito M, Saito T et al (1987) Purification and characterization of NADP-linked acetoacetyl-CoA reductase from *Zoogloea ramigera* I-16-M. Biochim Biophys Acta Biochim Biophys Acta 917(3):365–371

Gahlawat G, Srivastava AK (2012) Estimation of fundamental kinetic parameters of polyhydroxybutyrate fermentation process of *Azohydromonas australica* using statistical approach of media optimization. Appl Biochem Biotechnol 168(5):1051–1064

Gao D, Maehara A, Yamane T et al (2001) Identification of the intracellular polyhydroxyalkanoate depolymerase gene of *Paracoccus denitrificans* and some properties of the gene product. FEMS Microbiol Lett 196(2):159–164

Gao X, Yuan X-X, Shi Z-Y et al (2012) Production of copolyesters of 3-hydroxybutyrate and medium-chain-length 3-hydroxyalkanoates by *E. coli* containing an optimized PHA synthase gene. Microb Cell Factories 11(1):1–10. doi:10.1186/1475-2859-11-130

García A, Segura D, Espín G et al (2014) High production of poly-β-hydroxybutyrate (PHB) by an *Azotobacter vinelandii* mutant altered in PHB regulation using a fed-batch fermentation process. Biochem Eng J 82:117–123

Griebel R, Smith Z, Merrick J (1968) Metabolism of poly (β-hydroxybutyrate). I. Purification, composition, and properties of native poly (β-hydroxybutyrate) granules from *Bacillus megaterium*. Biochemistry 7(10):3676–3681

Grothe E, Chisti Y (2000) Poly (β-hydroxybutyric acid) thermoplastic production by *Alcaligenes latus*: behavior of fed-batch cultures. Bioprocess Eng 22(5):441–449

Gumel A, Annuar M, Chisti Y (2013) Recent advances in the production, recovery and applications of polyhydroxyalkanoates. J Polym Environ 21(2):580–605

Ha C-S, Cho W-J (2002) Miscibility, properties, and biodegradability of microbial polyester containing blends. Prog Polym Sci 27(4):759–809

Hahn SK, Chang YK, Lee SY (1995) Recovery and characterization of poly (3-hydroxybutyric acid) synthesized in *Alcaligenes eutrophus* and recombinant *Escherichia coli*. Appl Environ Microbiol 61(1):34–39

Haywood G, Anderson A, Dawes E (1989) The importance of PHB-synthase substrate specificity in polyhydroxyalkanoate synthesis by *Alcaligenes eutrophus*. FEMS Microbiol Lett 57(1):1–6
Hocking PJ, Marchessault RH (1994) Biopolyesters. In: Griffin GJ (ed) Chemistry and technology of biodegradable polymers. Blackie academic & professional, London, pp 48–96
Hoffmann N, Steinbüchel A, Rehm BH (2000) The *Pseudomonas aeruginosa* phaG gene product is involved in the synthesis of polyhydroxyalkanoic acid consisting of medium-chain-length constituents from non-related carbon sources. FEMS Microbiol Lett 184(2):253–259
Hoffmann N, Amara AA, Beermann BB et al (2002) Biochemical characterization of the *Pseudomonas putida* 3-hydroxyacyl ACP: CoA transacylase, which diverts intermediates of fatty acid de novo biosynthesis. J Biol Chem 277(45):42926–42936
Holmes P (1988) Biologically produced (R)-3-hydroxy-alkanoate polymers and copolymers. In: Developments in crystalline polymers. Springer, Dordrecht, pp 1–65
Horng YT, Chang KC, Chien CC et al (2010) Enhanced polyhydroxybutyrate (PHB) production via the coexpressed phaCAB and vgb genes controlled by arabinose PBAD promoter in *Escherichia coli*. Lett Appl Microbiol 50(2):158–167
Horng Y-T, Chien C-C, Huang C-T et al (2013) Biosynthesis of poly (3-hydroxybutyrate-co-3-hydroxyvalerate) with co-expressed propionate permease (prpP), beta-ketothiolase B (bktB), and propionate-CoA synthase (prpE) in *Escherichia coli*. Biochem Eng J 78:73–79
Huijberts G, de Rijk TC, de Waard P et al (1994) 13C nuclear magnetic resonance studies of *Pseudomonas putida* fatty acid metabolic routes involved in poly (3-hydroxyalkanoate) synthesis. J Bacteriol 176(6):1661–1666
Hume AR, Nikodinovic-Runic J, O'Connor KE (2009) FadD from Pseudomonas putida CA-3 is a true long-chain fatty acyl coenzyme A synthetase that activates phenylalkanoic and alkanoic acids. J Bacteriol 191(24):7554–7565
Ibrahim MH, Steinbüchel A (2010) High-cell-density cyclic fed-batch fermentation of a poly (3-hydroxybutyrate)-accumulating thermophile, *Chelatococcus* sp. strain MW10. Appl Environ Microbiol 76(23):7890–7895
Insomphun C, Kobayashi S, Fujiki T et al (2016) Biosynthesis of polyhydroxyalkanoates containing hydroxyl group from glycolate in *Escherichia coli*. AMB Express 6(1):1–8. doi:10.1186/s13568-016-0200-5
Jendrossek D, Handrick R (2002) Microbial degradation of Polyhydroxyalkanoates. Annu Rev Microbiol 56(1):403–432
Jiang X, Luo X, Zhou N-Y (2015) Two Polyhydroxyalkanoate synthases from distinct classes from the aromatic degrader *Cupriavidus pinatubonensis* JMP134 exhibit the same substrate preference. PLoS One 10 (11):e0142332. doi:http://dx.doi.org/10.1371/journal.pone.0142332
Kadouri D, Jurkevitch E, Okon Y et al (2005) Ecological and agricultural significance of bacterial polyhydroxyalkanoates. Crit Rev Microbiol 31(2):55–67
Kang Z, Wang Q, Zhang H et al (2008) Construction of a stress-induced system in *Escherichia coli* for efficient polyhydroxyalkanoates production. Appl Microbiol Biotechnol 79(2):203–208
Kang Z, Du L, Kang J et al (2011) Production of succinate and polyhydroxyalkanoate from substrate mixture by metabolically engineered *Escherichia coli*. Bioresour Technol 102(11):6600–6604
Kanjanachumpol P, Kulpreecha S, Tolieng V et al (2013) Enhancing polyhydroxybutyrate production from high cell density fed-batch fermentation of *Bacillus megaterium* BA-019. Bioprocess Biosyst Eng 36(10):1463–1474
Kaur G, Roy I (2015) Strategies for large-scale production of polyhydroxyalkanoates. Chem Biochem Eng Q 29(2):157–172
Kaur G, Srivastava A, Chand S (2012) Advances in biotechnological production of 1, 3-propanediol. Biochem Eng J 64:106–118
Keenan TM, Tanenbaum SW, Stipanovic AJ et al (2004) Production and characterization of poly-β-hydroxyalkanoate copolymers from *Burkholderia cepacia* utilizing xylose and Levulinic acid. Biotechnol Prog 20(6):1697–1704
Keshavarz T, Roy I (2010) Polyhydroxyalkanoates: bioplastics with a green agenda. Curr Opin Microbiol 13(3):321–326

Khanna S, Srivastava AK (2005) Recent advances in microbial polyhydroxyalkanoates. Process Biochem 40(2):607–619

Kılıçay E, Demirbilek M, Türk M et al (2011) Preparation and characterization of poly (3-hydroxy butyrate-co-3-hydroxyhexanoate)(PHBHHX) based nanoparticles for targeted cancer therapy. Eur J Pharm Sci 44(3):310–320

Kim BS, Lee SY, Chang HN (1992) Production of poly-β-hydroxybutyrate by fed-batch culture of recombinant *Escherichia coli*. Biotechnol Lett 14(9):811–816

Kim E-J, Son HF, Kim S et al (2014a) Crystal structure and biochemical characterization of beta-keto thiolase B from polyhydroxyalkanoate-producing bacterium *Ralstonia eutropha* H16. Biochem Biophys Res Commun 444(3):365–369

Kim J, Chang JH, Kim E-J et al (2014b) Crystal structure of (R)-3-hydroxybutyryl-CoA dehydrogenase PhaB from *Ralstonia eutropha*. Biochem Biophys Res Commun 443(3):783–788

Kim HS, Oh YH, Jang Y-A et al (2016) Recombinant *Ralstonia eutropha* engineered to utilize xylose and its use for the production of poly (3-hydroxybutyrate) from sunflower stalk hydrolysate solution. Microb Cell Factories 15(1):1–13. doi:10.1186/s12934-016-0495-6

Kirk RG, Ginzburg M (1972) Ultrastructure of two species of *Halobacterium*. J Ultrastruct Res 41(1–2):80–94

Kshirsagar PR, Suttar R, Nilegaonkar SS et al (2013) Scale up production of polyhydroxyalkanoate (PHA) at different aeration, agitation and controlled dissolved oxygen levels in fermenter using *Halomonas campisalis* MCM B-1027. J Biochem Technol 4(1):512–517

Kulkarni S, Kanekar P, Nilegaonkar S et al (2010) Production and characterization of a biodegradable poly (hydroxybutyrate-co-hydroxyvalerate)(PHB-co-PHV) copolymer by moderately haloalkalitolerant *Halomonas campisalis* MCM B-1027 isolated from Lonar Lake, India. Bioresour Technol 101(24):9765–9771

Łabużek S, Radecka I (2001) Biosynthesis of PHB tercopolymer by *Bacillus cereus* UW85. J Appl Microbiol 90(3):353–357

Lau N-S, Chee J-Y, Tsuge T et al (2010) Biosynthesis and mobilization of a novel polyhydroxyalkanoate containing 3-hydroxy-4-methylvalerate monomer produced by *Burkholderia* sp. USM (JCM15050). Bioresour Technol 101(20):7916–7923

Le Meur S, Zinn M, Egli T et al (2013) Poly (4-hydroxybutyrate)(P4HB) production in recombinant *Escherichia coli*: P4HB synthesis is uncoupled with cell growth. Microb Cell Factories 12(1):1–11. doi:10.1186/1475-2859-12-123

Lee SY (1996) Bacterial polyhydroxyalkanoates. Biotechnol Bioeng 49(1):1–14

Lee S, Jeon E, Yun HS et al (2011) Improvement of fatty acid biosynthesis by engineered recombinant *Escherichia coli*. Biotechnol Bioprocess Eng 16(4):706–713

Lemoigne M (1926) Products of dehydration and of polymerization of β-hydroxybutyric acid. Bull Soc Chem Biol 8:770–782

Leong YK, Show PL, Ooi CW et al (2014) Current trends in polyhydroxyalkanoates (PHAs) biosynthesis: insights from the recombinant *Escherichia coli*. J Biotechnol 180:52–65

Li Z, Loh XJ (2015) Water soluble polyhydroxyalkanoates: future materials for therapeutic applications. Chem Soc Rev 44(10):2865–2879

Li R, Chen Q, Wang PG et al (2007) A novel-designed *Escherichia coli* for the production of various polyhydroxyalkanoates from inexpensive substrate mixture. Appl Microbiol Biotechnol 75(5):1103–1109

Li XT, Sun J, Chen S et al (2008) In vitro investigation of maleated poly (3-hydroxybutyrate-co-3-hydroxyhexanoate) for its biocompatibility to mouse fibroblast L929 and human microvascular endothelial cells. J Biomed Mater Res A 87(3):832–842

Li Q, Chen Q, Li M-J et al (2011) Pathway engineering results the altered polyhydroxyalkanoates composition in recombinant *Escherichia coli*. New Biotechnol 28(1):92–95

Li X, Chang H, Luo H et al (2015) Poly (3-hydroxybutyrate-co-3-hydroxyhexanoate) scaffolds coated with PhaP-RGD fusion protein promotes the proliferation and chondrogenic differentiation of human umbilical cord mesenchymal stem cells in vitro. J Biomed Mater Res A 103(3):1169–1175

Li T, Ye J, Shen R et al (2016) Semi-rational approach for ultra-high poly (3-hydroxybutyrate) accumulation in *Escherichia coli* by combining one-step library construction and high-throughput screening. ACS Synth Biol. doi:10.1021/acssynbio.6b00083

Liebergesell M, Steinbüchel A (1992) Cloning and nucleotide sequences of genes relevant for biosynthesis of poly (3-hydroxybutyric acid) in *Chromatium vinosum* strain D. Eur J Biochem 209(1):135–150

Lillo JG, Rodriguez-Valera F (1990) Effects of culture conditions on poly (β-hydroxybutyric acid) production by *Haloferax mediterranei*. Appl Environ Microbiol 56(8):2517–2521

Liu Y, Huang S, Zhang Y et al (2014) Isolation and characterization of a thermophilic *Bacillus shackletonii* K5 from a biotrickling filter for the production of polyhydroxybutyrate. J Environ Sci 26(7):1453–1462

Lomas AJ, Webb WR, Han J et al (2013) Poly (3-hydroxybutyrate-co-3-hydroxyhexanoate)/collagen hybrid scaffolds for tissue engineering applications. Tissue Eng Part C Methods 19(8):577–585

López-Cuellar M, Alba-Flores J, Rodríguez JG et al (2011) Production of polyhydroxyalkanoates (PHAs) with canola oil as carbon source. Int J Biol Macromol 48(1):74–80

Lu X, Zhang J, Wu Q et al (2003) Enhanced production of poly (3-hydroxybutyrate-co-3-hydroxyhexanoate) via manipulating the fatty acid β-oxidation pathway in *E. coli*. FEMS Microbiol Lett 221(1):97–101

Lu XY, Wu Q, Zhang WJ et al (2004) Molecular cloning of polyhydroxyalkanoate synthesis operon from *Aeromonas hydrophila* and its expression in *Escherichia coli*. Biotechnol Prog 20(5):1332–1336

Lu XY, Zhang Y, Wang L (2010) Preparation and in vitro drug-release behavior of 5-fluorouracil-loaded poly (hydroxybutyrate-co-hydroxyhexanoate) nanoparticles and microparticles. J Appl Polym Sci 116(5):2944–2950

Madison LL, Huisman GW (1999) Metabolic engineering of poly (3-hydroxyalkanoates): from DNA to plastic. Microbiol Mol Biol Rev 63(1):21–53

Manchak J, Page WJ (1994) Control of polyhydroxyalkanoate synthesis in *Azotobacter vinelandii* strain UWD. Microbiology 140(4):953–963

Mansfield DA, Anderson AJ, Naylor LA (1995) Regulation of PHB metabolism in *Alcaligenes eutrophus*. Can J Microbiol 41(13):44–49

Masaeli E, Wieringa PA, Morshed M et al (2014) Peptide functionalized polyhydroxyalkanoate nanofibrous scaffolds enhance Schwann cells activity. Nanomed Nanotechnol Biol Med 10(7):1559–1569

Masamune S, Palmer MA, Gamboni R et al (1989a) Bio-Claisen condensation catalyzed by thiolase from *Zoogloea ramigera*. Active site cysteine residues. J Am Chem Soc 111(5):1879–1881

Masamune S, Walsh C, Sinskey A et al (1989b) Poly-(R)-3-hydroxybutyrate (PHB) biosynthesis: mechanistic studies on the biological Claisen condensation catalyzed by β-ketoacyl thiolase. Pure Appl Chem 61(3):303–312

Masood F, Chen P, Yasin T et al (2013) Encapsulation of Ellipticine in poly-(3-hydroxybutyrate-co-3-hydroxyvalerate) based nanoparticles and its in vitro application. Mater Sci Eng C Mater Biol Appl 33(3):1054–1060

Masood F, Yasin T, Hameed A (2015) Polyhydroxyalkanoates–what are the uses? Current challenges and perspectives. Crit Rev Biotechnol 35(4):514–521

Matsumoto K, Matsusaki H, Taguchi S et al (2001) Cloning and characterization of the *Pseudomonas* sp. 61-3 phaG gene involved in polyhydroxyalkanoate biosynthesis. Biomacromolecules 2(1):142–147

Matsumoto K, Tanaka Y, Watanabe T et al (2013) Directed evolution and structural analysis of NADPH-dependent Acetoacetyl coenzyme A (Acetoacetyl-CoA) reductase from *Ralstonia eutropha* reveals two mutations responsible for enhanced kinetics. Appl Environ Microbiol 79(19):6134–6139

Matsusaki H, Abe H, Doi Y (2000) Biosynthesis and properties of poly (3-hydroxybutyrate-co-3-hydroxyalkanoates) by recombinant strains of *Pseudomonas* sp. 61-3. Biomacromolecules 1(1):17–22

Mauclaire L, Brombacher E, Bünger J et al (2010) Factors controlling bacterial attachment and biofilm formation on medium-chain-length polyhydroxyalkanoates (mcl-PHAs). Colloids Surf B Biointerfaces 76(1):104–111
McCool GJ, Cannon MC (2001) PhaC and PhaR are required for polyhydroxyalkanoic acid synthase activity in *Bacillus megaterium*. J Bacteriol 183(14):4235–4243
Mejía M, Segura D, Espín G et al (2010) Two-stage fermentation process for alginate production by *Azotobacter vinelandii* mutant altered in poly-β-hydroxybutyrate (PHB) synthesis. J Appl Microbiol 108(1):55–61
Mendes JBE, Riekes MK, de Oliveira VM et al (2012) PHBV/PCL microparticles for controlled release of resveratrol: physicochemical characterization, antioxidant potential, and effect on hemolysis of human erythrocytes. Sci World J 2012. doi:http://dx.doi.org/10.1100/2012/542937
Meng D-C, Shi Z-Y, Wu L-P et al (2012) Production and characterization of poly (3-hydroxypro pionate-co-4-hydroxybutyrate) with fully controllable structures by recombinant *Escherichia coli* containing an engineered pathway. Metab Eng 14(4):317–324
Mercan N, Beyatli Y (2005) Production of poly-beta-hydroxybutyrate (PHB) by *Rhizobium meliloti*, *R. viciae* and *Bradyrhizobium japonicum* with different carbon and nitrogen sources, and inexpensive substrates. Zuckerindustrie 130(5):410–415
Merrick J, Doudoroff M (1964) Depolymerization of poly-β-hydroxybutyrate by an intracellular enzyme system. J Bacteriol 88(1):60–71
Miková G, Chodák I (2006) Properties and modification of poly(3-hydroxybutanoate). Chem List 100(12):1075–1083
Mittendorf V, Robertson EJ, Leech RM et al (1998) Synthesis of medium-chain-length polyhydroxyalkanoates in *Arabidopsis thaliana* using intermediates of peroxisomal fatty acid β-oxidation. Proc Natl Acad Sci U S A 95(23):13397–13402
Modis Y, Wierenga RK (1999) A biosynthetic thiolase in complex with a reaction intermediate: the crystal structure provides new insights into the catalytic mechanism. Structure 7(10):1279–1290
Modis Y, Wierenga RK (2000) Crystallographic analysis of the reaction pathway of *Zoogloea ramigera* biosynthetic thiolase. J Mol Biol 297(5):1171–1182
Mumtaz T, Yahaya NA, Abd-Aziz S et al (2010) Turning waste to wealth-biodegradable plastics polyhydroxyalkanoates from palm oil mill effluent–a Malaysian perspective. J Clean Prod 18(14):1393–1402
Narayanan A, Ramana KV (2012) Polyhydroxybutyrate production in *Bacillus mycoides* DFC1 using response surface optimization for physico-chemical process parameters. 3. Biotech 2(4):287–296
Nath A, Dixit M, Bandiya A et al (2008) Enhanced PHB production and scale up studies using cheese whey in fed batch culture of *Methylobacterium* sp. ZP24. Bioresour Technol 99(13):5749–5755
Ng K-S, Wong Y-M, Tsuge T et al (2011) Biosynthesis and characterization of poly (3-hydroxyb utyrate-co-3-hydroxyvalerate) and poly (3-hydroxybutyrate-co-3-hydroxyhexanoate) copolymers using jatropha oil as the main carbon source. Process Biochem 46(8):1572–1578
Nikel PI, de Almeida A, Melillo EC et al (2006) New recombinant *Escherichia coli* strain tailored for the production of poly (3-hydroxybutyrate) from agroindustrial by-products. Appl Environ Microbiol 72(6):3949–3954
Nikel PI, Giordano AM, de Almeida A et al (2010) Elimination of D-lactate synthesis increases poly (3-hydroxybutyrate) and ethanol synthesis from glycerol and affects cofactor distribution in recombinant *Escherichia coli*. Appl Environ Microbiol 76(22):7400–7406
Nishioka M, Nakai K, Miyake M et al (2001) Production of poly-β-hydroxybutyrate by thermophilic cyanobacterium, *Synechococcus* sp. MA19, under phosphate-limited conditions. Biotechnol Lett 23(14):1095–1099
Nomura CT, Taguchi K, Taguchi S et al (2004a) Coexpression of genetically engineered 3-ketoacyl-ACP synthase III (fabH) and polyhydroxyalkanoate synthase (phaC) genes leads to short-chain-length-medium-chain-length polyhydroxyalkanoate copolymer production from glucose in *Escherichia coli* JM109. Appl Environ Microbiol 70(2):999–1007

Nomura CT, Tanaka T, Gan Z et al (2004b) Effective enhancement of short-chain-length-medium-chain-length polyhydroxyalkanoate copolymer production by coexpression of genetically engineered 3-Ketoacyl-acyl-carrier-protein synthase III (f abH) and polyhydroxyalkanoate synthesis genes. Biomacromolecules 5(4):1457–1464

Nomura CT, Taguchi K, Gan Z et al (2005) Expression of 3-ketoacyl-acyl carrier protein reductase (fabG) genes enhances production of polyhydroxyalkanoate copolymer from glucose in recombinant *Escherichia coli* JM109. Appl Environ Microbiol 71(8):4297–4306

Normi YM, Hiraishi T, Taguchi S et al (2005) Characterization and properties of G4X mutants of *Ralstonia eutropha* PHA synthase for poly (3-hydroxybutyrate) biosynthesis in *Escherichia coli*. Macromol Biosci 5(3):197–206

Novikov LN, Novikova LN, Mosahebi A et al (2002) A novel biodegradable implant for neuronal rescue and regeneration after spinal cord injury. Biomaterials 23(16):3369–3376

Obruca S, Marova I, Melusova S et al (2011) Production of polyhydroxyalkanoates from cheese whey employing *Bacillus megaterium* CCM 2037. Ann Microbiol 61(4):947–953

Ojumu T, Yu J, Solomon B (2004) Production of polyhydroxyalkanoates, a bacterial biodegradable polymers. Afr J Biotechnol 3(1):18–24

Okamura K, Marchessault R (1967) X-ray structure of poly-β-hydroxybutyrate. In: Ramachandran G (ed) Conformation of biopolymers, vol 2, pp 709–720. doi:10.1016/B978-1-4832-2843-3.50023-6

Ouyang S-P, Luo RC, Chen S-S et al (2007) Production of polyhydroxyalkanoates with high 3-hydroxydodecanoate monomer content by fadB and fadA knockout mutant of *Pseudomonas putida* KT2442. Biomacromolecules 8(8):2504–2511

Palmer M, Differding E, Gamboni R et al (1991) Biosynthetic thiolase from *Zoogloea ramigera*. Evidence for a mechanism involving Cys-378 as the active site base. J Biol Chem 266(13):8369–8375

Panda B, Sharma L, Mallick N (2005) Poly-β-hydroxybutyrate accumulation in *Nostoc muscorum* and *Spirulina platensis* under phosphate limitation. J Plant Physiol 162(12):1376–1379

Park SJ, Lee SY (2004) Biosynthesis of poly (3-hydroxybutyrate-co-3-hydroxyalkanoates) by metabolically engineered *Escherichia coli* strains. Appl Biochem Biotechnol 114(1–3):335–346

Park SJ, J-i C, Lee SY (2005) Engineering of *Escherichia coli* fatty acid metabolism for the production of polyhydroxyalkanoates. Enzym Microb Technol 36(4):579–588

Peña C, Castillo T, García A et al (2014) Biotechnological strategies to improve production of microbial poly-(3-hydroxybutyrate): a review of recent research work. Microb Biotechnol 7(4):278–293

Peng S-W, Guo X-Y, Shang G-G et al (2011) An assessment of the risks of carcinogenicity associated with polyhydroxyalkanoates through an analysis of DNA aneuploid and telomerase activity. Biomaterials 32(10):2546–2555

Peoples OP, Sinskey AJ (1989) Poly-beta-hydroxybutyrate (PHB) biosynthesis in *Alcaligenes eutrophus* H16. Identification and characterization of the PHB polymerase gene (phbC). J Biol Chem 264(26):15298–15303

Philip S, Keshavarz T, Roy I (2007) Polyhydroxyalkanoates: biodegradable polymers with a range of applications. J Chem Technol Biotechnol 82(3):233–247

Phithakrotchanakoon C, Champreda V, S-i A et al (2013) Engineered *Escherichia coli* for short-chain-length medium-chain-length polyhydroxyalkanoate copolymer biosynthesis from glycerol and dodecanoate. Biosci Biotechnol Biochem 77(6):1262–1268

Qi Q, Rehm BH (2001) Polyhydroxybutyrate biosynthesis in *Caulobacter crescentus*: molecular characterization of the polyhydroxybutyrate synthase. Microbiology 147(12):3353–3358

Quillaguaman J, Hashim S, Bento F et al (2005) Poly (β-hydroxybutyrate) production by a moderate halophile, *Halomonas boliviensis* LC1 using starch hydrolysate as substrate. J Appl Microbiol 99(1):151–157

Raberg M, Bechmann J, Brandt U et al (2011) Versatile metabolic adaptations of *Ralstonia eutropha* H16 to a loss of PdhL, the E3 component of the pyruvate dehydrogenase complex. Appl Environ Microbiol 77(7):2254–2263

Rai R, Keshavarz T, Roether J et al (2011a) Medium chain length polyhydroxyalkanoates, promising new biomedical materials for the future. Mater Sci Eng R 72(3):29–47
Rai R, Yunos DM, Boccaccini AR et al (2011b) Poly-3-hydroxyoctanoate P (3HO), a medium chain length polyhydroxyalkanoate homopolymer from *Pseudomonas mendocina*. Biomacromolecules 12(6):2126–2136
Rao U, Sridhar R, Sehgal P (2010) Biosynthesis and biocompatibility of poly (3-hydroxybutyrate-co-4-hydroxybutyrate) produced by *Cupriavidus necator* from spent palm oil. Biochem Eng J 49(1):13–20
Rathbone S, Furrer P, Lübben J et al (2010) Biocompatibility of polyhydroxyalkanoate as a potential material for ligament and tendon scaffold material. J Biomed Mater Res A 93(4):1391–1403
Reddy MV, Nikhil G, Mohan SV et al (2012) *Pseudomonas otitidis* as a potential biocatalyst for polyhydroxyalkanoates (PHA) synthesis using synthetic wastewater and acidogenic effluents. Bioresour Technol 123:471–479
Rehm BH (2003) Polyester synthases: natural catalysts for plastics. Biochem J 376(1):15–33
Rehm BH (2015) Polyhydroxyalkanoates. http://lipidlibrary.aocs.org/Biochemistry/content.cfm?Item Number=41298. Accessed 25 Feb 2016
Rehm BH, Steinbüchel A (1999) Biochemical and genetic analysis of PHA synthases and other proteins required for PHA synthesis. Int J Biol Macromol 25(1):3–19
Rehm BH, Qingsheng Q, Beermann BB et al (2001) Matrix-assisted in vitro refolding of *Pseudomonas aeruginosa* class II polyhydroxyalkanoate synthase from inclusion bodies produced in recombinant *Escherichia coli*. Biochem J 358(1):263–268
Rehm BH, Antonio RV, Spiekermann P et al (2002) Molecular characterization of the poly (3-hydroxybutyrate)(PHB) synthase from *Ralstonia eutropha*: in vitro evolution, site-specific mutagenesis and development of a PHB synthase protein model. Biochim Biophys Acta 1594(1):178–190
Ren Q, Sierro N, Witholt B et al (2000) FabG, an NADPH-dependent 3-ketoacyl reductase of *Pseudomonas aeruginosa*, provides precursors for medium-chain-length poly-3-hydroxyalkanoate biosynthesis in *Escherichia coli*. J Bacteriol 182(10):2978–2981
Ren Q, De Roo G, Van Beilen JB et al (2005) Poly (3-hydroxyalkanoate) polymerase synthesis and in vitro activity in recombinant *Escherichia coli* and *Pseudomonas putida*. Appl Microbiol Biotechnol 69(3):286–292
Reusch RN (1995) Low molecular weight complexed poly (3-hydroxybutyrate): a dynamic and versatile molecule in vivo. Can J Microbiol 41(13):50–54
Ritchie G, Senior P, Dawes E (1971) The purification and characterization of acetoacetyl-coenzyme A reductase from *Azotobacter beijerinckii*. Biochem J 121(2):309–316
Rocha RC, da Silva LF, Taciro MK et al (2008) Production of poly (3-hydroxybutyrate-co-3-hydroxyvalerate) P (3HB-co-3HV) with a broad range of 3HV content at high yields by *Burkholderia sacchari* IPT 189. World J Microbiol Biotechnol 24(3):427–431
Ruan W, Chen J, Lun S (2003) Production of biodegradable polymer by *A. eutrophus* using volatile fatty acids from acidified wastewater. Process Biochem 39(3):295–299
Saito T, Fukui T, Ikeda F et al (1977) An NADP-linked acetoacetyl CoA reductase from *Zoogloea ramigera*. Arch Microbiol 114(3):211–217
Sathiyanarayanan G, Kiran GS, Selvin J et al (2013) Optimization of polyhydroxybutyrate production by marine *Bacillus megaterium* MSBN04 under solid state culture. Int J Biol Macromol 60:253–261
Sato S, Kanazawa H, Tsuge T (2011) Expression and characterization of (R)-specific enoyl coenzyme A hydratases making a channeling route to polyhydroxyalkanoate biosynthesis in *Pseudomonas putida*. Appl Microbiol Biotechnol 90(3):951–959
Satoh H, Mino T, Matsuo T (1999) PHA production by activated sludge. Int J Biol Macromol 25(1):105–109
Schubert P, Steinbüchel A, Schlegel HG (1988) Cloning of the *Alcaligenes eutrophus* genes for synthesis of poly-beta-hydroxybutyric acid (PHB) and synthesis of PHB in *Escherichia coli*. J Bacteriol 170(12):5837–5847

Schubert P, Krüger N, Steinbüchel A (1991) Molecular analysis of the *Alcaligenes eutrophus* poly (3-hydroxybutyrate) biosynthetic operon: identification of the N terminus of poly (3-hydroxybutyrate) synthase and identification of the promoter. J Bacteriol 173(1):168–175

Shabna A, Saranya V, Malathi J et al (2014) Indigenously produced polyhydroxyalkanoate based co-polymer as cellular supportive biomaterial. J Biomed Mater Res A 102(10):3470–3476

Shah M, Naseer MI, Choi MH et al (2010) Amphiphilic PHA–mPEG copolymeric nanocontainers for drug delivery: preparation, characterization and in vitro evaluation. Int J Pharm 400(1):165–175

Shah M, Ullah N, Choi MH et al (2012) Amorphous amphiphilic P (3HV-co-4HB)-b-mPEG block copolymer synthesized from bacterial copolyester via melt transesterification: nanoparticle preparation, cisplatin-loading for cancer therapy and in vitro evaluation. Eur J Pharm Biopharm 80(3):518–527

Shimamura E, Kasuya K, Kobayashi G et al (1994a) Physical properties and biodegradability of microbial poly (3-hydroxybutyrate-co-3-hydroxyhexanoate). Macromolecules 27(3):878–880

Shimamura E, Scandola M, Doi Y (1994b) Microbial synthesis and characterization of poly (3-hydroxybutyrate-co-3-hydroxypropionate). Macromolecules 27(16):4429–4435

Shuto H, Fukul T, Saito T et al (1981) An NAD-linked acetoacetyl-CoA reductase from *Zoogloea ramigera* I-16-M. Eur J Biochem 118(1):53–59

Sindhu R, Silviya N, Binod P et al (2013) Pentose-rich hydrolysate from acid pretreated rice straw as a carbon source for the production of poly-3-hydroxybutyrate. Biochem Eng J 78:67–72

Singh A, Mallick N (2008) Enhanced production of SCL-LCL-PHA co-polymer by sludge-isolated *Pseudomonas aeruginosa* MTCC 7925. Lett Appl Microbiol 46(3):350–357

Slater SC, Voige W, Dennis D (1988) Cloning and expression in *Escherichia coli* of the *Alcaligenes eutrophus* H16 poly-beta-hydroxybutyrate biosynthetic pathway. J Bacteriol 170(10):4431–4436

Slater S, Houmiel KL, Tran M et al (1998) Multiple β-ketothiolases mediate poly (β-hydroxyalkanoate) copolymer synthesis in *Ralstonia eutropha*. J Bacteriol 180(8):1979–1987

Steinbüchel A (1991) Polyhydroxyalkanoic acids. In: Biomaterials. Springer, pp 123–213

Steinbüchel A, Valentin HE (1995) Diversity of bacterial polyhydroxyalkanoic acids. FEMS Microbiol Lett 128(3):219–228

Steinbüchel A, Lütke-Eversloh T (2003) Metabolic engineering and pathway construction for biotechnological production of relevant polyhydroxyalkanoates in microorganisms. Biochem Eng J 16(2):81–96

Su Z, Li P, Wu B et al (2014) PHBVHHx scaffolds loaded with umbilical cord-derived mesenchymal stem cells or hepatocyte-like cells differentiated from these cells for liver tissue engineering. Mater Sci Eng C Mater Biol Appl 45:374–382

Sudesh K, Abe H, Doi Y (2000) Synthesis, structure and properties of polyhydroxyalkanoates: biological polyesters. Prog Polym Sci 25(10):1503–1555

Suriyamongkol P, Weselake R, Narine S et al (2007) Biotechnological approaches for the production of polyhydroxyalkanoates in microorganisms and plants-a review. Biotechnol Adv 25(2):148–175

Tan G-YA, Chen C-L, Li L et al (2014) Start a research on biopolymer polyhydroxyalkanoate (PHA): a review. Polymers 6(3):706–754

Thompson S, Mayerl F, Peoples OP et al (1989) Mechanistic studies on. Beta.-ketoacyl thiolase from *Zoogloea ramigera*: identification of the active-site nucleophile as Cys89, its mutation to Ser89, and kinetic and thermodynamic characterization of wild-type and mutant enzymes. Biochemistry 28(14):5735–5742

Timm A, Steinbüchel A (1990) Formation of polyesters consisting of medium-chain-length 3-hydroxyalkanoic acids from gluconate by *Pseudomonas aeruginosa* and other fluorescent pseudomonads. Appl Environ Microbiol 56(11):3360–3367

Tomizawa S, Hyakutake M, Saito Y et al (2011) Molecular weight change of polyhydroxyalkanoate (PHA) caused by the PhaC subunit of PHA synthase from *Bacillus cereus* YB-4 in recombinant *Escherichia coli*. Biomacromolecules 12(7):2660–2666

Tripathi AD, Srivastava SK, Singh RP (2013) Statistical optimization of physical process variables for bio-plastic (PHB) production by *Alcaligenes* sp. Biomass Bioenergy 55:243–250
Tsuge T, Taguchi K, Doi Y (2003) Molecular characterization and properties of (R)-specific enoyl-CoA hydratases from *Pseudomonas aeruginosa*: metabolic tools for synthesis of polyhydroxy-alkanoates via fatty acid ß-oxidation. Int J Biol Macromol 31(4):195–205
Tsuge T, Hyakutake M, Mizuno K (2015) Class IV polyhydroxyalkanoate (PHA) synthases and PHA-producing *Bacillus*. Appl Microbiol Biotechnol 99(15):6231–6240
Urtuvia V, Villegas P, González M et al (2014) Bacterial production of the biodegradable plastics polyhydroxyalkanoates. Int J Biol Macromol 70:208–213
Valentin H, Dennis D (1996) Metabolic pathway for poly (3-hydroxybutyrate-co-3-hydroxyvalerate) formation in *Nocardia corallina*: inactivation of mutB by chromosomal integration of a kanamycin resistance gene. Appl Environ Microbiol 62(2):372–379
Valentin HE, Dennis D (1997) Production of poly (3-hydroxybutyrate-co-4-hydroxybutyrate) in recombinant *Escherichia coli* grown on glucose. J Biotechnol 58(1):33–38
Valentin HE, Steinbüchel A (1995) Accumulation of poly (3-hydroxybutyric acid-co-3-hydroxyvaleric acid-co-4-hydroxyvaleric acid) by mutants and recombinant strains of *Alcaligenes eutrophus*. J Environ Polymer Degradation 3(3):169–175
Vigneswari S, Nik LA, Majid M et al (2010) Improved production of poly (3-hydroxybutyrate-co-4-hydroxbutyrate) copolymer using a combination of 1, 4-butanediol and γ-butyrolactone. World J Microbiol Biotechnol 26(4):743–746
Wang F, Lee SY (1997) Production of poly (3-hydroxybutyrate) by fed-batch culture of filamentation-suppressed recombinant *Escherichia coli*. Appl Environ Microbiol 63(12):4765–4769
Wang Y, Bian Y-Z, Wu Q et al (2008) Evaluation of three-dimensional scaffolds prepared from poly (3-hydroxybutyrate-co-3-hydroxyhexanoate) for growth of allogeneic chondrocytes for cartilage repair in rabbits. Biomaterials 29(19):2858–2868
Wang H-h, X-t L, Chen G-Q (2009) Production and characterization of homopolymer polyhydroxyheptanoate (P3HHp) by a fadBA knockout mutant *Pseudomonas putida* KTOY06 derived from *P. putida* KT2442. Process Biochem 44(1):106–111
Wang L, Wang Z-H, Shen C-Y et al (2010) Differentiation of human bone marrow mesenchymal stem cells grown in terpolyesters of 3-hydroxyalkanoates scaffolds into nerve cells. Biomaterials 31(7):1691–1698
Wang A, Gan Y, Qu J et al (2012a) Application of Electrospun poly[(R)-3-hydroxybutyrate-co-(R)-3-hydroxyvalerate]-Ecoflex Mats in periodontal regeneration: a primary study. Paper presented at the Proceedings of the 2012 International Conference on Biomedical Engineering and Biotechnology
Wang Q, Tappel RC, Zhu C et al (2012b) Development of a new strategy for production of medium-chain-length polyhydroxyalkanoates by recombinant *Escherichia coli* via inexpensive non-fatty acid feedstocks. Appl Environ Microbiol 78(2):519–527
Wang B, Sharma-Shivappa RR, Olson JW et al (2013a) Production of polyhydroxybutyrate (PHB) by *Alcaligenes latus* using sugarbeet juice. Ind Crop Prod 43:802–811
Wang Y, Chen R, Cai J et al (2013b) Biosynthesis and thermal properties of PHBV produced from levulinic acid by *Ralstonia eutropha*. PLoS One 8(4):e60318. doi:http://dx.doi.org/10.1371/journal.pone.0060318
Wang H, He X-Q, Jin T et al (2016) Wnt11 plays an important role in the osteogenesis of human mesenchymal stem cells in a PHA/FN/ALG composite scaffold: possible treatment for infected bone defect. Stem Cell Res Ther 7(1):1–13. doi:10.1186/s13287-016-0277-4
Webb WR, Dale TP, Lomas AJ et al (2013) The application of poly (3-hydroxybutyrate-co-3-hydroxyhexanoate) scaffolds for tendon repair in the rat model. Biomaterials 34(28):6683–6694
Witkowski A, Joshi AK, Smith S (1997) Characterization of the interthiol acyltransferase reaction catalyzed by the β-ketoacyl synthase domain of the animal fatty acid synthase. Biochemistry 36(51):16338–16344
Wodzinska J, Snell K, Rhomberg A et al (1996) Polyhydroxybutyrate synthase: evidence for covalent catalysis. J Am Chem Soc 118(26):6319–6320

Xie WP, Chen G-Q (2008) Production and characterization of terpolyester poly (3-hydroxybutyrate-co-4-hydroxybutyrate-co-3-hydroxyhexanoate) by recombinant *Aeromonas hydrophila* 4AK4 harboring genes phaPCJ. Biochem Eng J 38(3):384–389

Xie H, Li J, Li L et al (2013) Enhanced proliferation and differentiation of neural stem cells grown on PHA films coated with recombinant fusion proteins. Acta Biomater 9(8):7845–7854

Xiong Y-C, Yao Y-C, Zhan X-Y et al (2010) Application of polyhydroxyalkanoates nanoparticles as intracellular sustained drug-release vectors. J Biomater Sci Polym Ed 21(1):127–140

Xu X-Y, Li X-T, Peng S-W et al (2010) The behaviour of neural stem cells on polyhydroxyalkanoate nanofiber scaffolds. Biomaterials 31(14):3967–3975

Xu F, Huang S, Liu Y et al (2014) Comparative study on the production of poly (3-hydroxybutyrate) by thermophilic *Chelatococcus daeguensis* TAD1: a good candidate for large-scale production. Appl Microbiol Biotechnol 98(9):3965–3974

Yabutani T, Maehara A, Ueda S et al (1995) Analysis of β-ketothiolase and acetoacetyl-CoA reductase genes of a methylotrophic bacterium, *Paracoccus denitrifleans*, and their expression in *Escherichia coli*. FEMS Microbiol Lett 133(1–2):85–90

Ying TH, Ishii D, Mahara A et al (2008) Scaffolds from electrospun polyhydroxyalkanoate copolymers: fabrication, characterization, bioabsorption and tissue response. Biomaterials 29(10):1307–1317

You M, Peng G, Li J et al (2011) Chondrogenic differentiation of human bone marrow mesenchymal stem cells on polyhydroxyalkanoate (PHA) scaffolds coated with PHA granule binding protein PhaP fused with RGD peptide. Biomaterials 32(9):2305–2313

Yuan M-Q, Shi Z-Y, Wei X-X et al (2008) Microbial production of medium-chain-length 3-hydroxyalkanoic acids by recombinant *Pseudomonas putida* KT2442 harboring genes fadL, fadD and phaZ. FEMS Microbiol Lett 283(2):167–175

Zhang C, Zhao L, Dong Y et al (2010) Folate-mediated poly (3-hydroxybutyrate-co-3-hydroxyoctanoate) nanoparticles for targeting drug delivery. Eur J Pharm Biopharm 76(1):10–16

Zhang Y, Sun W, Wang H et al (2013) Polyhydroxybutyrate production from oil palm empty fruit bunch using *Bacillus megaterium* R11. Bioresour Technol 147:307–314

Zhang C, Zhang Z, Zhao L (2015a) Folate-decorated poly (3-hydroxybutyrate-co-3-hydroxyoctanoate) nanoparticles for targeting delivery: optimization and in vivo antitumor activity. Drug Deliv:1–8

Zhang W, Chen C, Cao R et al (2015b) Inhibitors of polyhydroxyalkanoate (PHA) synthases: synthesis, molecular docking, and implications. Chembiochem 16(1):156–166

Zheng Z, Zhang M-J, Zhang G et al (2004) Production of 3-hydroxydecanoic acid by recombinant *Escherichia coli* HB101 harboring phaG gene. Antonie Van Leeuwenhoek 85(2):93–101

Zheng LZ, Li Z, Tian H-L et al (2005) Molecular cloning and functional analysis of (R)-3-hydroxyacyl-acyl carrier protein: coenzyme A transacylase from *Pseudomonas mendocina* LZ. FEMS Microbiol Lett 252(2):299–307

Zhou Q, Shi Z-Y, Meng D-C et al (2011) Production of 3-hydroxypropionate homopolymer and poly (3-hydroxypropionate-co-4-hydroxybutyrate) copolymer by recombinant *Escherichia coli*. Metab Eng 13(6):777–785

Zhu XH, Wang CH, Tong YW (2007) Growing tissue-like constructs with Hep3B/HepG2 liver cells on PHBV microspheres of different sizes. J Biomed Mater Res B Appl Biomater 82(1):7–16

Zhu C, Chiu S, Nakas JP et al (2013) Bioplastics from waste glycerol derived from biodiesel industry. J Appl Polym Sci 130(1):1–13

Zhuang Q, Wang Q, Liang Q et al (2014) Synthesis of polyhydroxyalkanoates from glucose that contain medium-chain-length monomers via the reversed fatty acid β-oxidation cycle in *Escherichia coli*. Metab Eng 24:78–86

Zinn M, Witholt B, Egli T (2001) Occurrence, synthesis and medical application of bacterial polyhydroxyalkanoate. Adv Drug Del Rev 53(1):5–21

Aneesh Balakrishna Pillai completed his B.Sc. (Botany & Biotechnology) from University of Kerala, Thiruvananthapuram and M.Sc. (Biotechnology) from Mahatma Gandhi University, Kottayam, Kerala. Presently, he is working for his Ph.D. at Rajiv Gandhi Centre for Biotechnology with Dr. H.K. Kumarapillai in the area of bacterial polyhydroxyalkanoates.

Hari Krishnan Kumarapillai is M.Sc. (Zoology) from Kanpur University, Uttar Pradesh, India and Ph.D. (Aquatic Biology) from University of Kerala, India. His current research areas are metagenomics, microbial biotechnology and bacterial biopolymers.

Biosurfactant-Aided Bioprocessing: Industrial Applications and Environmental Impact

3

Reetika Sharma and Harinder Singh Oberoi

Abstract

Surfactants are classified as ionic, nonionic, and zwitterionic surfactants based on the ionic properties of the polar head group. Biosurfactants are surface-active compounds produced by microbes, possessing both hydrophilic and hydrophobic moieties. In biosurfactants, the lipophilic moiety is generally a protein or peptide with a high fraction of hydrophobic side chains or a hydrocarbon chain of a fatty acid with 10 to 18 carbon atoms, whereas the hydrophilic moiety is an ester; hydroxyl, phosphate, and carboxylate group; or sugar. Biosurfactants have specific advantages over chemical surfactants, such as biodegradable and environmental-friendly nature, production at lower temperatures, effectiveness at low concentrations, low toxicity, high selectivity because of the presence of specific functional groups, and efficiency to work at extreme environmental conditions of temperatures, pH, and salinity, rendering them suitable for different industrial applications. However, large-scale commercial application of biosurfactants is impeded because of their high production costs, ineffective bioprocessing methods, less efficient microbial strains, and the exorbitant downstream processing costs. Biosurfactants find potential industrial application in areas, such as disruption of cell biomass, hydrocarbon bioremediation, and heavy metal bioremediation. Different groups of microbes, such as bacteria, yeasts, fungi, and actinomycetes are capable of producing biosurfactants. Some of the extensively studied biosurfactant producing microbial genera include *Pseudomonas,*

R. Sharma
Department of Environmental Science, Dr YS Parmar University of Horticulture and Forestry, Solan 173230, Himachal Pradesh, India
e-mail: sharma_reetika@yahoo.com; reet_sharma33@ymail.com

H.S. Oberoi (✉)
Division of Post Harvest Technology and Agricultural Engineering, ICAR- Indian Institute of Horticultural Research, Bengaluru 560089, India
e-mail: hari_manu@yahoo.com; harinder@iihr.res.in

P. Shukla (ed.), *Recent Advances in Applied Microbiology*,
DOI 10.1007/978-981-10-5275-0_3

Acinetobacter, Bacillus, Candida and *Torulopsis*. Development of improved and cost-efficient application technologies coupled with genetic engineering and strain improvement techniques and improved production processes will help in large-scale application of biosurfactants in the near future.

Keywords
Biosurfactants • Bioprocessing • Bioremediation • Lignocellulosic biomass • Production cost • Rhamnolipids • Sophorolipids

3.1 Introduction

Bioprocess employs complete living cells (microbes) or their metabolites (enzymes) for getting the desired end product. Industrial biotechnology deals with the application of biotechnological tools in different industrial processes, including bioprocessing and production of value-added products from renewable feedstocks. Manipulation of microbes and their physiology through the application of genetic engineering tools facilitates the development of new and cost-effective, cleaner, and environmentally friendly industrial manufacturing processes. Substantial amount of biomass generated from different agro-processing industries is not being commercially exploited, because of lack of infrastructure for collection, handling, and management of such a biomass. Despite being rich in nutrients, this valuable biomass is burnt leading to environmental pollution problems and loss of important rich resource. Thus, this enormous biomass considered as "waste" possesses great potential to be converted into a great variety of value-added products, such as biofuels, animal feeds, and human nutrients, which besides mitigating the greenhouse effect and atmospheric pollution problems is likely to help in better biomass management (Pothiraj et al. 2006).

According to a recent survey, global consumption of oil rose from 1.1 in 2014 to 1.9 million barrels per day (b/d) in 2016 (BP Statistical Review of World Energy 2016). In addition to the rising global oil demand, there are growing concerns about diminishing known petroleum reserves and the adverse effects of atmospheric greenhouse gases (GHG). This has resulted in search for alternative sources of energy and renewed interest in the production of fuels from plants or organic wastes, termed as "biofuels."

Biofuels mainly derived from food crops, such as corn, sugarcane, soybean, vegetable oil, etc., are regarded as "first-generation biofuels." In addition to their potential of being used after blending with petroleum-based fuels, they face many challenges including food-versus-fuel debate resulting in hike in food prices, production of food crops leading to change in land use pattern, and potential increase in GHG emissions. Moreover, the economic feasibility of the processes implied for first-generation biofuel production is based on the type of feedstock and the area of cultivation of that particular feedstock (Naqvi and Yan 2015). Therefore,

Table 3.1 Comparison of merits and demerits of first- and second-generation biofuel and petroleum fuels

	Petroleum refinery	First-generation biofuels	Second-generation biofuels
Feedstocks	Crude petroleum	Food crops, vegetable oils, corn sugar, etc.	Agricultural and forest residues
Products	Diesel, kerosene, jet fuel	Biodiesel, corn ethanol, etc.	Bio-oil, ethanol, butanol, mixed alcohols, etc.
Benefits	Major fuel	Environment friendly	Environment friendly
		Economic and social security	Nonfood cheap, abundant biomass
			Grown on marginal lands
Demerits	Expensive technology	Expensive technology	Efficient technology, development is still under progress
	Nonrenewable sources	Significant of land requirement	Renewable source

second-generation biofuels largely produced from lignocellulosic biomass are cheap and abundantly available as nonfood materials which offer a potential alternative to meet the growing global fuel demands. In addition, second-generation biofuels also provide several benefits to the society, such that they (i) are renewable and sustainable, (ii) help to mitigate the greenhouse gases (GHG) emission in the atmosphere, (iii) facilitate development of local economy through creation of job opportunities, (iv) reduce air pollution caused by burning or rotting of biomass in fields, and (v) ensure energy security for countries dependent on oil being imported from the other countries (Greenwell et al. 2012; Lee and Lavoie 2013). Comparison on advantages and disadvantages of the first-generation biofuels, petroleum fuels, and second-generation biofuels is presented in Table 3.1 which corroborates the importance of the second-generation biofuels.

There are two routes for converting biomass to biofuels: "thermochemical" route, commonly known as biomass-to-liquid (BTL) conversion process, and the "biochemical" route. In the case of BTL, the biomass is subjected to pyrolysis or gasification to generate syngas (composed of carbon monoxide and hydrogen) which is subsequently converted to fuels using either a catalytic process, such as the Fischer-Tropsch reactions, or by a biological conversion method (Balan 2014). In "biochemical" route, holocellulose content (cellulose and hemicellulose) available in the biomass is converted to monomeric sugars, which can be further utilized for the production of valuable compounds, like fuels, organic acids, etc. It is estimated that in economic terms, second-generation biofuel production processes are two to three times more expensive than the petroleum fuels on an energy-equivalent basis and over five times that of similar capacity first-generation bioethanol plants (Wright and Brown 2007; Carriquiry et al. 2011). Several challenges in the areas of bioprocessing of feedstocks, such as feedstock production and logistics, development of energy-efficient technologies for pretreatment and enzymatic hydrolysis of biomass, upstream and downstream processing cost, biofuel distribution, its acceptance in the society, and environmental impacts, need to be addressed to alleviate the fears and to lower down the production costs of the second-generation biofuels (Luo et al. 2010; Menon and Rao 2012).

A consortium of hydrolytic enzymes known as molecular scissors is required to cleave the complex network of cellulose, hemicellulose, and lignin in the lignocellulosic biomass to produce specific monomeric sugars (Zhang et al. 2012). Lower enzyme quantity and enzyme efficiency and higher enzyme costs required to hydrolyze lignocellulosic biomass are the limiting factors in bioprocessing applications. In the recent years, companies manufacturing enzymes commercially have made significant progress in overcoming these problems using different biotechnological and process engineering approaches (Alvira et al. 2013). Development of an efficient and potent process for production of ethanol as a liquid fuel after conversion of cellulose in the lignocellulosic biomass depends on many factors, such as lignin content in biomass, pretreatment effectiveness, cellulose crystallinity, substrate concentration, and productive hydrolysis of cellulose into monomeric units (Jeoh et al. 2007; Hall et al. 2010; Zeng et al. 2014). Moreover, high enzyme concentrations are needed to achieve efficient hydrolysis of biomass, thereby increasing the processing costs. The major difficulties surface up during the recycling of enzymes adsorbed on the residual lignocellulosic material. Thus, different ways to enhance the enzyme efficacy are required to reduce their consumption during hydrolysis, maintaining higher sugar productivity.

Surfactants are amphiphilic surface-active agents having both hydrophilic as well as hydrophobic entities that lessen surface tension between two immiscible fluids after accumulating at their interface (Fig. 3.1). The term "Surfactants" was created and registered as a trademark for the first time by the General Aniline and Film Corp for their surface-active products (GAF 1950; Schramm et al. 2003) and was later released in the public domain (Stevens 1969).

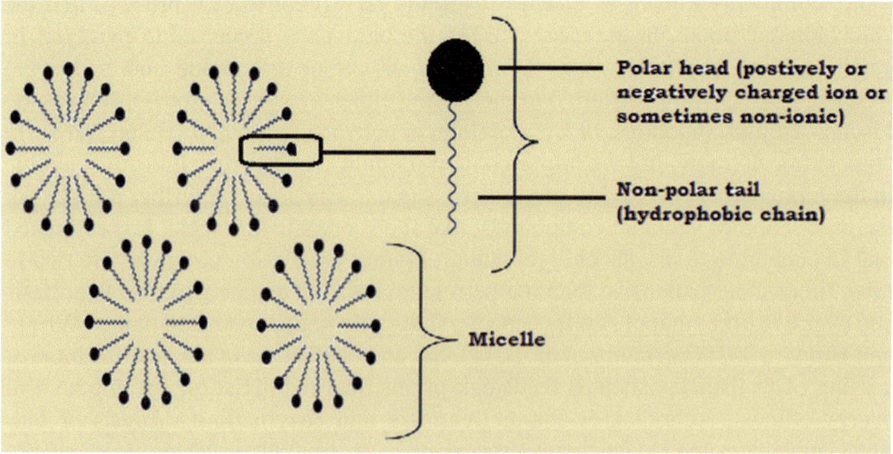

Fig. 3.1 Structure of several surfactant molecules forming micelles

Table 3.2 List of global biosurfactant manufacturers

S. No.	Name of the manufacturer	Biosurfactant manufactured
1	Kao Co Ltd., Japan	Sophorolipids
2	Iwata Chemical Co. Ltd., Japan	Rhamnolipids
3	Wako Pure Chemical Industries, USA	Surfactin
4	MG Intobio Co. Ltd., South Korea	Rhamnolipids
5	Jeneil Biosurfactant Company, USA	
6	Groupe Soliance, France	Sophorolipids
7	Ecover, Belgium	Sophorolipids
8	AGAE Technologies LLC, USA	Rhamnolipids
9	Apollo Biolife, India	Sophorolipids

Surfactants can either be of synthetic origin or biological origin (biosurfactants). Biosurfactants present an edge over chemical surfactants due to their numerous advantages, such as biodegradable and environment-friendly nature, production at lower temperatures, effectiveness at low concentrations, low toxicity, high selectivity because of the presence of specific functional groups, and efficiency to work at extreme environmental conditions of temperatures, pH, and salinity, which renders them suitable for different industrial applications (Kapadia and Yagnik 2013; Santos et al. 2013). However, their large-scale commercial application is hindered due to their high production costs, ineffective methods of bioprocessing, less efficient microbial strains, and the exorbitant downstream processing costs (Marchant and Banat 2012; Campos et al. 2013; Banat et al. 2014). Although, there is a restricted commercial production of biosurfactants due to huge production cost and low yields, in the recent times, new microbial strains and process interventions have helped in improving their productivity by 10–20-fold. Some of the companies manufacturing different types of biosurfactants are listed in Table 3.2.

Due to their important characteristics, surfactants are in high demand for bioprocessing applications (Saharan et al. 2011). This is largely because of the increasing interest in surfactant-aided hydrolysis of cellulose (Hseih et al. 2015; Min et al. 2015; Li et al. 2016). Different mechanisms, such as surfactant adsorption on air-liquid interface preventing enzyme denaturation, increase in available cellulose surface, and/or removal of inhibitory lignin, have been suggested for efficient enzymatic hydrolysis of cellulose (Min et al. 2015; Li et al. 2016). Based on kinetic analysis, surfactants can also increase the availability of reaction sites, leading to enhanced hydrolysis rate (Samiey et al. 2014).

Therefore, the aim of this chapter is to provide an insight into various possible mechanisms for action of biosurfactants and their application in bioprocessing along with their futuristic potential. Information presented in the chapter also provides an insight into the development of an efficient bioprocess for conversion of lignocellulose into biofuel and sets a platform for research on biosurfactant-aided pretreatment and hydrolysis for production of second-generation bioethanol from lignocellulosic biomass.

3.2 Surfactants and Their Classification

Surfactants are characteristic organic compounds, possessing both water-insoluble hydrophobic groups (tails) and water-soluble hydrophilic groups (heads) that decrease the surface tension at the interface of two liquids or liquid and solid (Saharan et al. 2011). Surfactants have diverse industrial applications, such as in food processing industries, agrochemical and pharmaceutical products, personal care and laundry products, petroleum, fuel additives, lubricants, paints, coatings and adhesives, photographic films, biological systems and various medical practices, soil remediation techniques, and also other environment-friendly methods (Schramm et al. 2003; Mishra et al. 2009).

It is estimated that during 2015–2020, world consumption of surfactants will continue to expand at an average annual rate of 1–5% (IHS 2016). Global surfactant market is calculated to grow by value at a compound annual growth rate (CAGR) of 5.5% and is predicted to reach a volume of 24,037.3 KT corresponding to $42,120.4 million in monetary terms by 2020. In terms of volumes, the anionic surfactants ruled the global market, with 7686.1 KT, followed by the nonionic surfactants at 6345.7 KT in 2014. Additionally, share of the amphoteric surfactants that represent 7.2% of the global surfactant market in 2014 is predicted to rise at the highest CAGR in the duration of 2015–2020 (www.marketsandmarkets.com/Market-Reports/biosurfactants-market-493.html).

On the basis of the ionic properties of the polar head group, surfactants are categorized into different groups, such as ionic (anionic and cationic), nonionic, and zwitterionic. Cationic surfactants carry a positive charge; on the other hand, a negative charge is present on the polar head groups of anionic surfactants. Zwitterionic or amphoteric surfactants possess both positive and negative charges depending on the environment in which they are present, whereas nonionic surfactants do not carry any charge on their head groups (Fig. 3.2). Different classes of surfactants with examples and their structures are presented in Table 3.3.

3.3 Ionic (Anionic and Cationic) Surfactants

Anionic surfactants get dissociated in water as an amphiphilic anion, which acts as surface-active portion of the molecule and a cation, which is generally an alkaline metal (Na^+, K^+) or a quaternary ammonium ion (Salager 2002). The commonly used anionic surfactants in various industrial applications are (a) carboxylates (alkyl carboxylates-fatty acid salts), (b) sulfates (alkyl sulfates, alkyl ether sulfates), (c) sulfonates: docusates (dioctyl sodium sulfosuccinate, alkyl benzene sulfonates), and (d) phosphate esters (alkyl-aryl ether phosphates; alkyl ether phosphates). These are also employed in pharmaceutical and cosmetic industries. For example, sodium lauryl sulfate BP containing sodium dodecyl sulfate with bacteriostatic action against gram-positive bacteria is applied as a skin cleaner and also a component in the medicated shampoos (Mishra et al. 2009; Sekhon 2013; Azarmi and Ashjaran 2015).

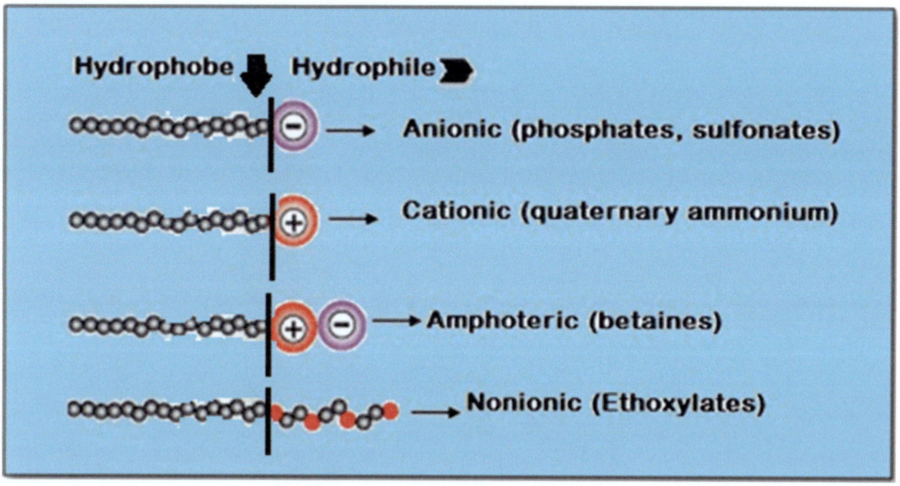

Fig. 3.2 Types of surfactants on the basis of polarity of head group

Cationic surfactants segregate in aqueous solution into a surface-active amphiphilic cation and an anion (mostly halogen type). Cationic surfactants majorly constitute a huge fraction of nitrogenous compounds, like amine and quaternary ammonium salts with one or many long alkyl chains originating from natural fatty acids (Salager 2002). Due to their positive charge, they adsorb strongly on negatively charged solid surfaces imparting special characteristics, like anti-caking, corrosion inhibition, dispersion, germicidal action, etc., to the substrates. These also have property to destroy a wide range of gram-positive and some gram-negative microbes. Such surfactants find application in cleansing wounds or burns on the skin (Mishra et al. 2009; Azarmi and Ashjaran 2015). The most widely used surfactants in this category are "Esterquats" with ester bonds which are generally more expensive than anionics and nonionics, because their synthesis involves high-pressure hydrogenation reactions. However, these surfactants show poor detergency (Sekhon 2013).

3.4 Nonionic Surfactants

Nonionic surfactants do not dissociate in water due to the non-dissociable nature of their hydrophilic group, such as alcohol, phenol, ether, ester, or amide, and are compatible with all other types of surfactants (Salager 2002). These can be described as polyolesters, polyoxyethylene esters, poloxamers polyolesters including glycol, glycerol esters, and sorbitan derivatives. Fatty acid esters of sorbitan (commonly known as Spans, e.g., Span 40, Span 60, Span 80, etc.) and their ethoxylated derivatives (frequently referred to as Tweens, e.g., Tween 20, Tween 40, Tween 80) are one of the most commonly used nonionic surfactants. Polyoxyethylenated mercaptans have slight unpleasant odor. The most repeatedly used surfactants in this group

Table 3.3 Classification of surfactants on the basis of the nature of the polar head group

Class	Examples	Structures
Anionic	Sodium stearate	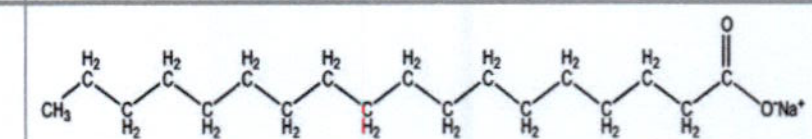
	Sodium dodecyl sulfate	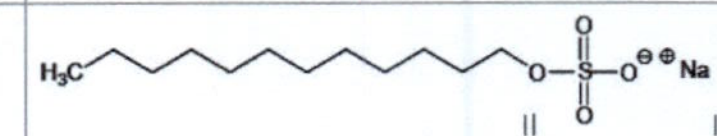
	Sodium dodecylbenzenesulfonate	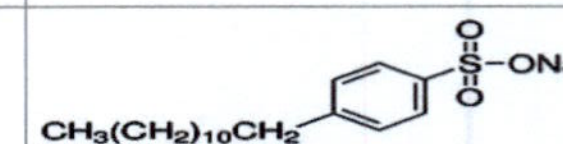
	Dialkyl sulfosuccinate	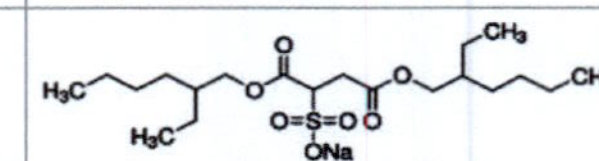
	Lignosulfonates	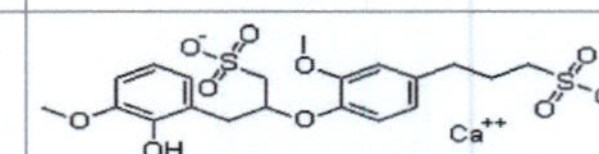
Cationic	Laurylamine hydrochloride	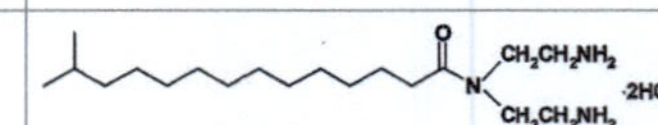
	Trimethyl dodecylammonium chloride	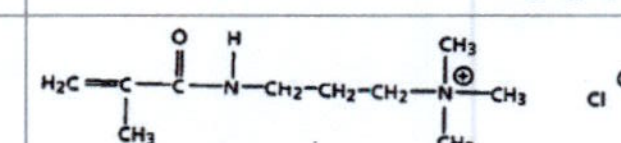
	Cetyltrimethylammonium bromide	
	Benzalkonium-ammonium chloride	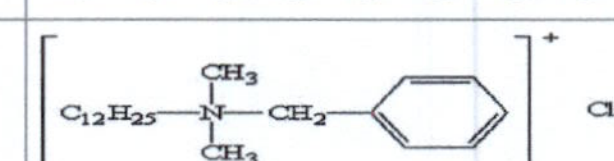
	N,N-cetylmethyl morpholinium cation	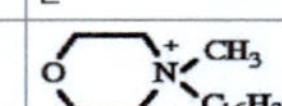

Nonionic	Polyoxyethylene alcohol	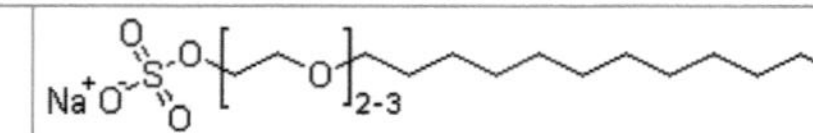
	Alkylphenol ethoxylate	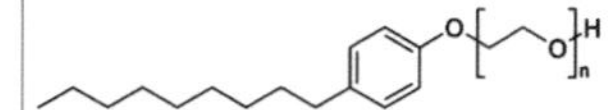
	Polysorbate 80	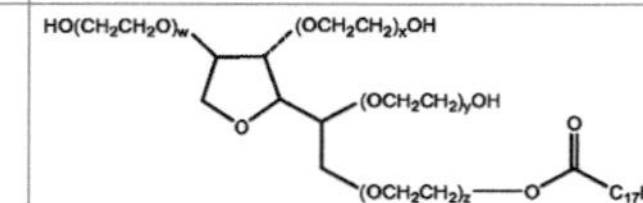
	Propylene oxide-modified polydimethylsiloxane	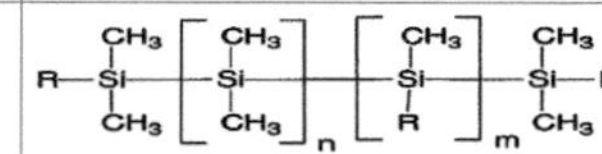
	1-5 sorbitan	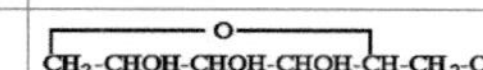
	1-4,3-6 isosorbitan	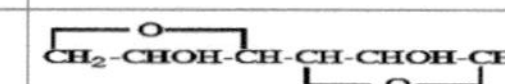
	Ethoxylated imide	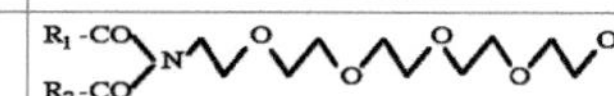
	Diacyl ethoxy urea	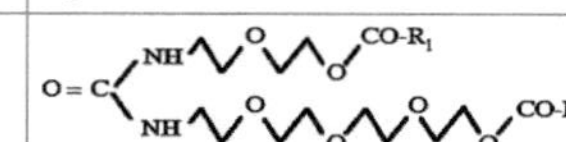

(continued)

Table 3.3 (continued)

Class	Examples	Structures
Zwitterionic	Lauramidopropyl betaine	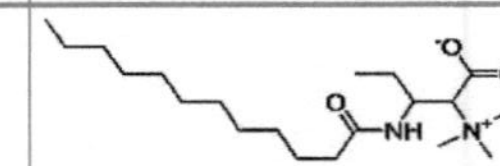
	Dodecyl betaine	
	Cocoamido-2-hydroxypropyl sulfobetaine	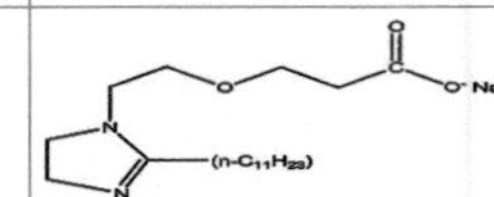
	Trimethylglycine betaine	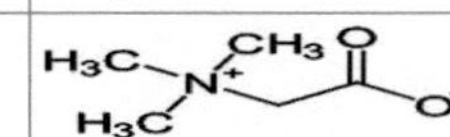

are Polysorbate 20, Polysorbate 80, and Poloxamer 188 which are generally applied in a concentration ranging between 0.001 and 0.1% and mainly find their application as emulsifying agents, dispersants, and solubilizers in pharmaceutical industry (Abraham 2003). Nonionic surfactants are mostly used as gelling and foaming agents and also in the fabrication of several drugs and nano-carriers for drug delivery systems (Mishra et al. 2009; Azarmi and Ashjaran 2015; Carter and Puig-Sellart 2016). In the USA, the neutral agent nonoxynol-9 is commonly used as vaginal spermicide. However, octoxynol has also been approved by the Food and Drug Administration (FDA) to be used in contraceptives and other vaginal drug products (Sekhon 2013).

3.5 Zwitterionic (Amphoteric) Surfactants

Zwitterionic or amphoteric surfactants exhibit both positive and negative charges on the surface-active portion, for example, betaines or sulfobetaines and natural substances, like amino acids, phosphatidylcholine (lecithin), and phospholipids. These surfactants are very mild, compatible with all the other types of surfactants, and less irritating to the skin and eyes than the other types revealing excellent dermatological properties. Moreover, due to their high foaming potential and insolubility in most organic solvents, amphoteric surfactants are largely used in shampoos, cosmetic products, and hand dishwashing liquids (Mishra et al. 2009; Azarmi and Ashjaran 2015).

3.6 Biosurfactants

Microbially produced surface-active compounds, possessing both hydrophilic and hydrophobic moieties, are commonly referred to as biosurfactants (Kugler et al. 2015). In biosurfactants, the lipophilic moiety is generally a protein or peptide with a high fraction of hydrophobic side chains or a hydrocarbon chain of a fatty acid with 10 to 18 carbon atoms, whereas the hydrophilic moiety is an ester; hydroxyl, phosphate, and carboxylate group; or sugar (Campos et al. 2013). Biosurfactants are structurally diverse group of secondary metabolites secreted in liquid culture media by aerobic microorganisms requiring a carbon source, such as carbohydrates, hydrocarbons, fats, and oils, to perform vital roles for their metabolic processes (Silva et al. 2014). Biological surfactants help in the microbial growth by facilitating the availability of hydrocarbons to the microbes with increase in the area at the aqueous-hydrocarbon interfaces across their cell membranes, thereby enhancing utilization by microorganisms, and also help in protection of microbes from harsh environmental conditions (Aulwar and Awasthi 2016). These compounds have amphipathic molecules which act between solutions of different polarities reducing surface tension, thereby allowing access to the hydrophobic substrates due to enhanced contact area between insoluble compounds (such as hydrocarbons) resulting in their enhanced mobility, availability to living forms, and, thus, biodegradation of such

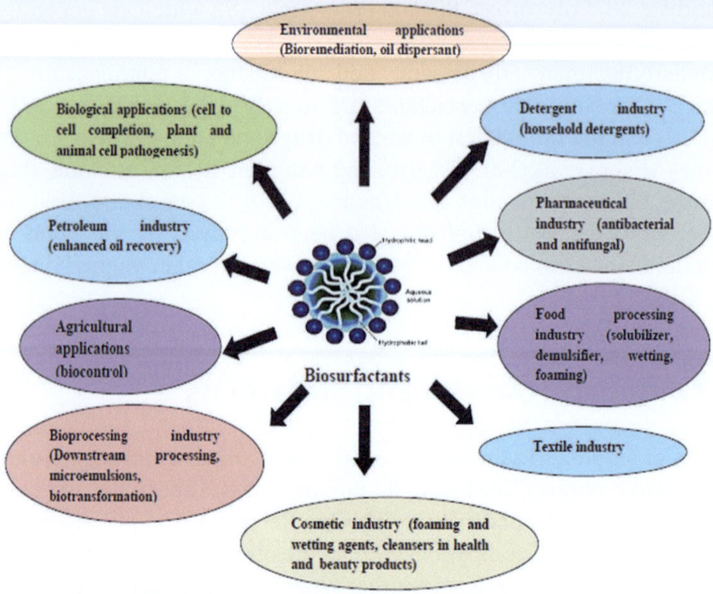

Fig. 3.3 Diverse industrial applications of biosurfactants

compounds (Arparna et al. 2011). Biosurfactants find diverse applications in a wide range of industries as mentioned previously (Banat et al. 2010; Marchant and Banat 2012; Campos et al. 2013; Lawniczak et al. 2013; Reis et al. 2013). A general outlay of various industrial applications of biosurfactants is presented in Fig. 3.3.

Due to the capabilities of biosurfactants for being used in diverse industrial processes, many patents have been granted in recent years involving biosurfactant production by different microbes, specifically *Pseudomonas* spp., *Acinetobacter* spp., *Bacillus* spp., *Candida* spp., and *Torulopsis* spp. (Sachdev and Cameotra 2013). Patents on the biosurfactant production by a consortium of microbes involving *Corynebacterium* spp., *Alcaligenes* spp., and *Methylomonas* spp. have also been filed (Shete et al. 2006; Rahman and Gakpe 2008).

3.7 Classification and Properties of Biosurfactants

Biosurfactants are generally divided into two classes, where class I contains low-molecular-mass molecules, such as glycolipids, lipopeptides, and phospholipids, which efficiently reduce surface and interfacial tensions, and class II comprises high-molecular-mass polymer agents including polymeric and particulate surfactants with effective emulsion-stabilizing properties (Kapadia and Yagnik 2013). Biosurfactants allow two immiscible phases to interact more readily by decreasing interfacial tension between the two dissimilar phases (Chavez and Maier 2011). Efficiency of biosurfactants is based on their concentration required to obtain the

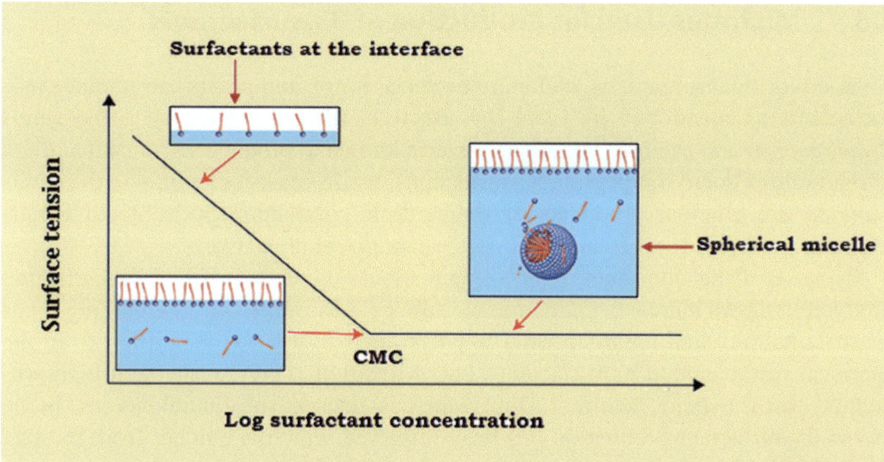

Fig. 3.4 Micelle formation in biosurfactants above critical micelle concentration (CMC)

critical micelle concentration (CMC). It is noteworthy to mention here that biosurfactant molecules aggregate at concentrations above the CMC to form micelles which further enable them to lessen the surface and interfacial tension resulting in enhanced solubility and bioavailability of otherwise reluctant hydrophobic organic compounds (Fig. 3.4). Therefore, critical micelle concentration is inversely related to the efficiency of biosurfactants, which implies that less biosurfactant concentration is needed to reduce the surface tension (Desai and Banat 1997). In standard terms, an effective biosurfactant can lower down the surface tension of water from 72 to 30 mN per meter and the interfacial tension between water and n-hexadecane from 40 to 1 mN per meter.

Biosurfactants are representative amphiphilic biomolecules that contain both hydrophilic and lipophilic groups. The hydrophile-lipophile balance (HLB) number is used to calculate the ratio of these groups, and its value between 0 and 60 defines the affinity of a surfactant for water or oil. Emulsifiers with low HLB are lipophilic and stabilize water-in-oil emulsification, while the emulsifiers with high HLB are hydrophilic (Desai and Banat 1997; Christofi and Ivshina 2002). HLB numbers calculated for nonionic surfactants are in the range from 0 to 20 wherein surfactants with HLB numbers > 10 possess affinity for water (hydrophilic) and HLB numbers < 10 show affinity toward oil (lipophilic). Recently, ionic surfactants have also been assigned HLB values extending above the value of 60 (Sajjadi et al. 2003). Thus, biosurfactant effectiveness can be estimated by its capability to minimize surface and interfacial tensions, by stabilization of emulsions, and also by measuring its hydrophilic-lipophilic balance (HLB).

3.8 Microbes Used for Production of Biosurfactants

A variety of microorganisms including bacteria, fungi, and yeasts can produce biosurfactants as mentioned in Table 3.4. Bacteria mainly belonging to the genus *Pseudomonas* and *Bacillus* are major bacteria known to produce significant amount of rhamnolipids and lipopeptide biosurfactants, while yeasts belonging to the genus *Candida* are efficient producers of rhamnolipids and lipopeptide biosurfactants (Arparna et al. 2011; Al-Bhary et al. 2013; Campos et al. 2013).

Because of the important diverse applications, the economics of biosurfactant production draws increasing attention worldwide. Moreover, due to emerging focus on sustainability and environmental impacts, industrial focus is shifting from the chemical surfactants to biosurfactants, but as mentioned previously, their high production cost is a major deterrent. During the past decade, substantial increase in the global biosurfactant production has been observed which is evident from the fact that in 2013, biosurfactant production was about 344,068 tons, but at a current compound annual growth rate (CAGR) of 4.3%, it is speculated to reach up to 476,512 tons by 2018 (Transparency Market Research 2014). Thus, there is an urgent need to develop new strategies for commercial production of biosurfactants through improved biotechnological processes. Two basic approaches are adopted globally for cost-efficient production of biosurfactants: (1) utilization of abundant, inexpensive, and waste biomass as substrate for the production media resulting in low initial raw material costs required for the process and (2) development and optimization of bioprocesses for maximizing biosurfactant production and recovery, leading to reduced operating costs (Saharan et al. 2011; Peirera et al. 2016). Therefore, it appears to be economically viable to produce biosurfactants using economical, renewable resources, such as agricultural wastes, fruit processing industries waste, molasses, vegetable oils, distillery waste, and dairy wastes as substrates (Makkar and Cameotra 2002; Krieger et al. 2010).

3.9 Application of Surfactants in Bioprocessing

Disruption of Cell Biomass In order to address current burning issues of energy security, food-versus-fuel debate and environmental-related issues due to the use of liquid fossil fuels, renewable biofuels are drawing a considerable attention worldwide as substitutes to petroleum-based transportation fuels (Lee 2011). Among various types of renewable fuels, microalgal biodiesel known as "third-generation biofuels" has been considered as one of the best alternatives because of the high photosynthetic efficiency, high growth rate, and high levels of extracted oil (Ahmad et al. 2011). Algal strains, like *Scenedesmus*, *Chlorella*, *Nannochloropsis*, and *Chlamydomonas*, are known to possess a large amount of high-density lipid inclusions (30–60% lipids as dry weight) which can serve as good feedstock for biofuel production (Xin et al. 2011; Bondioli et al. 2012; Seo et al. 2016). However, extraction of lipids from algal cells requires efficient cell disruption methods to allow penetration of solvents into the intracellular inclusions. Different approaches have been developed for microalgal biomass harvesting and cell disruption including

Table 3.4 Major types of biosurfactants produced by different microbial groups

Biosurfactant			
Group	Class	Microorganism	References
Glycolipids	Rhamnolipids	*Pseudomonas aeruginosa* spp. *Ustilago maydis*, *Serratia rubidaea*	Maier and Chavez (2000), Sifour et al. (2007), Teichmann et al. (2007) and Jadhav et al. (2011)
	Trehalolipids	*Mycobacterium* spp. *(tuberculosis, bovis, smegmatis, kansasii, malmoense, phlei), Rhodococcus (erythropolis, opacus, ruber), Arthrobacter paraffineus, Nocardia* spp., *Corynebacterium* spp. *(fasciens, pseudodiphtheria, matruchotii), Brevibacterium vitarumen*	Franzetti et al. (2010) and Kugler et al. (2015)
	Sophorolipids	*Torulopsis bombicola, Torulopsis petrophilum, Torulopsis apicola, Candida bombicola, C. antarctica, C. apicola, C. stellata, C. botisae*	Baviere et al. (1994) and Fesle et al. (2007)
Fatty acids, phospholipids, and neutral lipids	Corynomycolic acid	*Corynebacterium lepus, Clavibacter michiganensis*	Gerson and Zajic (1978) and Herman and Maier (2002)
	Spiculisporic acid	*Penicillium spiculisporum*	Ishigami et al. (2000)
	Phosphatidylethanolamine	*Acinetobacter* sp., *Rhodococcus erythropolis*	Appanna et al. (1995) and Santos et al. (2016)
Lipopeptides	Surfactin/iturin	*Bacillus subtilis, B. amyloliquefaciens*	Arguelles et al. (2009) and Liu et al. (2015)
	Lichenysin	*Bacillus licheniformis, Bacillus subtilis*	Yakimov et al. (1997) and Santos et al. (2016)
Polymeric biosurfactants	Emulsan	*Acinetobacter calcoaceticus* RAG-1	Zosim et al. (1982) and Santos et al. (2016)
	Alasan	*Acinetobacter radioresistens* KA-53	Toren et al. (2001) and Santos et al. (2016)
	Biodispersan	*Acinetobacter calcoaceticus* A2	Rosenberg et al. (1988) and Santos et al. (2016)
	Liposan	*Candida lipolytica, C. tropicalis*	Cirigliano and Carman (1984) and Santos et al. (2016)
	Mannoprotein	*Saccharomyces cerevisiae*	Cameron et al. (1988) and Santos et al. (2016)

centrifugation, flocculation, filtration, and flotation for harvesting and microwave heating, ultrasonic cavitation, bead milling, enzymatic, pulsed electric fields, and osmotic shock for cell disruption which are either energy or chemical extensive (Sheng et al. 2011; Liang et al. 2012; Halim et al. 2012).

Because of their proven ability to disrupt membranes, surfactants can play a crucial role in harvesting and disruption of microalgal cells in a cost-effective and energy-efficient manner. The hydrophobic components of surfactants have potential to insert themselves into outer membranes, thereby facilitating the lysis of the cells (Nasirpour et al. 2014). Existing literature suggests that the cationic surfactants could easily bind with a negatively charged microalgal membranes, resulting in effective cell disruption (Huang and Kim 2013; Lai et al. 2016; Salam et al. 2016). A recent study by Seo et al. (2016) described utilization of cationic surfactant-decorated Fe_3O_4 nanoparticles (CS-OTES-MNP) in microalgal cell harvesting, detachment, and cell disruption, schematic illustration of which is presented in Fig. 3.5. Despite the huge potential of surfactant-assisted lipid extraction, there are also certain limitations associated with it. The extent of surfactant binding does not just count on the charge but can also be associated with the hydrophilic-lipophilic interactions between microalgae and surfactant leading to the final cell disruption (Ulloa et al. 2012). Microalgal cell membrane composition not only varies with species but also with the physiological state of a single strain (Gerken et al. 2013; Lai et al. 2016). Thus, any deviation in the cell wall structure can impact the efficiency of surfactant-aided cell disruption method. Therefore, it is imperative to understand correlation between microbial growth cycle and surfactant-aided disruption process.

3.10 Hydrocarbon Bioremediation

Hydrocarbons are the hydrophobic organic chemicals which show restricted water solubility and are toxic and of persistent nature and have an adverse effect on the living forms. Moreover, excessive use of hydrocarbons nowadays has created numerous environmental contamination problems. Additionally, moderate to poor recovery of hydrocarbon contaminants by physicochemical treatments and limited accessibility to microbes and to oxidative and reductive chemicals during the in situ and/or ex situ applications make their removal difficult (Plociniczak et al. 2011). Therefore, biosurfactants hold a great potential in the biological remediation technologies.

Biosurfactant-aided hydrocarbon bioremediation is enhanced due to increase in substrate availability to microorganisms which involves enhancement of the hydrophobicity of the cell surface due to its interactions with the biosurfactant, leading to more effective and easy associations with the microbial cells (Mulligan and Gibbs 2004). Biosurfactants, therefore, consequently can help in enhancing biodegradation and removal of hydrocarbons by the bacteria capable of growing on hydrocarbon contaminants present in polluted soil (Urum and Pekdemir 2004; Nievas et al.

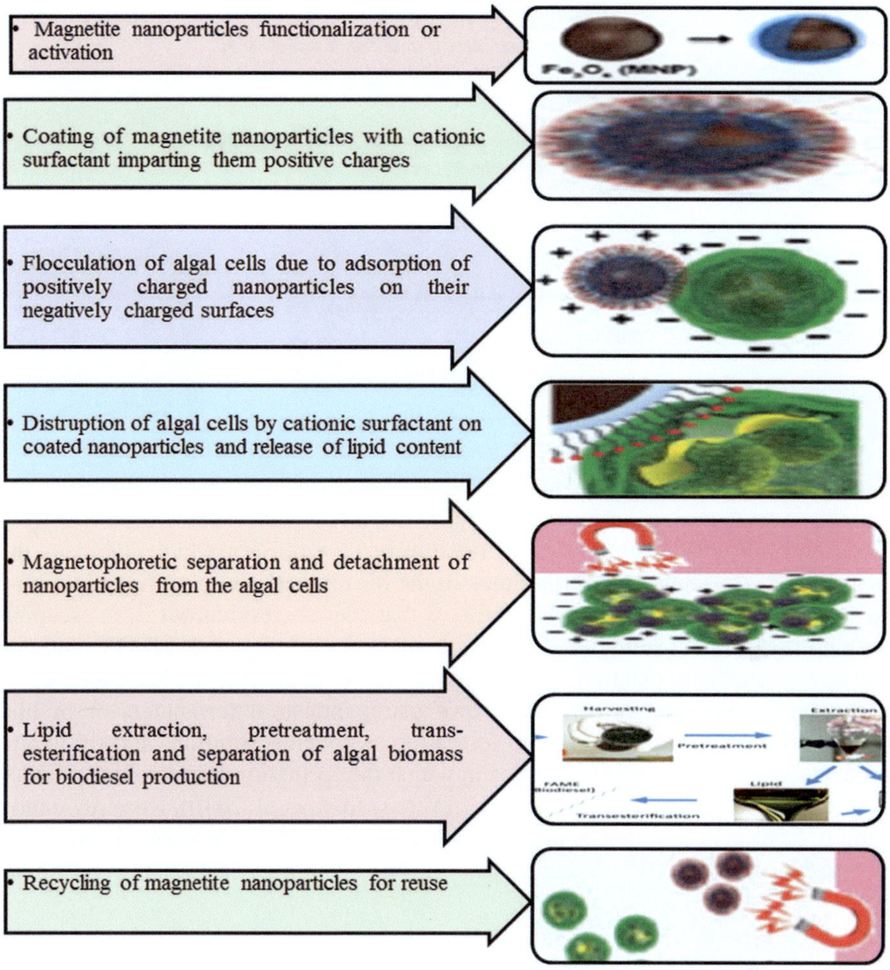

Fig. 3.5 Steps in the utilization of cationic surfactant-decorated magnetite nanoparticles (MNP) in microalgae harvesting, detachment, and cell disruption

2008; Chaprao et al. 2015; Liu et al. 2015; Adrion et al. 2016a; Adrion et al. 2016b; de la Cueva et al. 2016; Sawadogo et al. 2016).

One of the most important applications of biosurfactants is the "microbial enhanced oil recovery" (MEOR) which is generally implied to the recovery of a notable fraction of the residual oil remaining in reservoirs that otherwise is difficult to obtain even after exhausting all the physicomechanical recovery procedures (Banat et al. 2000; Sen 2008). In this process, microbes or their primary or secondary metabolites, such as biosurfactants, biopolymers, acids, solvents, and enzymes, are applied to enhance oil recovery from depleted reservoirs, where biosurfactants stick tightly to the oil/water interface and decrease interfacial tension between oil/

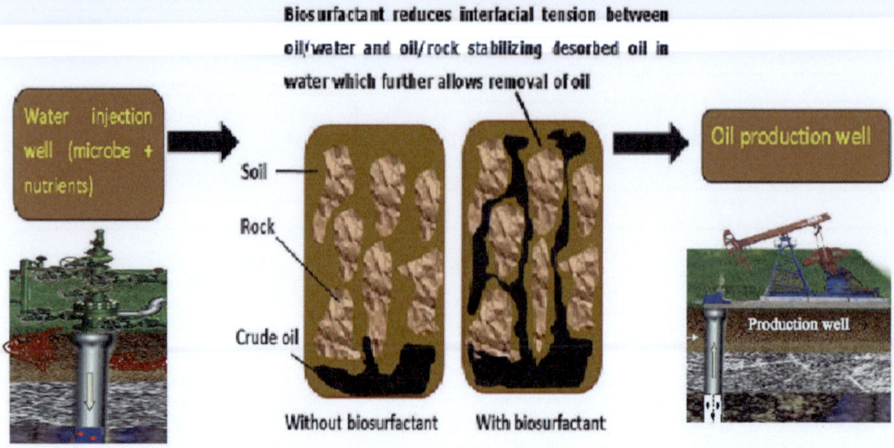

Fig. 3.6 Illustration of role of biosurfactants in microbial enhanced oil recovery (MEOR)

water and oil/rock, leading to the formation of an emulsion which stabilizes the desorbed oil in water. This further aids in the oil removal along with the injection water while decreasing the capillary forces that prevent residual oil from escaping through rock pores (Suthar et al. 2008) as illustrated in Fig. 3.6. MEOR methods can be classified into two main groups: (a) ex situ production of the MEOR metabolites wherein microorganisms are grown using industrial fermenters or mobile plants and then injected into the oil formation as aqueous solutions and (b) in situ production of the MEOR metabolites in which the formation of microbial metabolites takes place directly in the reservoir (Yernazarova et al. 2016). Recently, many authors have also reported applications of biosurfactants in MEOR (Amin 2010; El-Sheshtawy et al. 2015; Golabi 2016).

3.11 Heavy Metal Bioremediation

Due to toxic nature of heavy metals, their contamination in soil ecosystems has serious consequences as even low concentration of heavy metals is very hazardous to the living organisms (Plociniczak et al. 2011). Anionic biosurfactants form complexes with metal with the help of ionic bonds, stronger than the existing metal-soil bonds, and due to the reduced interfacial tension, metal-biosurfactant complexes are detached from the soil surfaces to the soil solution. The cationic biosurfactants replace the same charged metal ions with the help of ion exchange competing for negatively charged surfaces (Mulligan and Gibbs 2004; Juwarkar et al. 2007; Asci et al. 2008). In addition to this, metal ions can also be separated from the soil matrices by the biosurfactant micellar inclusions, wherein the polar head groups of micelles chelate metals, mobilizing them in water as shown in Fig. 3.7 (Mulligan 2005). Many authors have previously reported the utility of biosurfactants in

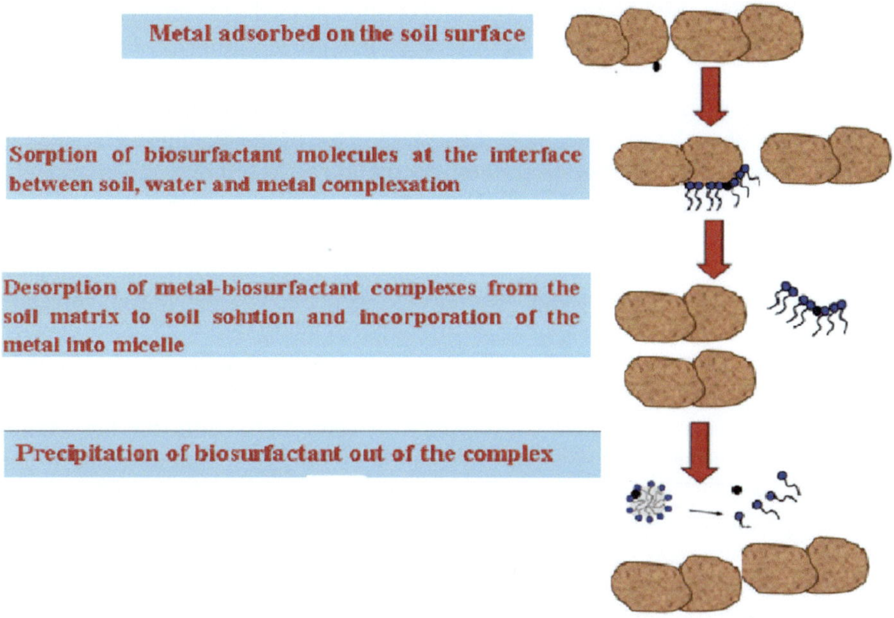

Fig. 3.7 Schematic presentation of biosurfactant-aided heavy metal removal from the soil

bioremediation of soils contaminated with heavy metals (Singh and Cameotra 2004; Juwarkar et al. 2008; Das et al. 2009; Peng et al. 2009; Fu and Wang 2011).

3.12 Bioconversion of Lignocellulosic Biomass

Enormous amount of lignocellulosic biomass in nature has the potential for bioconversion into a variety of high value-added products like biofuels, fine chemicals, and cheap energy sources (Anwar and Gulfraz 2014; Kumar et al. 2016). Lignocellulosic biomass mainly comprises cellulose (30–60%), hemicellulose (20–40%), and lignin (10–25%) which are interconnected in a hetero-matrix constituting roughly 90% of the dry matter, while the rest is composed of ash and other extractives (Rosta Estela and Luis 2013; Nanda et al. 2014). Cellulose and hemicellulose that account for more than 50% of total mass can be potentially converted to sugars for their subsequent conversion to ethanol through a series of processes (Oberoi et al. 2010). Enzymatic hydrolysis and fermentation processes have been combined into several process configurations, such as separate hydrolysis and fermentation (SHF), simultaneous saccharification and fermentation (SSF), simultaneous saccharification and co-fermentation (SSCF), and consolidated bioprocessing (CBP) that includes enzyme production, enzymatic saccharification, and fermentation in a single step (Kumar et al. 2016).

Lignocellulosic biomass-derived ethanol is often termed as "second generation" or "2G" offering several advantages, such as greenhouse gas mitigation, near carbon neutrality, lesser dependence on fossil fuels, and improvement in nation's energy security (Naik et al. 2010). The US Department of Energy (DOE) report of 2011 suggests that in the USA alone, more than a billion ton of lignocellulosic biomass is potentially available at ~$60/ton for conversion into > 20 billion gallons of cellulosic biofuels (Perlack and Stokes 2011). Previous reports suggest that the total crop residue available is more than one billion ton in the USA alone and more than nine billion ton worldwide (Lal 2005). Several authors have reported the bioethanol potential of various abundant major agro-wastes generated globally (Kim and Dale 2004; Perlack et al. 2005; Naik et al. 2010; Sarkar et al. 2012).

An immense amount of food waste, such as raw, cooked, edible, and nonedible portions of food crops, is generated during their production, storage, distribution, transportation, processing, and consumption of food stuffs from the household, commercial, and industrial sources. A recent study on global food loss and waste reported a food loss in the range of 27–32% for all the food produced in the world. According to a recent survey in the world, cereal losses at 19–32%, root and tuber losses at 33–60%, and fruit and vegetable losses at 37–55% have been estimated (Global Food Policy Report 2016). Reduction in the food loss and waste can lead to increased global food availability. This waste is a rich source of important biomolecules such as lipids, carbohydrates, amino acids, and phosphates which can be utilized as a substrate for the development of cost-effective biofuels (Pleissner et al. 2013). Carbohydrate-rich food hydrolysate produced after enzymatic hydrolysis of food waste can be transformed into bioethanol, whereas the lipid fraction from hydrolysate can be converted to biodiesel as shown in Fig. 3.8 (Karmee and Lin 2014). Across the globe, a lot of work has been carried out on the tremendous potential of food waste for the generation of biofuels, by different research groups (Kim et al. 2011; Yan et al. 2011; Yan et al. 2013; Yang et al. 2014; Matsakas et al. 2014; Pleissner et al. 2014).

Commercial production of biofuels from lignocellulosic biomass is still hindered by many factors such as (1) biomass recalcitrance requiring effective pretreatment process, (2) high enzyme concentrations to achieve high rate of cellulose hydrolysis, and (3) enzyme adsorption to the lignocellulosic material making enzyme recycling difficult (Gregg and Saddler 1996). The possible ways of inhibition of cellulases by lignin during hydrolysis include (1) nonproductive adsorption of cellulase onto lignin, (2) physical blockage of cellulase on lignocellulose chain structure, and (3) enzyme inhibition due to soluble lignin-derived compounds like ferulic acid, syringaldehyde, vanillin, etc. (Saini et al. 2016). Thus, nonproductive binding of cellulases with lignin results in decreased efficiency of lignocellulosic hydrolysis. Therefore, development of economically feasible cellulose hydrolysis process for ethanol production along with identification of methods to increase enzyme effectiveness is mandated.

Addition of surfactants, such as nonionic detergents and protein, has been reported to significantly increase the enzymatic conversion of cellulose into soluble sugars (Kaar and Holtzapple 1998; Erikssson et al. 2002; Kristensen et al. 2007;

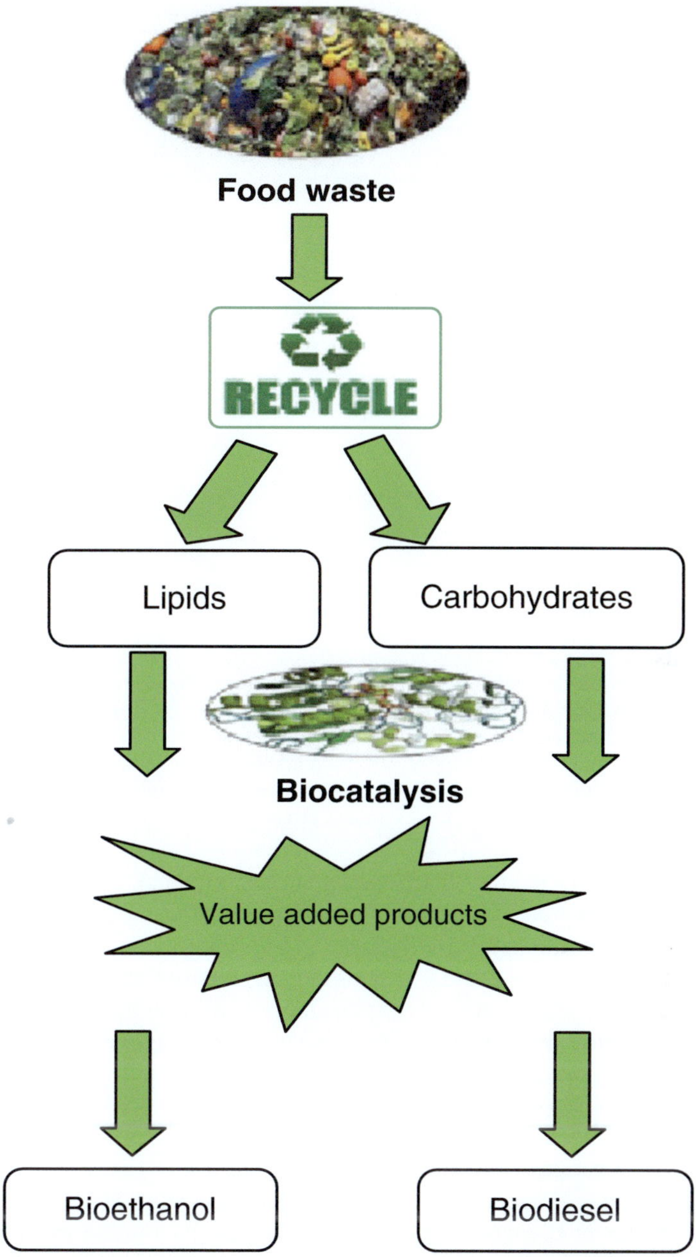

Fig. 3.8 Recycling of food waste into value-added products such as biofuels

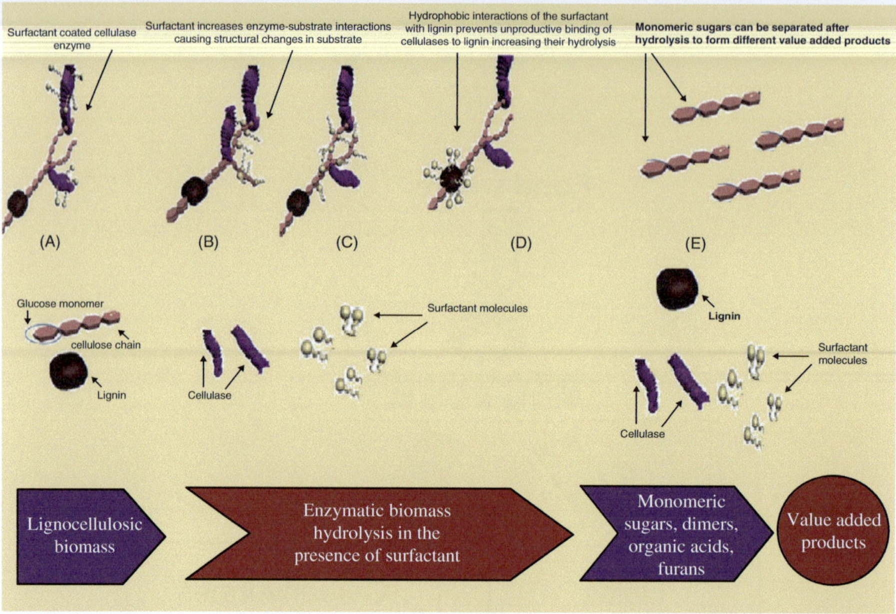

Fig. 3.9 Possible theories explaining enhanced cellulose hydrolysis in the presence of surfactant: (**a**) surfactant "protects" cellulase and increases its stability, (**b**) and (**c**) surfactant increases enzyme-substrate interactions and also causes structural changes in substrate making it more accessible for effective cellulose conversion, (**d**) hydrophobic part of the surfactant binds through hydrophobic interactions to lignin and prevents unproductive binding of cellulases to lignin increasing their efficiency, and (**e**) conversion of cellulose into monomeric units which can be further used for development of value-added products

Bardant et al. 2013; Hseih et al. 2015; Min et al. 2015; Li et al. 2016). On the basis of earlier studies, possible different explanations of surfactant effect on cellulose hydrolysis have been illustrated in Fig. 3.9.

The exact mechanism explaining how surfactants improve enzymatic hydrolysis is still unknown, but several possible explanations have been proposed to describe enhanced surfactant-aided enzymatic hydrolysis of lignocellulosic biomass, and these are:

1. Surfactants enhance removal of lignin by forming emulsions, thereby increasing the access of substrate's reaction sites to the cellulases.
2. Surfactants decrease irreversible, nonproductive adsorption of cellulase to nonproductive sites of biomass allowing its availability in solution to have higher activity, thus resulting in higher yields and better recycling of enzymes (Karr and Holtzapple 1998; Park et al. 1992; Eriksson et al. 2002).
3. Enhanced enzyme activity due to improved electrostatic interaction between surfactant and enzyme occurs either by activation of a certain amino acid in the

enzyme active site or by reforming enzyme secondary structure, especially the α-helixes (Eckard et al. 2013a).

4. Surfactants provide protection to enzymes from thermal deactivation and denaturation by reducing the surface tension and viscosity of hydrolysate even after extended incubation period (Yoon and Robyt 2005; Kim et al. 2006; Eckard et al. 2013b). Addition of proteins such as bovine serum albumin (BSA) has also been reported to prevent unproductive binding of cellulases to lignin after blocking its reactive sites (Wang et al. 2013).

Currently, various authors have reported the positive effect of surfactant addition on the enzymatic hydrolysis of lignocellulosics (Qing et al. 2010; Sipos et al. 2011; Parnthong and Kungsanant 2014; Hseih et al. 2015; Min et al. 2015; Li et al. 2016; Mesquita et al. 2016). Liang et al. (2016) measured the effect of molecular structure of lignosulfonate-based polyoxyethylene ether (LS-PEG) on the enzymatic hydrolysis of Avicel and corn stover. They showed that the glucose yield of corn stover increased from 16.7 to 51.9% with a respective increase in PEG content and molecular weight of LS-PEG during hydrolysis. In an unpublished study, we had observed that the addition of 0.2% Tween 20 increased the sugar yield by about 15–20% during hydrolysis of alkali-pretreated sweet sorghum bagasse with crude enzymes produced by *Aspergillus terreus* using optimized parameters, as compared to control, where no surfactant was used. In addition to the sugar concentration, higher productivity of sugars was observed as surfactant-aided hydrolysis was carried out at 60 °C, instead of 50 °C, generally used for industrial applications. Previous studies on enzymatic hydrolysis of pretreated lignocellulosic biomass have reported the use of thermostable crude cellulases and hemicellulases, which helped in achieving a higher sugar yield and productivity (Soni et al. 2010; Srivastava et al. 2014; Rawat et al. 2014; Sharma et al. 2015). It is a well-established fact that the rate of reaction increases with increase in temperature. Therefore, maintaining a temperature of 70–80 °C during hydrolysis with the use of thermostable enzymes and surfactants is likely to improve the sugar productivity manyfold, highly desired by the biofuel industry.

3.13 Future Perspectives

Surfactants represent a major class of industrial chemicals widely used in different industries. Petroleum-derived commercially available chemical surfactants have diverse industrial applications, such as applications in pharmaceuticals, including their role as enhancers for percutaneous absorption, in respiration distress therapy, in suspension aerosols, as emulsifying agents, and in influencing drug absorption (Mishra et al. 2009), in biological systems (Ikegami et al. 2000), in synthesis of nanostructured or mesostructured materials with diverse uses, in health and personal care products (Mainkar and Jolly 2001), in food industries (Kralova and Sjöblom 2009), and also in crop protection (Green and Beestman 2007). Notwithstanding these advantages, the toxicity and persistence of surfactants in different environmental situations have also been highlighted by several authors (Ying 2006; Ivankovic and Hrenovic 2010). Thus, there is an imminent requirement to realize the effects of surfactants during their normal applications as well as accidental spills in the environment.

Biosurfactants due to their diverse, less toxic, and biodegradable nature have gained importance in recent times as viable alternatives to the chemical surfactants. There is a wide range of potential commercial applications of biosurfactants in diverse industries, such as pulp and paper industry, food industry, textiles, pharmaceutical industry as therapeutic agents, paint industry, remediation processes, cell-biomass disruption, bioprocessing, and even uranium ore processing (Banat et al. 2010; Silva et al. 2014; Peireira et al. 2016). One of the emerging areas of research using biosurfactants is the conversion of agricultural biomass into sugars for their subsequent conversion to liquid fuels. There is very scanty information available on biosurfactant-aided enzymatic hydrolysis of pretreated biomass. The use of biosurfactants during pretreatment might not only yield good results for lignin deconstruction/removal but may also reduce the higher energy required to achieve the same level of efficiency. This coupled with high-temperature biosurfactant-aided enzymatic hydrolysis as discussed elsewhere is likely to revolutionize the research in second-generation bioethanol production. It is well recognized that pretreatment and hydrolysis are the most cost- and energy-intensive operations in conversion of lignocellulosic biomass to liquid fuels (Parnthong and Kungsanant 2014; Mesquita et al. 2016).

Similarly, the use of biosurfactants in developing valuable compounds from fruit and vegetable processing waste holds promise for the future. Since the fruit and vegetable processing waste is high in moisture, biosurfactants in combination with enzymes or other extraction methods may help in efficient extraction of pigments, carotenoids, and other nutritionally important compounds for their use in the development of functional foods. Because of the presence of lower lignin and high carbohydrates, fruit and vegetable processing wastes offer good opportunities for production of bioethanol and other value-added products. Use of surfactants/biosurfactants in improving ethanol productivity from the fruit and vegetable processing residues is likely to draw attention of many researchers involved in biofuel research.

Commercial production and development of biosurfactants is dictated by their safety issues and high production cost. Biosurfactants, currently under investigation, such as sophorolipids and rhamnolipids have not been reported to pose any safety or health issue (Marchant and Banat 2012). However, in certain isolated cases, rhamnolipids can act as immune modulators or as virulence factors in *P. aeruginosa* infections known as an opportunistic pathogen (McClure and Schiller 1996; Zulianello et al. 2006). Commercial production of rhamnolipid has been initiated by a company, Jeneil Biotech, Milwaukee, USA (www.jenielbiotech.com), which has reported no health issues associated with its use. Furthermore, large-scale production of sophorolipids, generally produced by yeasts, is also already underway in Asia with no reported health/safety issues (Marchant and Banat 2012).

One of the ways to reduce the overall costs is through improving the stability of biosurfactants. Another way to reduce the high production costs is to use a wide variety of microorganisms, such as *Bacillus*, *Candida*, *Pseudomonas*, *Thiobacillus*, etc., for producing biosurfactants using various renewable substrates, such as sugars, oils, alkanes, and agro-industrial wastes including molasses, potato processing wastes, olive oil mill effluent, plant oil extracts and waste, distillery and whey wastes, by-products of vegetable industries, dairy and sugar industry wastes, and cassava wastewater (Makkar and Cameotra 1999; Youssef et al. 2004; Makkar et al. 2011). However, these microbes cannot be used for commercial production of biosurfactants on a large scale due to their low productivity. Thus, the microbial strain producing biosurfactants should be carefully selected and engineered for enhanced productivity. Moreover, the production process should also be engineered in such a way so as to minimize overall capital, operating, maintenance, and product recovery costs (Reis et al. 2013).

Keeping in mind the demand and need for green technology, biosurfactants can also have a potential application in the synthesis of nanoparticles, which is an emerging promising environment-compatible method. Use of biosurfactants reduces the formation of aggregates facilitating homogeneous and uniform morphology of the nanoparticles during synthesis (Kiran et al. 2010; Mujumdar et al. 2016). Nano-emulsions are advantageous because of the Brownian motion of very small droplets causing a significant deduction in the gravity force. This further inhibits their coalescence, since these droplets are non-deformable, and hence, surface fluctuations are avoided. Consequently, low surface tension of the whole system results in high penetration of active and the small-sized droplets allowing their enhanced uniform rapid deposition on substrates (Morsy 2014). Reddy et al. (2009) reported stabilization of silver nanoparticles for 2 months using surfactin, known as a biodegradable, less toxic stabilizing agent. These surfactant-aided nano-emulsions have diverse applications in agrochemicals, as lubricants and cutting oils and corrosion inhibitors, in remediation technology, in pharmaceutics, in cosmetics, in foods, and in personal care products (Oliveira et al. 2014; Morsy 2014; Jaiswal et al. 2015; Mujumdar et al. 2016; Santos et al. 2016). Therefore, there is an imperative need to focus research efforts on the biosurfactant-mediated stabilization of the nanoparticles for diverse industrial applications.

3.14 Conclusions

- Biosurfactants are potential replacements for synthetic surfactants in several industrial processes due to their biodegradability and lower toxicity.
- Presently, despite of their many advantages, high production costs and shortage of detailed information on toxicity testing of biosurfactants make them economically incompetent in comparison to chemically produced surfactants available in the market.
- Measures selected to target simplification and optimization of the kinds of the products for specific applications, such as use of sterilized or pasteurized fermentation broth without the requirement for extraction, concentration, or purifi-

cation of the biosurfactant, may remarkably decrease the process production cost.

- Optimization of upstream and downstream approaches during the production processes may also have a significant influence on the overall cost reduction. Success of biosurfactants in bioremediation will be in need of specific targeting and complete information of the physicochemical nature of the pollutant-affected areas.
- Role of biosurfactants in MEOR has numerous utilities with respect to environment, but extensive research is still required for ex situ production and commercial application.
- With increased efforts on the development of improved and cost-efficient application technologies, genetic engineering, and strain improvement techniques and production processes, biosurfactants are predicted to be one of the most multifaceted and valued compounds for use in various processes in the coming time.

References

Abraham M (2003) Wetting of hydrophobic rough surfaces: to be heterogeneous or not to be. Langmuir 4:8343–8348

Adrion AC, Nakamura J, Shea D, Aitken MD (2016a) Screening nonionic surfactants for enhanced biodegradation of polycyclic aromatic hydrocarbons remaining in oil after conventional biological treatment. Environ Sci Technol 50(7):3838–3845. doi:10.1021/acs.est.5b05243

Adrion AC, Singleton DR, Jun N, Damian S, Aitken MD (2016b) Improving polycyclic aromatic hydrocarbon biodegradation in contaminated soil through low-level surfactant addition after conventional bioremediation. Environ Eng Sci 33(9):659–670. doi:10.1089/ees.2016.0128

Ahmad AL, Yasin NHM, Derek CJC, Lim JK (2011) Microalgae as a sustainable energy source for biodiesel production: a review. Sust Energ Rev 15:584–593

Al-Bhary SN, Al-Wahaibi YM, Elshafie AE, Al-Bemani AS, Joshi SJ, Al-akhmari HS, Al-Sulaimani HS (2013) Biosurfactant production by *Bacillus subtilis* B20 using date molasses and its possible application in enhanced oil recovery. Int Biodeterior Biodegrad 81:141–146

Alvira P, Ballesteros M, Negro MJ (2013) Progress on enzymatic saccharification technologies for biofuels production. In: Gupta VK, Tuohy MG (eds) Biofuel technologies: recent developments. Springer, Berlin, Germany, pp 145–169

Amin GA (2010) A potent biosurfactant producing bacterial strain for application in enhanced oil recovery applications. J Pet Environ Biotechnol 1:104–111. doi:10.4172/2157-7463.1000104

Anwar Z, Gulfraz IM (2014) Agro-industrial lignocellulosic biomass a key to unlock the future bio-energy: a brief review. J Radiation Res Appl Sci 7:163–173

Appanna VD, Finn H, Pierre M (1995) Exocellular phosphatidylethanolamine production and multiple-metal tolerance in *Pseudomonas fluorescens*. FEMS Microbiol Lett 131:53–56

Arguelles-Arias A, Ongena M, Halimi B (2009) *Bacillus amyloliquefaciens* GA1 as a source of potent antibiotics and other secondary metabolites for biocontrol of plant pathogens. Microb Cell Factories 8:63–71

Arparna A, Srinikethan G, Hedge S (2011) Effect of addition of biosurfactant produced by *Pseudomonas* ssp. on biodegradation of crude oil. In: 2nd International proceedings of chemical, biological and environmental engineering, Singapore, 26–28 February 2011, vol 6, p 71–75

Asci Y, Nurbaş M, Acikel YS (2008) A comparative study for the sorption of Cd(II) by soils with different clay contents and mineralogy and the recovery of Cd(II) using rhamnolipid biosurfactant. J Hazd Mater 154:663–673

Aulwar U, Awasthi RS (2016) Production of biosurfactant and their role in bioremediation. J Ecosys Ecograph 6:202. doi:10.4172/2157-7625.1000202

Azarmi R, Ashjaran A (2015) Type and application of some common surfactants. J Chem Pharm Res 7(2):632–640

Balan V (2014) Current challenges in commercially producing biofuels from lignocellulosic biomass. Hindawi Publishing Corporation ISRN Biotechnology 463074. doi:10.1155/2014/463074

Banat IM, Makkar RS, Cameotra SS (2000) Potential commercial applications of microbial surfactants. Appl Environ Microbiol 53:495–508

Banat IM, Franzetti A, Gandolfi I, Bestetti G, Martinotti MG, Fracchia L, Smyth TJ, Marchant R (2010) Microbial biosurfactants production, applications. Appl Microbiol Biotechnol 87:427–444

Banat IM, Satpute SK, Cameotra SS (2014) Cost effective technologies and renewable substrates for biosurfactants production. Front Microbiol 5:1–18

Bardant TB, Abimanyu SH, Hanum AK (2013) Effect of non-ionic surfactant addition to cellulase performance in high substrate loading hydrolysis of palm oil EFB and water hyacinth. Indo J Chem 13(1):53–58

Baviere M, Degouy D, Lecourtier J (1994) Process for washing solid particles comprising a sophoroside solution. US Patent 5:32–407

Bondioli P, Bella LD, Rivolta G, Zittelli GC, Bassi N, Rodolfi L, Casini D, Prussi M, Chiaramonti D, Tredici MR (2012) Oil production by the marine microalgae *Nannochloropsis* sp. F and M-M24 and *Tetraselmis suecica* F and M-M33. Bioresour Technol 114:567–572

BP Statistical Review of World Energy (2016) Centre for Energy Economics Research and Policy, Heriot-watt university, 65th edn. Whitehouse Associates/Pureprint Group Limited, London, pp 1–48

Cameron DR, Cooper DG, Neufeld RJ (1988) The mannoprotein of *Saccharomyces cerevisiae* is an effective bioemulsifier. Appl Environ Microbiol 54:1420–1425

Campos JM, Stamford TLM, Sarubbo LA, Luna JM, Rufino RD, Banat IM (2013) Microbial biosurfactants as additives for food industries. Biotechnol Prog 29:1097–1108

Carriquiry MA, Du X, Timilsina GR (2011) Second generation biofuels: economics and policies. Energ Pol 39(7):4222–4234

Carter KC, Puig-Sellart M (2016) Nanocarriers made from non-ionic surfactants or natural polymers for pulmonary drug delivery. Curr Pharm Des 22(22):3324–3331

Chaprao MJ, Ferreira INS, Correa PF, Rufino RD, Luna JM, Silva EJ, Sarubbo LA (2015) Application of bacterial and yeast biosurfactants for enhanced removal and biodegradation of motor oil from contaminated sand. Electron J Biotechnol 18:471–479

Chavez SG, Maier RM (2011) Biosurfactants: a general overview. In: Chavez SG (ed) Biosurfactants. Springer-Verlag, Berlin, pp 1–11

Christofi N, Ivshina IB (2002) Microbial surfactants and their use in field studies of soil remediation. J Appl Microbiol 93:915–929

Cirigliano MC, Carman GM (1984) Purification and characterization of liposan, a bioemulsifier from *Candida lipolytica*. Appl Environ Microbiol 50:846–850

Das P, Mukherjee S, Sen R (2009) Biosurfactant of marine origin exhibiting heavy metal remediation properties. Bioresour Technol 100:4887–4890

Desai JD, Banat IM (1997) Microbial production of surfactants and their commercial potential. Microbiol Mol Biol Res 61:47–64

de la Cueva SC, Rodríguez CH, Cruz NOS (2016) Changes in bacterial populations during bioremediation of soil contaminated with petroleum hydrocarbons. Water Air Soil Pollut 227:91. doi:10.1007/s11270-016-2789-z

Eckard AD, Muthukumarappan K, Gibbons W (2013a) A review of the role of amphiphiles in biomass to ethanol conversion. Appl Sci 3:396–419. doi:10.3390/app3020396

Eckard AD, Muthukumarappan K, Gibbons W (2013b) Enzyme recycling in a simultaneous and separate saccharification and fermentation of corn stover: comparing polymeric micelles of surfactants and polypeptides. Bioresour Technol 132:202–209

El-Sheshtawy HS, Aiada I, Osmanb ME, Abo-Elnasr AA, Kobisya AS (2015) Production of biosurfactant from *Bacillus licheniformis* for microbial enhanced oil recovery and inhibition the growth of sulfate reducing bacteria. Egypt J Petr 24(2):155–162

Eriksson T, Borjesson J, Tjerneld F (2002) Mechanism of surfactant effect in enzymatic hydrolysis of lignocellulose. Enzym Microb Technol 31:353–364

Felse PA, Shah V, Chan J (2007) Sophorolipid biosynthesis by *Candida bombicola* from industrial fatty acid residues. Enz Microbiol Technol 40:316–323

Franzetti A, Gandolfi I, Bestetti G, Smyth TJ, Banat IM (2010) Production and applications of trehalose lipid biosurfactants. Eur J Lipid Sci Tech 112:617–627

Fu F, Wang Q (2011) Removal of heavy metal ions from wastewaters: a review. J Env Manag 92(3):407–418

GAF (1950) General Aniline and Film Corp. for their surface active products. For an example of one of GAF Corp's. early advertisements promoting their trademarked surfactants. Business Week, March 11, p 42

Gerken HG, Donohoe B, Knoshaug EP (2013) Enzymatic cell wall degradation of *Chlorella* vulgaris and other microalgae for biofuels production. Planta 237:239–253. doi:10.1007/s00425-012-1765-0

Gerson OF, Zajic JE (1978) Surfactant production from hydrocarbons by *Corynebacterium lepus,* sp. nov. and *Pseudomonas asphaltenicus*, sp. nov. dev. Ind J Microbiol 19:577–599

Global Food Policy Report (2016) International Food Policy Research Institute (IFPRI), Washington, DC. ISBN: 978-0-89629-582-7, pp 23–34. doi:10.2499/9780896295827

Golabi E (2016) Experimental study of effect of microbial enhanced oil recovery on rag Sefid reservoir. Int J Chem Stud 4(1):43–45

Green JM, Beestman GB (2007) Recently patented and commercialized formulation and adjuvant technology. Crop Protec 26(3):320–327

Greenwell HC, Loyd-Evans M, Wenner C (2012) Biofuels, science and society. Interface Focus 3:1–4

Gregg D, Saddler JN (1996) Factors affecting cellulose hydrolysis and the potential of enzyme recycle to enhance the efficiency of an integrated wood to ethanol process. Biotechnol Bioeng 51:375–383

Halim R, Danquah MK, Webley PA (2012) Extraction of oil from microalgae for biodiesel production: a review. Biotechnol Adv 30(3):709–732. doi:10.1016/j.biotechadv.2012.01.001

Hall M, Bansal P, Lee JH, Realff MJ, Bommarius AS (2010) Cellulose crystallinity-a key predictor of the enzymatic hydrolysis rate. FEBS J 277(6):1571–1582

Herman DC, Maier RM (2002) Biosynthesis and applications of glycolipid and lipopeptide biosurfactants. In: Kuo TM, Gardner HW (eds) Lipid biotechnology. Marcel Dekker, New York, pp 629–654

Hsieh CC, Cannella D, Jørgensen H, Felby C, Thygesen LG (2015) Cellobiohydrolase and endoglucanase respond differently to surfactants during the hydrolysis of cellulose. Biotechnol Biofuels 8(52):1–10. doi:10.1186/s13068-015-0242-y

Huang WC, Kim JD (2013) Cationic surfactant-based method for simultaneous harvesting and cell disruption of a microalgal biomass. Bioresour Technol 149:579–581

Information Handling Services (2016) Chemical Economics Handbook: Surfactants, household detergents and their raw materials, p 16–19. (www.ihs.com/products/surfactants-household-detergents-chemical-economics-handbook.html)

Ikegami M, Whitsett JA, Jobe A, Ross G, Fisher J, Korfhagen T (2000) Surfactant metabolism in SP-D gene-targeted mice. Am J Physiol Lung Cell Mol Physiol 279(3):468–476

Ishigami Y, Zhang Y, Ji F (2000) Spiculisporic acid. Functional development of biosurfactants. Chim Oggi 18:32–34

Ivankovic T, Hrenovic J (2010) Surfactants in the environment: a review. Arh Hig Rada Toksikol 61:95–110. doi:10.2478/10004-1254-61-2010-1943

Jadhav M, Kalme S, Tamboli D (2011) Rhamnolipid from *Pseudomonas desmolyticum* NCIM-2112 and its role in the degradation of brown 3REL. J Basic Microbiol 51:1–12

Jaiswal M, Dudhe R, Sharma PK (2015) Nanoemulsion: an advanced mode of drug delivery system. 3. Biotech 5(2):123–127. doi:10.1007/s13205-014-0214-0
Jeoh T, Ishizawa CI, Davis MF, Himmel ME, Adney WS, Johnson DK (2007) Cellulase digestibility of pretreated biomass is limited by cellulose accessibility. Biotechnol Bioeng 98(1):112–122
Juwarkar AA, Nair A, Dubey KV, Singh SK, Devotta S (2007) Biosurfactant technology for remediation of cadmium and lead contaminated soils. Chemosphere 68:1996–2002
Juwarkar AA, Dubey KV, Nair A, Singh SK (2008) Bioremediation of multi-metal contaminated soil using biosurfactant—a novel approach. Ind J Microbiol 48:142–146
Kaar WE, Holtzapple M (1998) Benefits from tween during enzymatic hydrolysis of corn stover. Biotechnol Bioeng 59:419–427
Kapadia SG, Yagnik BN (2013) Current trend and potential for microbial biosurfactants. Asian J Exp Biol Sci 4:1–8
Karmee SK, Lin CSK (2014) Valorisation of food waste to biofuel: current trends and technological challenges. Sustainable Chem Proc 2:22–32
Kim S, Dale BE (2004) Global potential bioethanol production from wasted crops and crop residues. Biomass Bioenergy 26:361–375
Kim J, Grate JW, Wang P (2006) Nanostructures for enzyme stabilization. Chem Eng Sci 61(3):1017–1026
Kim JH, Lee JC, Pak D (2011) Feasibility of producing ethanol from food waste. Waste Manag 31:2121–2125
Kiran GS, Sabu A, Selvin J (2010) Synthesis of silver nanoparticles by glycolipid biosurfactant produced from marine *Brevibacterium casei* MSA19. J Biotechnol 148:221–225
Kralova I, Sjoblom J (2009) Surfactants used in food industry: a review. J Dispers Sci Technol 30:1363–1383
Krieger N, Doumit C, David AM (2010) Production of microbial biosurfactants by solid-state cultivation. Adv Exp Med Biol 672:203–210
Kristensen JB, Borjesson J, Maria H, Tjerneld BF, Jorgensen H (2007) Use of surface active additives in enzymatic hydrolysis of wheat straw lignocellulose. Enz Microb Technol 40:888–895
Kugler JH, Le Roes-Hill M, Syldatk C, Hausmann R (2015) Surfactants tailored by the class Actinobacteria. Front Microbiol 6:212–219. doi:10.3389/fmicb.2015.00212
Kumar R, Tabatabaei M, Karimi K, Horváth IS (2016) Recent updates on lignocellulosic biomass derived ethanol – a review. Biofuel Res J 9:347–356
Lai YJS, De Francesco F, Aguinaga A, Parameswaran P, Rittmanna BE (2016) Improving lipid recovery from *Scenedesmus* wet biomass by surfactant-assisted disruption. Green Chem 18:1319–1326
Lal R (2005) World crop residues production and implications of its use as a biofuel. Environ Int 31(4):575–584
Ławniczak L, Marecik R, Chrzanowski L (2013) Contributions of biosurfactants to natural or induced bioremediation. Appl Microbiol Biotechnol 97:2327–2339
Lee DH (2011) Algal biodiesel economy and competition among biofuels. Bioresour Technol 102:43–49
Lee RA, Lavoie JM (2013) From first – to third-generation biofuels: challenges of producing a commodity from a biomass of increasing complexity. Animal Front 3(2):6–11
Li Y, Sun Z, Ge X, Zhang J (2016) Effects of lignin and surfactant on adsorption and hydrolysis of cellulases on cellulose. Biotechnol Biofuels 9(20):1–10. doi:10.1186/s13068-016-0434-0
Liang K, Zhang Q, Cong W (2012) Enzyme-assisted aqueous extraction of lipid from microalgae. J Agric Food Chem 60(47):11771–11776. doi:10.1021/jf302836v
Liang LX, Qing QX, Ming LH, Hao LZ, Xin ZN, Hao HJ, Xia PY (2016) Enhancement of lignosulfonate-based polyoxyethylene ether on enzymatic hydrolysis of lignocelluloses. Indus Crops Prod 80:86–92
Liu JF, Mbadinga SM, Yang SZ, Gu JD, Mu BM (2015) Chemical structure, property and potential applications of biosurfactants produced by *Bacillus subtilis* in petroleum recovery and spill mitigation. Int J Mol Sci 16:4814–4837. doi:10.3390/ijms16034814

Luo L, van der Voet E, Huppes G (2010) Biorefining of lignocellulosic feedstock—technical, economic and environmental considerations. Bioresour Technol 101(13):5023–5032
Maier RM, Chávez SG (2000) *Pseudomonas aeruginosa* rhamnolipids: biosynthesis and potential applications. Appl Microbiol Biotechnol 54:625–633
Mainkar AR, Jolly CI (2001) Formulation of natural shampoos. Int J Cosmetic Sci 23:59–62
Makkar RS, Cameotra SS (1999) Biosurfactant production by microorganisms on unconventional carbon sources. J Surfact Deterg 2:2–16
Makkar RS, Cameotra SS (2002) An update on use of unconventional substrates for biosurfactants production and their new applications. Appl Microbiol Biotechnol 58:428–434
Makkar RS, Banat IM, Cameotra SS (2011) Advances in utilization of renewable substrates for biosurfactant production. AMB Express 1:5–17
Marchant R, Banat IM (2012) Microbial biosurfactants: challenges and opportunities for future exploitation. Trends Biotechnol 30:558–565. doi:10.1016/j.tibtech.2012.07.003
Matsakas L, Kekos D, Loizidou M, Christakopoulos P (2014) Utilization of household food waste for the production of ethanol at high dry material content. Biotechnol Biofuels 7:4–12
McClure CD, Schiller NL (1996) Inhibition of macrophage phagocytosis by *Pseudomonas aeruginosa* rhamnolipids in vitro and in vivo. Curr Microbiol 33:109–117
Menon V, Rao M (2012) Trends in bioconversion of lignocellulose: biofuels, platform chemicals and biorefinery concept. Prog Energ Combustion Sci 38(4):522–550
Mesquita JF, Ferraz A, Aguiar A (2016) Alkaline-sulfite pretreatment and use of surfactants during enzymatic hydrolysis to enhance ethanol production from sugarcane bagasse. Bioprocess Biosyst Eng 39:441–448. doi:10.1007/s00449-015-1527-z
Min BC, Bhayani BV, Jampana V, Ramarao BV (2015) Enhancement of the enzymatic hydrolysis of fines from recycled paper mill waste rejects. Bioresour Bioproc 2(40):1–10. doi:10.1186/s40643-015-0068-2
Mishra M, Muthuprasanna P, Surya prabha K, Rani PS, Babu IAS, Chandiran IS, Arunachalam G, Shalini S (2009) Basics and potential applications of surfactants – a review. Int J PharmTech Research, 1: 1354-1365, ISSN:0974-4304
Morsy SMI (2014) Review article: role of surfactants in nanotechnology and their applications. Int J Curr Microbiol App Sci 3(5):237–260
Mujumdar S, Bashetti S, Pardeshi S, Thombre RS (2016) Industrial applications of biosurfactants. In: Thangadurai D, Sangeetha J (eds) Industrial biotechnology: sustainable production and bioresource utilization. CRC Press, Boca Raton, pp 61–90. ISBN 177188262X, 9781771882620
Mulligan CN (2005) Environmental applications for biosurfactants. Environ Pollut 133:183–198
Mulligan CN, Gibbs BF (2004) Types, production and applications of biosurfactants. Proc Ind Nat Sci Acad 1:31–55
Naik SN, Goud VV, Rout PK, Dalai AK (2010) Production of first and second generation biofuels: a comprehensive review. Renew Sust Energ Rev 14:578–597
Nanda S, Mohammad J, Reddy S, Kozinski J, Dalai A (2014) Pathways of lignocellulosic biomass conversion to renewable fuels. Biomass Conver Bioref 4:157–191. doi:10.1007/s13399-013-0097-z
Naqvi M, Yan J (2015) First-generation biofuels. Handbook of clean energy systems. Wiley, Chichester, pp 1–18. doi:10.1002/9781118991978.hces207
Nasirpour N, Mousavi SM, Shojaosadati SA (2014) A novel surfactant-assisted ionic liquid pretreatment of sugarcane bagasse for enhanced enzymatic hydrolysis. Bioresour Technol 169:33–37
Nievas ML, Commendatore MG, Estevas JL, Bucala V (2008) Biodegradation pattern of hydrocarbons from a fuel oil-type complex residue by an emulsifier-producing microbial consortium. J Hazard Matter 154:96–104
Oberoi HS, Vadlani PV, Madl RL, Saida L, Abeykoon JP (2010) Ethanol production from orange peels: two-stage hydrolysis and fermentation studies using optimized parameters through experimental design. J Agric Food Chem 58:3422–3429
Park JW, Takahata Y, Kajiuchi T, Akehata T (1992) Effects of nonionic surfactant on enzymatic hydrolysis of used newspaper. Biotechnol Bioeng 39:117–120

Parnthong J, Kungsanant S (2014) Statistical optimization for application of nonionic surfactants in enzymatic hydrolysis of palm fiber for ethanol production. Int J Chem Eng App 5:23–25
Peng JF, Song YH, Yuan P, Cui XY, Qiu GL (2009) The remediation of heavy metals contaminated sediment. J Hazard Mat 161(30):633–640
Pereira BL, Francisco SM, da Silva SS (2016) Recent advances in sustainable production and application of biosurfactants in Brazil and Latin America. Indus Biotechnol 12(1):31–39. doi:10.1089/ind.2015.0027
Perlack RD, Wright LL, Turhollow AF, Graham RL, Stokes BJ, Erbach DC (2005) Biomass as a feedstock for a bioenergy and bioproducts industry: the technical feasibility of a billion-ton annual supply. Oak Ridge National laboratory, US Department of Agriculture (USDA), pp 1–54. Available electronically at: http://www.osti.gov/bridge
Perlack RD, Stokes BJ (2011) US billion-ton update: biomass supply for a bioenergy and bioproducts industry. US Department of Energy, Oak Ridge National Laboratory, Oak Ridge
Pleissner D, Lam WC, Sun Z, Lin CSK (2013) Food waste as nutrient source in heterotrophic microalgae cultivation. Bioresour Technol 137:139–146
Pleissner D, Kwan TH, Lin CSK (2014) Fungal hydrolysis in submerged fermentation for food waste treatment and fermentation feedstock preparation. Bioresour Technol 158:48–54
Płociniczak MP, Płaza GA, Seget ZP, Cameotra SS (2011) Environmental applications of biosurfactants: recent advances. Int J Mol Sci 12:633–654. doi:10.3390/ijms12010633
Pothiraj C, Kanmani P, Balaji P (2006) Bioconversion of lignocellulose materials. Mycobiol 34(4):159–165
Qing Q, Yang B, Wyman CE (2010) Impact of surfactants on pretreatment of corn stover. Bioresour Technol 101:5941–5951
Rahman PKSM, Gakpe E (2008) Production, characterization and applications of biosurfactants-review. Biotechnol 7:360–370. doi:10.3923/biotech.2008.360.370
Rawat R, Srivastava N, Chadha BS, Oberoi HS (2014) Generating fermentable sugars from rice straw using functionally active cellulolytic enzymes from *Aspergillus niger* HO. Energ Fuels 28:5067–5075. doi:10.1021/ef500891g
Reddy AS, Chen CY, Baker SC, Chen CC, Jean JS, Fan CW, Chen HR, Wang JC (2009) Synthesis of silver nanoparticles using surfactin: a biosurfactant stabilizing agent. Mater Lett 63:1227–1230
Reis RS, Pacheco GJ, Pereira AG, Freire DMG (2013) Biosurfactants: production and applications. Chapter 2 in biodegradation-life of science, pp 31–63. http://dx.doi.org/10.5772/56144
Rosa Estela QCE, Luis FMJ (2013) Hydrolysis of biomass mediated by cellulases for the production of sugars. In: Chandel AK, da- Silva SS (eds) Sustainable degradation of lignocellulosic biomass-techniques, applications and commercialization. Rijeka, Croatia, pp 119–155. doi:10.5772/53719
Rosenberg E, Rubinovitz C, Legmann R, Ron EZ (1988) Purification and chemical properties of *Acinetobacter calcoaceticus* A2 Biodispersan. Appl Environ Microbiol 54:323–326
Sachdev DP, Cameotra SS (2013) Biosurfactants in agriculture. Appl Microbiol Biotechnol 97:1005–1016
Saharan BS, Sahu RK, Sharma D (2011) A review on biosurfactants: fermentation, current developments and perspectives. Genetic Eng Biotechnol J 29:1–14
Saini JK, Patel AK, Adsul M, Singhania RR (2016) Cellulase adsorption on lignin: a roadblock for economic hydrolysis of biomass. Renew Energy 98:29–42
Sajjadi S, Jahanzad F, Yianneskis M, Brooks BW (2003) Phase inversion in abnormal O/W/O emulsions: effect of surfactant hydrophilic–lipophilic balance. Ind Eng Chem Res 42(15):3571–3577. doi:10.1021/ie021044e
Salager JL (2002) Surfactants: types and uses. Laboratory of formulation, interfaces rheology and processes, FIRP booklet E300:1–48
Salam KA, Velasquez-Orta SB, Harvey AP (2016) Surfactant-assisted direct biodiesel production from wet *Nannochloropsis oculata* by in situ transesterification/reactive extraction. Biofuel Res J 9:366–371

Samiey B, Cheng CH, Wu J (2014) Effects of surfactants on the rate of chemical reactions. J Chem Article ID 908476:1–14. http://dx.doi.org/10.1155/2014/908476

Santos DKF, Rufino RD, Luna JM, Santos VA, Salgueiro AA, Sarubbo LA (2013) Synthesis and evaluation of biosurfactant produced by *Candida lipolytica* using animal fat and corn steep liquor. J Pet Sci Eng 105:43–50

Santos DKF, Rufino RD, Luna JM, Santos VA, Sarubbo LA (2016) Review: biosurfactants multifunctional biomolecules of the 21st century. Int J Mol Sci 17:401–432. doi:10.3390/ijms17030401

Sarkar N, Ghosh SK, Bannerjee S, Aikat K (2012) Bioethanol production from agricultural wastes: an overview. Renew Energ 37:19–27

Sawadogo A, Otoidobiga HC, Nitiema LW, Traore AS, Dianou D (2016) Optimization of hydrocarbons biodegradation by bacterial strains isolated from wastewaters in Ouagadougou, Burkina Faso: case study of SAE 40/50 used oils and diesel. J Agric Chem Environ 5:1–11. doi.org/10.4236/jacen.2016.51001

Schramm LL, Stasiuk EN, Marangoni GD (2003) Surfactants and their applications. Ann Rep Prog Chem 99:30–48. doi:10.1039/b208499f

Sekhon BS (2013) Surfactants: pharmaceutical and medicinal aspects. J Pharma Technol Res Manage 1:11–36

Sen R (2008) Biotechnology in petroleum recovery: the microbial EOR. Prog Energ Combust 34:714–724

Seo JY, Kumar RP, Kim B, Seo JC, Park JY, Na JG, Jeon SG, Park SB, Lee K, Oh YK (2016) Downstream integration of microalgae harvesting and cell disruption by means of cationic surfactant-decorated Fe_3O_4 nanoparticles. Green Chemi 18:1–9. doi:10.1039/c6gc00904b

Sharma R, Rawat R, Bhogal RS, Oberoi HS (2015) Multi-component thermostable cellulolytic enzyme production by *Aspergillus niger* HN-1 using pea pod waste: appraisal of hydrolytic potential with lignocellulosic biomass. Process Biochem 50:696–704

Sheng J, Vannela R, Rittmann BE (2011) Evaluation of cell-disruption effects of pulsed-electric-field treatment of *Synechocystis* PCC 6803. J Env Sci Technol 8(8):3795–3802

Shete AM, Wadhawa G, Banat IM, Chopade BA (2006) Mapping of patents on bioemulsifier and biosurfactant: a review. J Scient Indus Res 65:91–115

Sifour M, Al-Jilawi MH, Aziz GM (2007) Emulsification properties of biosurfactant produced from *Pseudomonas aeruginosa* RB 28. Pak J Biol Sci 10:1331–1335

Silva RC, Almeida DG, Rufino RD, Luna JM, Santos VA, Sarubbo LA (2014) Applications of biosurfactants in the petroleum industry and the remediation of oil spills. Int J Mol Sci 15:12523–12542

Singh P, Cameotra SS (2004) Enhancement of metal bioremediation by use of microbial surfactants. Biochem Biophy Res Commun 319:291–297

Sipos B, Szilagyi M, Sebestyen Z, Perazzini R, Dienes D, Jakab E, Crestini C, Reczey K (2011) Mechanism of the positive effect of poly(ethylene glycol) addition in enzymatic hydrolysis of steam pretreated lignocelluloses. C R Biol 334:812–823

Soni SK, Batra N, Bansal N, Soni R (2010) Bioconversion of sugarcane bagasse into second generation bioethanol after enzymatic hydrolysis with-in house produced cellulases from *Aspergillus* sp. S_4B_2F. Bioresources 5(2):741–758

Srivastava N, Rawat R, Sharma R, Oberoi HS, Srivastava M, Singh J (2014) Effect of nickel–cobaltite nanoparticles on production and thermostability of cellulases from newly isolated thermotolerant *Aspergillus fumigatus* NS (class: Eurotiomycetes). Appl Biochem Biotechnol 174:1092–1103. doi:10.1007/s12010-014-0940-0

Stevens CE (1969) In Kirk-Othmer encyclopedia of chemical technology, vol 19, 2nd edn. Wiley, New York, pp 507–593

Suthar H, Hingurao K, Desai A, Nerurkar A (2008) Evaluation of bioemulsifier mediated microbial oil recovery using sand pack column. J Microbiol Methods 75:225–230

Teichmann B, Linne U, Hewald S (2007) A biosynthetic gene cluster for a secreted cellobiose lipid with antifungal activity from *Ustilago maydis*. Mol Microbiol 66:525–533
Toren A, Navon-Venezia S, Ron EZ, Rosenberg E (2001) Emulsifying activity of purified alas an proteins from *Acinetobacter radioresistens*. Appl Environ Microbiol 67:110–1106
Transparency Market Research (2014) Microbial biosurfactants market (rhamnolipids, sophorolipids, mannosylerythritol lipids for household detergents, industrial & institutional cleaners, personal care, oilfield chemicals, agricultural chemicals, food processing, textile and other applications – global industry analysis, size, share, growth, trends and forecast, 2014–2020. Available at: www.transparencymarketresearch.com/microbial-biosurfactants-market.html
Ulloa G, Coutens C, Sánchez M, Jineiro J, Fábregas J, Deive FJ, Rodríguez A, Nuneza MJ (2012) On the double role of surfactants as microalga cell lysis agents and antioxidants extractants. Green Chem 14:1044–1051
Urum K, Pekdemir T (2004) Evaluation of biosurfactants for crude oil contaminated soil washing. Chemosphere 57:1139–1150
Wang H, Mochidzuki K, Kobayashi S (2013) Effect of bovine serum albumin (BSA) on enzymatic cellulose hydrolysis. Appl Biochem Biotechnol 170:541–551. doi:10.1007/s12010-013-0208-0
Wright M, Brown R (2007) Comparative economics of biorefineries based on the biochemical and thermochemical platforms. Biofuels Bioprod Biorefin 1:49–56
Xin L, Hong-ying H, Yu-ping Z (2011) Growth and lipid accumulation properties of a freshwater microalga *Scenedesmus* sp. under different cultivation temperature. Bioresour Technol 102:3098–3102
Yakimov M, Amro M, Bock M (1997) The potential of *Bacillus licheniformis* strains for in situ enhanced oil recovery. J Pet Sci Eng 18:147–160
Yan S, Li J, Chen X, Wu J, Wang P, Ye J, Yao J (2011) Enzymatical hydrolysis of food waste and ethanol production from the hydrolysate. Renew Energ 36:1259–1265
Yan S, Chen X, Wu J, Wang P (2013) Pilot scale production of fuel ethanol from concentrated food waste hydrolysates using *Saccharomyces cerevisiae* H058. Bioprocess Biosyst Eng 36:937–946
Yang X, Lee JH, Yoo HY, Shin HY, Thapa LP, Park C, Kim SW (2014) Production of bioethanol and biodiesel using instant noodle waste. Bioprocess Biosyst Eng. doi:10.1007/s00449-014-1135-3
Yernazarova A, Kayirmanova G, Baubekova A, Zhubanova A (2016) Chapter 5: Microbial enhanced oil recovery. "chemical enhanced oil recovery (cEOR) – a practical overview",. Ed. Laura RZ, InTech, Rijeka ISBN 978-953-51-2701-7, doi: 10.5772/64805
Ying GG (2006) Fate, behavior and effects of surfactants and their degradation products in the environment. Environ Int 32:417–431
Yoon SH, Robyt JF (2005) Activation and stabilization of 10 starch-degrading enzymes by Triton X-100, polyethylene glycols, and polyvinyl alcohols. Enzyme Microb Technol 37:556–562
Zeng Y, Zhao S, Yang S, Ding SY (2014) Lignin plays a negative role in the biochemical process for producing lignocellulosic biofuels. Curr Opin Biotechnol 27:38–45
Zhang Z, Donaldson H, Ma X (2012) Advancements and future directions in enzyme technology for biomass conversion. Biotechnol Adv 30(4):913–919
Zosim Z, Gutnick DL, Rosenberg E (1982) Properties of hydrocarbon-in-water emulsions stabilized by *Acinetobacter* RAG-1 emulsan. Biotechnol Bioeng 24:281–292
Zulianello L, Canard C, Köhler T, Caille D, Lacroix JS, Meda P (2006) Rhamnolipids are virulence factors that promote early infiltration of primary human airway epithelia by *Pseudomonas aeruginosa*. Infect Immun 74:3134–3147

Reetika Sharma has completed her M.Sc. (Microbiology) from Chaudhary Sarwan Kumar Himachal Pradesh Agricultural University, Palampur, Himachal Pradesh, India and Ph.D. (Microbiology) from Punjab Agricultural University, Ludhiana, India. Her current research interests are: fermentation and microbial biotechnology, enzyme technology, biofuels and microbial diversity analysis in the environment.

Harinder Singh Oberoi is M.Sc. and PhD (Microbiology) from Punjab Agricultural University, Ludhiana, India and Post-Doc from Kansas State University, Manhattan, KS, USA. His current research interests are biovalorization of agricultural/ horticultural residues, fermentation, food safety and development of functional foods.

Part II

Microbes in Health

Autophagy and Bacterial Pathogenesis: An Interactive Overview

4

Madhu Puri, Trinad Chakraborty, and Helena Pillich

Abstract

Autophagy is a cellular homeostasis-essential mechanism in which damaged organelles, protein aggregates, or pathogens are enclosed in double-membraned autophagosomes and are subsequently degraded by lysosomal enzymes. Autophagic control of bacterial replication promotes bacterial clearance during infection. However, several pathogenic bacteria have devised strategies to escape/inhibit autophagy, so as to enable their growth. This review discusses the role of autophagy in the pathogenesis of some intracellular bacteria: recent mechanisms by which bacteria are targeted by autophagy and also the strategies employed by bacteria to counter autophagy.

Keywords

Autophagy • Intracellular bacteria • Pathogenesis

4.1 Introduction

Autophagy is an evolutionarily conserved eukaryotic cellular degradation system wherein cargo molecules are enclosed in double-membraned vacuoles known as autophagosomes and are subsequently degraded by lysosomal hydrolases. Damaged cellular organelles, protein aggregates, or pathogens constitute autophagic cargo (Lin and Baehrecke 2015). The term autophagy, coined by the Belgian biochemist Christian de Duve in 1963, is derived from the Greek words "auto" meaning self and "phagy" meaning eating (deReuck and Cameron 1963). Several factors can

M. Puri (✉) • T. Chakraborty • H. Pillich
Institute of Medical Microbiology, Justus-Liebig University, Schubertstrasse 81, 35392 Giessen, Germany
e-mail: madhu.006.2009@gmail.com

P. Shukla (ed.), *Recent Advances in Applied Microbiology*,
DOI 10.1007/978-981-10-5275-0_4

trigger the induction of autophagy in cells, which include amino acid starvation, low cellular energy levels, withdrawal of growth factors, hypoxia, oxidative stress, endoplasmic reticulum (ER) stress, damaged cellular organelles, and infection. Autophagy is an essential part of cellular homeostasis, and defects in autophagy are associated with many diseases, including neurodegenerative diseases, diabetes, cardiomyopathy, tumorigenesis, fatty liver, and Crohn's disease (Burman and Ktistakis 2010). Moreover, autophagy is an indispensable cellular defense mechanism against intracellular pathogens (Lin and Baehrecke 2015).

Autophagy can be of three types: macro-autophagy, micro-autophagy, and chaperone-mediated autophagy. Macro-autophagy involves the entrapment of cytoplasmic cargo into autophagosomes and fusion with lysosomes where the cargo is degraded. Micro-autophagy comprises the direct lysosomal uptake of cytosolic components by the invagination of the lysosomal membrane. In chaperone-mediated autophagy, chaperone proteins recognized by the lysosomal membrane receptor lysosome-associated membrane protein 2A form a complex with cargo and are translocated across the lysosomal membrane (Glick et al. 2010). In this article, the term "autophagy" refers to the process of macro-autophagy. Autophagy can also be classified as selective and nonselective. Selective autophagy is mediated by autophagy receptors or cargo receptors which specifically recognize cargo for degradation, whereas in nonselective autophagy, cargo is indiscriminately cloistered into developing autophagosomes (Moy and Cherry 2013).

Autophagy is mediated by autophagy-related genes (Atg), first identified in *Saccharomyces cerevisiae* (Tsukada and Ohsumi 1993); subsequently, orthologs of Atg have been discovered in many species (Mizushima et al. 2011). Autophagy induction leads to the activation of the Unc-51 like autophagy-activating kinase 1 (ULK1), which in turn activates Beclin-1 (mammalian homologue of Atg6) (Di Bartolomeo et al. 2010; Chan 2012; Russell et al. 2013). The class III phosphatidylinositol 3-phosphate kinase Vps34 phosphorylates phosphatidylinositol to yield phosphatidylinositol 3-phosphate (PtdIns(3)P) which provides a docking site for WD-repeat protein which interacts with phosphoinositides (WIPI) protein family. The WIPI proteins then promote the formation of autophagy isolation membranes (Proikas-Cezanne et al. 2015). Atg12 binds to Atg5 and subsequently to Atg16L1 to form a complex which binds and activates Atg3 (Hanada et al. 2007). Atg3 attaches with mammalian homologues of the ubiquitin-like yeast protein Atg8 (microtubule-associated protein 1 light chain 3 [LC3]) to phosphatidylethanolamine (PE) on the surface of autophagosomes and thereby contributes to the closure of autophagosomes (Fujita et al. 2008). Thus, the inactive cytosolic form of LC3 (LC3-I) is converted to the active membrane-bound form (LC3-II). Lysosomes subsequently fuse with closed autophagosomes, thus leading to the degradation of cargo (Fig. 4.1).

Selective autophagy is mediated by autophagy receptors which specifically recognize ubiquitinated cargo and deliver it to autophagosomes. Autophagy receptors are, therefore, characterized by the presence of an ubiquitin-binding domain which binds to ubiquitin molecules on the cargo surface and an LC3-interacting region which interacts with LC3 present on autophagosomal membranes (Johansen and Lamark 2011). Recent advances in the field of selective autophagy have led to the

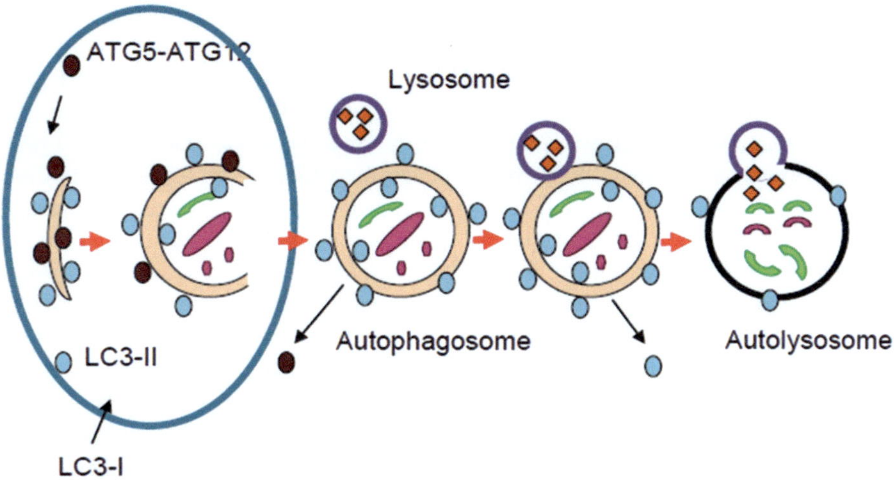

Fig. 4.1 Stages in autophagy

discovery of the presence of seven autophagy receptors: these include sequestosome 1 (SQSTM1, also known as p62), nuclear dot protein 52 (NDP52), optineurin (OPTN), neighbor of BRCA1 gene 1 (NBR1), TAX1-binding protein (TAX1BP1), Toll-interacting protein (Tollip), and nuclear receptor subfamily 1, group D, member 1 (NR1D1) (Bjørkøy et al. 2005; Thurston et al. 2009; Osawa et al. 2011; Kirkin et al. 2009; Newman et al. 2012; Lu et al. 2014; Chandra et al. 2015a). The role of autophagy receptors in sensing both Gram-positive and Gram-negative intracellular bacteria is also known (Zheng et al. 2009; Dupont et al. 2009; Ogawa et al. 2011; Khweek et al. 2013).

Over the years, autophagy has been recognized both as a pro- and anti-infection mechanism for pathogenic bacteria. The aim of this review is to compile and compare the contribution of autophagy in cellular defense against pathogenic bacteria, as well as the survival strategies used by these bacteria in exploiting autophagy for their intracellular survival and growth.

4.2 Interaction of Bacteria with the Autophagy System of Host Cells

4.2.1 *Listeria monocytogenes*

L. monocytogenes is a Gram-positive, facultative, intracellular bacterium and the causative agent of listeriosis (Vázquez-Boland et al. 2001). Cytoplasmic *L. monocytogenes* induces autophagy in cells (Rich et al. 2003). Expression of the pore-forming toxin listeriolysin O (LLO) by *L. monocytogenes* activates autophagy, while phospholipases A and B are not essential for autophagy induction (Py et al. 2007). The peptidoglycan recognition protein LE (PGRP-LE) in *Drosophila* recognizes diaminopimelic acid-type peptidoglycans on *L. monocytogenes*, thereby leading to autophagy induction and intracellular growth restriction of *L. monocytogenes* (Yano et al. 2008). Another study has identified a novel antibacterial gene called Listericin in *Drosophila*, which is expressed in a PGRP-LE-dependent manner, thus leading to the growth inhibition of *L. monocytogenes* (Goto et al. 2010). The involvement of other PRRs in autophagy activation upon *L. monocytogenes* infection has also been reported: Toll-like receptor 2 and Nod-like receptors 1 and 2, acting via the downstream extracellular signal-regulated kinases, are involved in the activation of the autophagic response during infection with *L. monocytogenes* (Anand et al. 2011). LLO-dependent phagosomal lysis during *L. monocytogenes* infection triggers amino acid starvation, leading to autophagy induction (Tattoli et al. 2013). Another phenomenon, termed LC3-associated phagocytosis (LAP), is induced by LLO and is known to aid the formation of spacious *Listeria*-containing phagosomes (SLAPs: LC3-positive *L. monocytogenes*-containing phagosomes) which are LC3-positive single-membrane *L. monocytogenes*-containing phagosomes (Lam et al. 2013). SLAPs provide a safe niche for slow-growing *L. monocytogenes* to establish persistent infection (Birmingham et al. 2008). The autophagy receptors, SQSTM1 and NDP52, are recruited to intracellular *L. monocytogenes* (Yoshikawa et al. 2009; Mostowy et al. 2011). NDP52 interacts with galectin 8 which binds to host glycans exposed on damaged vacuole remnants and colocalizes with intracellular *L. monocytogenes* (Thurston et al. 2012). Thus, the induction of autophagy during *L. monocytogenes* infection is an essential event in controlling infection.

L. monocytogenes has developed certain strategies to evade autophagic recognition. The ActA protein (ubiquitously is distributed on the surface of *L. monocytogenes* and aids intracellular movement for the bacterium) recruits host cell actin machinery components, thus enabling *L. monocytogenes* to disguise itself as a host cell organelle and successfully evade autophagy. On the other hand, *L. monocytogenes* which lacks ActA (LmΔ*actA*) is ubiquitinated and consequently leads to SQSTM1 binding and entrapment by autophagosomes where it is degraded (Yoshikawa et al. 2009). Another report by Dortet et al. (2011) has shown that in the absence of ActA, internalin K (InlK) present in *L. monocytogenes* interacts with the major vault protein (MVP) to decorate its surface with MVP (in a manner similar to that of actin complex recruitment by ActA) and evades autophagy.

4.2.2 *Salmonella typhimurium*

S. typhimurium is a Gram-negative, facultative intracellular bacterium which can cause gastroenteritis (Holden 2002). Infection with *S. typhimurium* induces autophagy. The SipB protein of *S. typhimurium* leads to macrophage death by autophagy induction (Hernandez et al. 2003). The type III secretion system (TTSS) of *S. typhimurium* ruptures *Salmonella*-containing vacuoles (SCV), and the bacteria in these damaged SCVs are subsequently targeted by autophagy. These autophagy-targeted bacteria are ubiquitinated (Birmingham et al. 2006). The infection of epithelial cells with *S. typhimurium* triggers amino acid starvation due to SCV membrane damage, which, in turn, activates autophagy (Tattoli et al. 2012). The autophagy adaptor SQSTM1 was first found to be associated with bacterial autophagy when it was shown to colocalize with intracellular *S. typhimurium* (Zheng et al. 2009). Thurston et al. (2009) have reported the involvement of another adaptor, NDP52, in the autophagy of *S. typhimurium*. NDP52 binds the adaptor proteins Nap1 and Sintbad and subsequently recruits the TANK-binding kinase 1 (TBK1) to ubiquitinated *S. typhimurium* and thereby elicits an autonomous innate immune response. Both SQSTM1 and NDP52 are recruited to *S. typhimurium* independently of each other, with similar kinetics, and the depletion of either adaptor cripples autophagy. Additionally, immunofluorescence assays revealed that SQSTM1 and NDP52 do not colocalize but form non-overlapping microdomains surrounding *S. typhimurium*. The authors suggest that SQSTM1 and NDP52 work in conjunction to drive efficient antibacterial autophagy (Cemma et al. 2011). The danger receptor galectin 8 recruits NDP52 to damaged SCVs and restricts the proliferation of *S. typhimurium* by autophagy (Thurston et al. 2012). The phosphorylation of the autophagy adaptor OPTN by TBK1 enhances the LC3-binding affinity of OPTN, which consequently leads to increased autophagic clearance of *S. typhimurium* in HeLa cells (Wild et al. 2011). OPTN deficiency in zebrafish increases its mortality as a result of *Salmonella* infection (Chew et al. 2015). Recently, TAX1BP1, which was previously not known to be associated with xenophagy (bacterial autophagy), was found to be associated with the autophagy of *Salmonella*. This study has reported that the knockdown of TAX1BP1 results in the accumulation of ubiquitinated *S. typhimurium* in the cytosol of infected cells and that TAX1BP1 is recruited to ubiquitinated *S. typhimurium*. TAX1BP1 mediates the autophagic clearance of *S. typhimurium* via the actin-based motor protein myosin VI (Tumbarello et al. 2015). Additionally, NDP52 plays a double role in *S. typhimurium* autophagy: it targets bacteria to autophagosomes and subsequently promotes the maturation of *Salmonella*-containing autophagosomes by binding to LC3A, LC3B, GABARAPL2, and myosin VI (Verlhac et al. 2015). Shahnazari et al. (2010) have reported the involvement of diacylglycerol (DAG) in *Salmonella* autophagy. They have shown that DAG colocalizes with *Salmonella*-containing autophagosomes. Phosphatidic acid phosphatase, phospholipase D, and protein kinase C are required for the autophagy of *S. typhimurium*. Moreover, this study also states that DAG and SQSTM1 act in independent signaling pathways to mediate *S. typhimurium* autophagy. Autophagy as a consequence of *S. typhimurium* infection has been documented not just in vitro but also in vivo. The genetic inactivation of autophagy in *Caenorhabditis elegans* and *Dictyostelium discoideum* increases the intracellular replication of *S. typhimurium* and decreases animal lifespan, which subsequently leads to its death (Jia et al. 2009).

Like *L. monocytogenes*, *S. typhimurium* has also devised mechanisms to escape autophagy. *S. typhimurium* induces the formation of ubiquitinated aggregates in cultured cells, but the deubiquitinase activity of the *S. typhimurium* virulence protein SseL deubiquitinates SQSTM1-bound proteins found in *S. typhimurium*-induced aggregates and thereby reduces autophagic flux and favors bacterial replication (Thomas et al. 2012). The non-receptor tyrosine kinase focal adhesion kinase promotes the intracellular survival of *S. typhimurium* in macrophages by suppressing autophagy *via* the Akt-mTOR signaling pathway (Owen et al. 2014). Curiously, in contrast to previous reports, a recent study has reported that autophagy facilitates *S. typhimurium* replication in HeLa cells. This study states that a part of the cytosolic *S. typhimurium* population associates with SQSTM1 and/or LC3 and replicates quickly, thereby facilitating cell detachment and dissemination of *S. typhimurium* to neighboring cells. These contradictory results were proposed to be due to different infection times, cell lysis protocols, and cell types used which suggests cell tropism (Yu et al. 2014).

4.2.3 *Mycobacterium tuberculosis*

M. tuberculosis is an obligate pathogenic bacterium which causes human tuberculosis. Many significant developments have been made in the study of autophagy in mycobacterial infections. Autophagy is induced upon infection with *M. tuberculosis* and functions as an important defense mechanism against infection (Gutierrez et al. 2004). Autophagy has also been shown to protect mice from active tuberculosis by suppressing *M. tuberculosis* burden and inflammation. *Atg5* knockout mice exhibit increased bacterial load and extensive pulmonary inflammation with heightened IL-17 and IL-1α levels (Castillo et al. 2012). The apoptosis inhibitor of macrophages (AIM), a scavenger protein secreted by macrophages, enhances the mycobactericidal activity of macrophages by increasing autophagy as demonstrated by increased LC3II and Beclin 1 levels (Sanjurjo et al. 2013). TBK1 coordinates the assembly and function of the autophagic machinery and also phosphorylates SQSTM1 in response to *M. tuberculosis* infection. In addition, TBK1 induces the proinflammatory cytokine IL-1β, which leads to the autophagic elimination of *M. tuberculosis* (Pilli et al. 2012). Another pioneering study in the same year has reported that the transmembrane protein STING recognizes *M. tuberculosis* extracellular DNA which leads to its ubiquitination and subsequent recruitment of SQSTM1 and NDP52 (Watson et al. 2012). The involvement of SQSTM1 in the autophagy of *M. tuberculosis* has also been demonstrated in macrophages by Seto et al. (2012). Recently, a novel autophagy adaptor, NR1D1, has also been linked to mycobacterial autophagy. This study has reported that NR1D1 expression increases the number of acidic vacuoles and LC3-II levels in a time and concentration-dependent manner, as well as modulates lysosome biogenesis upon infection with *M. tuberculosis* (Chandra et al. 2015a). Another recent report by Watson et al. (2015) has stated that the cytosolic DNA sensor cyclic GMP-AMP synthase is essential for the activation of interferon production via the STING/TBK1/IRF3 pathway during infection of macrophages with *M. tuberculosis*. Micro-RNAs (miRNAs) are small, noncoding RNA molecules involved in RNA silencing and post-transcriptional regulation of gene expression (Ambros 2004). Increased expression of miR-155 enhances the autophagic activity of macrophages and thus promotes the maturation of mycobacterial phagosomes and decreases the survival rate of intracellular mycobacteria. Furthermore, miR-155 binds to the Ras homologue enriched in the brain (Rheb), a negative regulator of autophagy, accelerates autophagy, and eliminates the intracellular mycobacteria by the suppression of Rheb expression (Wang et al. 2013).

Numerous reports have highlighted the suppression of autophagy by *M. tuberculosis*. IL-27 inhibits IFN-γ- and starvation-induced autophagy, which blocks phagosome maturation and promotes the intracellular growth of *M. tuberculosis* (Sharma et al. 2014). The early secretory antigenic target 6 (ESAT-6) system 1 (ESX-1) secretion-associated protein B (EspB) of *M. tuberculosis* suppresses autophagosome formation and LC3B expression induced by IFN-γ stimulation by reducing the IFN-γ receptor 1 expression (Huang and Bao 2014). miRNAs, in addition to their ability to enhance autophagy, inhibit autophagy during *M. tuberculosis* infection. The overexpression of miR-30A inhibits autophagy, thereby promoting the proliferation of *M. tuberculosis* (Chen et al. 2015). miR-125a targets the UV radiation resistance-associated gene (UVRAG) to inhibit autophagy activation and antimicrobial responses to *M. tuberculosis* (Kim et al. 2015). A very recent report by Duan et al. (2016) has reported that the enhanced intracellular survival (EIS) gene of *M. tuberculosis* upregulates IL-10 *via* increased acetylation of histone H3 and thus activates the mTOR pathway to suppress autophagy. *M. tuberculosis* inhibits the recruitment of the late endosome marker RAB7 to form intermediate autophagic compartments called amphisomes and thus selectively modulates autophagic flux in macrophages (Chandra et al. 2015b). Another interesting study has recently revealed that *M. tuberculosis* escapes from macrophage phagosomes to the cytoplasm in a phospholipase A2-dependent manner. Phagosome escape by some strains of *M. tuberculosis* occurs quite early in some cases, taking place either soon after or concurrently with phagocytic uptake of the bacteria; this early escape is vital for these strains because of their reduced capacity to tolerate phagosomal stresses. Thus, escape from phagosomes to the cytoplasm serves as a "virulence-rescue" mechanism because cytoplasmic localization favors suppression of autophagy in macrophages (Jamwal et al. 2016).

4.2.4 *Shigella flexneri*

S. flexneri is a Gram-negative facultative intracellular bacterium which causes diarrhea. *S. flexneri* infection induces autophagy, both in vitro and in vivo (Suzuki et al. 2007; Chang et al. 2013). The TTSS IcsB mutant of *S. flexneri*, which is incapable of intracellular dissemination, is trapped within autophagosomes and undergoes autophagy, which implies that *S. flexneri* can escape autophagy by the expression of IcsB. Moreover, IcsA (VirG), a surface protein which is required for the intracellular actin-based motility of *S. flexneri*, binds to Atg5 to induce autophagy when bacteria are in the cytoplasm (Ogawa et al. 2005). IcsB binds cholesterol to facilitate *Shigella* escape from autophagy (Kayath et al. 2010). To get access into the cytoplasm, the pathogen lyses phagosomal vacuoles which are formed during host cell entry. *Shigella*-lysed vacuole membrane remnants are ubiquitinated, recognized by

SQSTM1 and LC3, and, subsequently, are degraded by autophagy (Dupont et al. 2009). Mostowy et al. (2010) have demonstrated that cytoskeletal components called septins are recruited with autophagy proteins to enclose *S. flexneri* in cage-like structures in the cytosol of host cells and, thus, restrict *Shigella* dissemination. In another study, the same group has reported the involvement of the adaptors SQSTM1 and NDP52 in the autophagy of cytoplasmic *Shigella*. Recruitment of SQSTM1 and NDP52 to *S. flexneri* is interdependent, and dependent on IcsA and IcsB. In addition, the depletion of NBR1 decreases SQSTM1 and NDP52 recruitment to *S. flexneri* (Mostowy et al. 2011). Mitochondrial proteins associated with *S. flexneri* promote the formation of septin cages to entrap bacteria for autophagy. It has also been shown that actin-polymerizing *S. flexneri* fragment mitochondria to escape from septin caging and eventual autophagic degradation (Sirianni et al. 2016).The pathogen evades autophagic recognition via competitive binding of IcsB to IcsA when bacteria are in the cytoplasm and thus prevent Atg5-binding to IcsA (Ogawa et al. 2005). In addition, the host factor formin binding protein 1 like (FNBB1L, also known as Toca-1) is recruited by cytoplasmic *Shigella via* IcsB which prevents LC3 association so as to restrict LAP and the recruitment of LC3 to vacuolar membrane remnants during early infection with *Shigella* and also inhibits autophagy late during infection (Baxt and Goldberg 2014).

4.2.5 *Legionella pneumophila*

L. pneumophila is an aerobic, pleomorphic, nonspore-forming, facultative intracellular Gram-negative bacterium which is the causative agent of legionellosis. *L. pneumophila* was first reported to be associated with autophagy by Swanson and Isberg (1995). They have shown that *L. pneumophila* replicates within nascent autophagosomal vacuoles which originate from the ER, and starvation-induced autophagy in macrophages promotes growth of *L. pneumophila*. Thus, their results support the hypothesis that *L. pneumophila* utilizes autophagy to establish a favorable intracellular niche for its replication in mouse macrophages. In contrast, autophagy is dispensable for the intracellular replication of *L. pneumophila* in *Dictyostelium discoideum* (Otto et al. 2004). Entry-triggered formation of *Legionella*-containing phagosomes is accompanied by their envelopment via ER membranes, Atg7 translocation, and subsequent LC3 binding (Amer and Swanson 2005). Khweek et al. (2013) have shown that *L. pneumophila*-containing vacuoles undergo ubiquitination followed by SQSTM1 recruitment prior to their eventual targeting for autophagy. Autophagy induced by 2-deoxy-D-glucose suppresses the intracellular growth of *L. pneumophila* in A/J mouse peritoneal macrophages (Matsuda et al. 2009).

L. pneumophila interferes with autophagy by using its effector protein RavZ to irreversibly inactivate LC3 attached to phosphatidylethanolamine on autophagosome membranes (Choy et al. 2012). The mechanism of RavZ-mediated LC3 inactivation by *L. pneumophila* has been elucidated by Horenkamp et al. (2015). They demonstrated that the PI3P-binding module and a catalytic domain helix of RavZ enable it to bind high-curvature membranes of autophagosomes, thereby maintaining localization to highly curved domains in autophagosome-intermediate membranes. These interactions enhance substrate affinity of RavZ and thus facilitate effective interference of *L. pneumophila* with host autophagy. Another recent study has reported that *L. pneumophila* disrupts host sphingolipid biosynthesis to inhibit autophagy *via* its effector protein sphingosine-1 phosphate lyase (LpSpl),which decreases sphingolipid levels essential for macrophage function. LpSpl is critical for efficient *L. pneumophila* infection, in vivo (Rolando et al. 2016).

4.2.6 *Streptococcus pyogenes*

S. pyogenes or group A *Streptococcus* (GAS) is a beta-hemolytic bacterium which causes a wide variety of diseases in humans, which include acute pharyngitis, rheumatic fever, acute glomerulonephritis, and toxic shock syndrome. The first study to report the induction of autophagy following GAS infection was published in 2004 (Nakagawa et al. 2004). This study showed that in HeLa cells, autophagy acts as a defense mechanism against GAS infection, as cytoplasmic GAS bacilli are enclosed in autophagosome-like compartments and are eventually degraded. Streptolysin O (SLO), the pore-forming toxin which allows GAS to escape from the endosome into the cytoplasm, is required for the autophagic process which suggests that cytosolic bacteria are detected by autophagosomal machinery (Sakurai et al. 2010). Indeed, the autophagy adaptor NDP52 is recruited to ubiquitinated GAS and delivers it to autophagosomes (von Muhlinen et al. 2010). Several studies have highlighted the contribution of Ras-related proteins (Rab) in the autophagy of GAS. Rab7 is necessary for the early phase of GAS-containing autophagosome-like vacuoles (GcAVs; Yamaguchi et al. 2009). Rab9A is recruited to GcAVs after the maturation of autophagosomes and is essential for GcAV enlargement and lysosomal fusion. Rab23 is required for the formation of GcAVs and for targeting GAS to autophagic vacuoles. Additionally, both Rab9A and Rab23 do not colocalize with autophagosomes under starvation conditions, which suggests that they function in different stages of autophagy during GAS infection (Nozawa et al. 2012). Another study has reported that Rab5 is involved in bacterial invasion and endosome fusion during GAS infection, whereas Rab7 seems to be multifunctional and is involved in bacterial invasion, endosome maturation, and autophagosome formation (Sakurai et al. 2010). Rab17 mediates the supply of membrane from recycling endosomes to GcAVs (Haobam et al. 2014).

GAS has evolved mechanisms for the subversion of autophagy. SLO stimulates its autophagy in pharyngeal keratinocytes, but the combined activity of SLO and its co-toxin NADase prevents the maturation of GAS-containing autophagosomes and thus prolongs the intracellular growth and survival of GAS (O'Seaghdha and Wessels 2013). Lu et al. (2015) have reported that the infection of endothelial cells with GAS results in the formation of GAS-containing vesicles with defective acidification, which results in their failure to maintain low pH of GAS-containing autophagosomes. This failure allows GAS to proliferate inside LAMP-1- and

LC3-positive vesicles. Furthermore, GAS produces a protease, SpeB, which degrades autophagic host proteins (SQSTM1, NDP52, NBR1) which allow autophagosomal evasion (Barnett et al. 2013).

4.2.7 *Streptococcus pneumoniae*

S. pneumoniae is a Gram-positive bacterium which is responsible for >1 million deaths in children annually (Tai 2016). Though often considered to be an extracellular pathogen (Henriques-Normark and Tuomanen 2013), an increasing number of studies have demonstrated that *S. pneumoniae* can invade non-phagocytic cells (Agarwal and Hammerschmidt 2009; Gradstedt et al. 2013). *S. pneumoniae* produces the pore-forming toxin pneumolysin (PLY) which induces autophagy in the epithelial cell line A549 via increased production of reactive-oxygen species (ROS) and mTOR inhibition. Interestingly, autophagy enables *S. pneumoniae* clearance (Li et al. 2015). Whether *S. pneumonia*, like *S. pyogenes*, is also able to evade or subvert autophagosomal degradation described for other intracellular bacteria requires further investigation.

4.3 Conclusions

The autophagy of bacteria, termed as xenophagy, is a defense mechanism of host cells aimed at the elimination of bacteria and subsequent infection. Intracellular bacteria are targeted for autophagy by various mechanisms: detection of degraded phagosomes formed during entry, recognition by PRRs or bacterial cytosolic ubiquitination, and autophagy adaptor binding. Pathogenic bacteria have also evolved various mechanisms to inhibit/evade autophagy, in order to stimulate their growth or establish a niche to enable persistent infection. Therefore, it can be concluded from these studies that the process of autophagy can be compared to a double-edged sword: on the one hand, it is a crucial defense mechanism against pathogenic infection, whereas on the other hand, pathogenic bacteria exploit it to favor their own growth and survival (Fig. 4.2).

However, on the brighter side of the picture, a few recent studies have suggested that autophagy induction by pathogenic bacteria can be targeted as a novel therapeutic strategy to treat infections. The induction of autophagy by isoniazid treatment decreases the proinflammatory responses induced by *M. tuberculosis* in macrophages (Kim et al. 2012). The antiprotozoan drug, nitazoxanide, and its metabolite, tizoxanide, strongly stimulate autophagy and thus inhibit mTORC1 signaling and the intracellular proliferation of *M. tuberculosis* (Lam et al. 2012). Autophagy induction by vitamin D inhibits the intracellular replication of both *M. tuberculosis* and human immunodeficiency virus type 1 (HIV1) (Campbell and Spector 2012). The peptide Tat-beclin 1, derived from the autophagy protein beclin 1, induces autophagy and inhibits infection with HIV-1, chikungunya, and West Nile virus, both in vitro and in vivo (Shoji-Kawata et al. 2013). These few studies

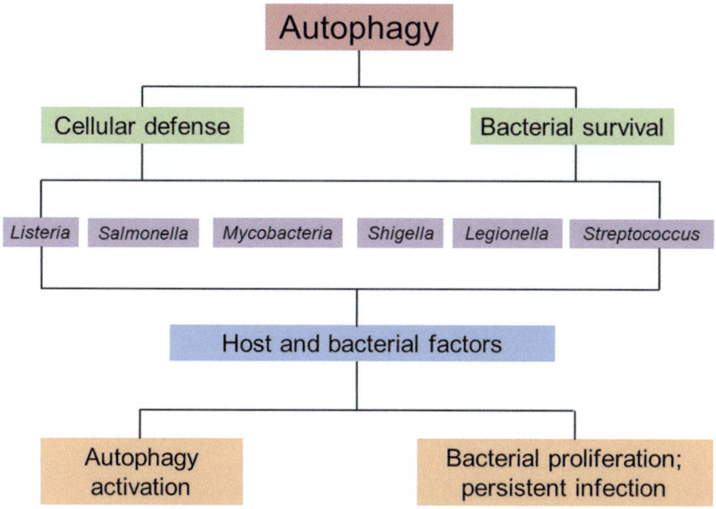

Fig. 4.2 The dual role of autophagy in bacterial pathogenesis. Autophagy can function both as a cellular defense mechanism and a pro-survival mechanism for bacterial pathogens. Pathogenic bacteria, viz., *Listeria monocytogenes*, *Salmonella typhimurium*, *Mycobacterium tuberculosis*, *Shigella flexneri*, *Legionella pneumophila*, and *Streptococcus pyogenes*, possess virulence factors or recruit host cell factors, which can trigger autophagy activation, and also enable bacterial proliferation in phagosomal/autophagosomal vacuoles to establish a safe niche for persistent infection

have been conducted on a handful of organisms; further studies with other microorganisms are warranted. Future research should explore targeting autophagic mechanisms for the treatment of bacterial infections.

Acknowledgments This work was funded by the Bundesministerium für Bildung und Forschung (ERA-NET PathoGenoMics LISTRESS and Infect-ERA PROANTILIS to T.C.) and by the Deutsche Forschungsgemeinschaft (SFB-TR84, project A04 to T.C.).

References

Agarwal V, Hammerschmidt S (2009) Cdc42 and the phosphatidylinositol3-kinase-Akt pathway are essential for PspC-mediated internalization of pneumococci by respiratory epithelial cells. J Biol Chem 284(29):19427–19436. doi:10.1074/jbc.M109.003442

Ambros V (2004) The functions of animal microRNAs. Nature 431(7006):350–355

Amer AO, Swanson MS (2005) Autophagy is an immediate macrophage response to Legionella pneumophila. Cell Microbiol 7(6):765–778

Anand PK, Tait SW, Lamkanfi M, Amer AO, Nunez G, Pagès G, Pouysségur J et al (2011) TLR2 and RIP2 pathways mediate autophagy of Listeria monocytogenes via extracellular signal-regulated kinase (ERK) activation. J BiolChem 286(50):42981–42991

Barnett TC, Liebl D, Seymour LM, Gillen CM, Lim JY, Larock CN, Davies MR, Schulz BL, Nizet V, Teasdale RD (2013) Walker MJ 2013 the globally disseminated M1T1 clone of group A

Streptococcus evades autophagy for intracellular replication. Cell Host Microbe 14(6):675–682. doi:10.1016/j.chom.2013.11.003
Baxt LA, Goldberg MB (2014) Host and bacterial proteins that repress recruitment of LC3 to Shigella early during infection. PLoS One 9(4):e94653. doi:10.1371/journal.pone.0094653. eCollection
Birmingham CL, Smith AC, Bakowski MA, Yoshimori T, Brumell JH (2006) Autophagy controls Salmonella infection in response to damage to the Salmonella-containing vacuole. J BiolChem 281(16):11374–11383
Birmingham CL, Canadien V, Kaniuk NA, Steinberg BE, Higgins DE, Brumell JH (2008) Listeriolysin O allows Listeria monocytogenes replication in macrophage vacuoles. Nature 451(7176):350–354
Bjørkøy G, Lamark T, Brech A, Outzen H, Perander M, Overvatn A, Stenmark H, Johansen T (2005) p62/SQSTM1 forms protein aggregates degraded by autophagy and has a protective effect on huntingtin-induced cell death. J Cell Biol 171(4):603–614
Burman C, Ktistakis NT (2010) Autophagosome formation in mammalian cells. Semin Immunopathol 32(4):397–413
Campbell GR, Spector SA (2012) Vitamin D inhibits human immunodeficiency virus type 1 and Mycobacterium tuberculosis infection in macrophages through the induction of autophagy. PLoS Pathog 8(5):e1002689. doi:10.1371/journal.ppat.1002689.Epub
Castillo EF, Dekonenko A, Arko-Mensah J, Mandell MA, Dupont N, Jiang S et al (2012) Autophagy protects against active tuberculosis by suppressing bacterial burden and inflammation. Proc Natl Acad Sci U S A 109(46):E3168–E3176. doi:10.1073/pnas.1210500109. Epub 2012 Oct 23
Cemma M, Kim PK, Brumell JH (2011) The ubiquitin-binding adaptor proteins p62/SQSTM1 and NDP52 are recruited independently to bacteria-associated microdomains to target Salmonella to the autophagy pathway. Autophagy 7(3):341–345
Chan EY (2012) Regulation and function of uncoordinated-51 like kinase proteins. Antioxid Redox Signal 17(95):775–785
Chandra P, Ghanwat S, Matta SK, Yadav SS, Mehta M, Siddiqui Z et al (2015a) Mycobacterium tuberculosis inhibits RAB7 recruitment to selectively modulate autophagy flux in macrophages. Sci Rep 5:16320. doi:10.1038/srep16320
Chandra V, Bhagyaraj E, Nanduri R, Ahuja N, Gupta P (2015b) NR1D1 ameliorates Mycobacterium tuberculosis clearance through regulation of autophagy. Autophagy 11(11):1987–1997
Chang SY, Lee SN, Yang JY, Kim DW, Yoon JH, Ko HJ et al (2013) Autophagy controls an intrinsic host defense to bacteria by promoting epithelial cell survival: a murine model. PLoS One 8(11):e81095. doi:10.1371/journal.pone.0081095.eCollection
Chen Z, Wang T, Liu Z, Zhang G, Wang J, Feng S et al (2015) Inhibition of autophagy by miR-30A induced by Mycobacteria tuberculosis as a possible mechanism of immune escape in human macrophages. Jpn J Infect Dis 68(5):420–424. doi:10.7883/yoken.JJID.2014.466.Epub
Chew TS, O'Shea NR, Sewell GW, Oehlers SH, Mulvey CM, Crosier PS et al (2015) Optineurin deficiency in mice contributes to impaired cytokine secretion and neutrophil recruitment in bacteria-driven colitis. Dis Model Mech 8(8):817–829
Choy A, Dancourt J, Mugo B, O'Connor TJ, Isberg RR, Melia TJ et al (2012) The Legionella effector RavZ inhibits host autophagy through irreversible Atg8 deconjugation. Science 338(6110):1072–1076. doi:10.1126/science.1227026.Epub
deReuck AVS, Cameron MP (1963) Ciba foundation symposium on lysosomes. JA Churchill Ltd, London
Di Bartolomeo S, Corazzari M, Nazio F et al (2010) The dynamic interaction of AMBRA1 with the dynein motor complex regulates mammalian autophagy. J Cell Biol 191(1):155–168
Dortet L, Mostowy S, Samba-Louaka A, Gouin E, Nahori MA, Wiemer EA, Dussurget O, Cossart P (2011) Recruitment of the major vault protein by InlK: a Listeria monocytogenes strategy to avoid autophagy. PLoS Pathogens 7(8):e1002168

Duan L, Yi M, Chen J, Li S, Chen W (2016) Mycobacterium tuberculosis EIS gene inhibits macrophage autophagy through up-regulation of IL-10 by increasing the acetylation of histone H3. Biochem Biophys Res Commun 473(4):1229–1234. doi:10.1016/j.bbrc.2016.04.045.Epub

Dupont N, Lacas-Gervais S, Bertout J, Paz I, Freche B, Van Nhieu GT, van der Goot FG, Sansonetti PJ, Lafont F (2009) Shigella phagocytic vacuolar membrane remnants participate in the cellular response to pathogen invasion and are regulated by autophagy. Cell Host Microbe 6(2):137–149

Fujita N, Itoh H, Omori H et al (2008) The Atg16L complex specifies the site of LC3 lipidation for membrane biogenesis in autophagy. Mol Biol Cell 19(5):2092–2100

Glick D, Barth S, Macleod KF (2010) Autophagy: cellular and molecular mechanisms. J Pathol 221(1):3–12

Goto A, Yano T, Terashima J, Iwashita S, Oshima Y, Kurata S (2010) Cooperative regulation of the induction of the novel antibacterial listericin by peptidoglycan recognition protein LE and the JAK-STAT pathway. J Biol Chem 285(21):15731–15738

Gradstedt H, Iovino F, Bijlsma JJ (2013) Streptococcus pneumoniae invades endothelial host cells via multiple pathways and is killed in a lysosome dependent manner. PLoS One 8(6):e65626. doi:10.1371/journal.pone.0065626

Gutierrez MG, Master SS, Singh SB, Taylor GA, Colombo MI, Deretic V (2004) Autophagy is a defense mechanism inhibiting BCG and Mycobacterium tuberculosis survival in infected macrophages. Cell 119(6):753–766

Hanada T, Noda NN, Satomi Y et al (2007) The Atg12-Atg5 conjugate has a novel E3-like activity for protein lipidation in autophagy. J Biol Chem 282(52):37298–37302

Haobam B, Nozawa T, Minowa-Nozawa A, Tanaka M, Oda S, Watanabe T et al (2014) Rab17-mediated recycling endosomes contribute to autophagosome formation in response to Group A Streptococcus invasion. Cell Micorbioll 6(12):1806–1821. doi:10.1111/cmi.12329.Epub

Henriques-Normark B, Tuomanen EI (2013) The pneumococcus: epidemiology, microbiology, and pathogenesis. Cold Spring Harb Perspect Med 3(7):pii: a010215. doi:10.1101/cshperspect.a010215

Hernandez LD, Pypaert M, Flavell RA, Galán JE (2003) A Salmonella protein causes macrophage cell death by inducing autophagy. J Cell Biol 163(5):1123–1131

Holden DW (2002) Trafficking of the Salmonella vacuole in macrophages. Traffic 3(3):161–169

Horenkamp FA, Kaufmann KJ, Kohler LJ, Sherwood RK, Krueger KP, Shteyn V et al (2015) The Legionella anti-autophagy effector RavZ targets the autophagosome via PI3P- and curvature-sensing motifs. Dev Cell 34(5):569–576. doi:10.1016/j.devcel.2015.08.010.Epub

Huang D, Bao L (2014) Mycobacterium tuberculosis EspB protein suppresses interferon-γ-induced autophagy in murine macrophages. J Microbiol Immunol Infect 49:859–865. doi:10.1016/j.jmii.2014.11.008. pii: S1684–1182(14)00238–2

Jamwal SV, Mehrotra P, Singh A, Siddiqui Z, Basu A, Rao KV (2016) Mycobacterial escape from macrophage phagosomes to the cytoplasm represents an alternate adaptation mechanism. Sci Rep 6:23089. doi:10.1038/srep23089

Jia K, Thomas C, Akbar M, Sun Q, Adams-Huet B, Gilpin C et al (2009) Autophagy genes protect against Salmonella typhimurium infection and mediate insulin signaling-regulated pathogen resistance. Proc Natl Acad Sci U S A 106(34):14564–14569. doi:10.1073/pnas.0813319106. Epub 2009 Aug 10

Johansen T, Lamark T (2011) Selective autophagy mediated by autophagic adaptor proteins. Autophagy 7(3):279–296

Kayath CA, Hussey S, El Hajjami N, Nagra K, Philpott D, Allaoui A (2010) Escape of intracellular Shigella from autophagy requires binding to cholesterol through the type III effector, IcsB. Microbes Infect 12(12–13):956–966. doi:10.1016/j.micinf.2010.06.006.Epub

Khweek AA, Caution K, Akhter A, Abdulrahman BA, Tazi M, Hassan H, Majumdar N et al (2013) A bacterial protein promotes the recognition of the Legionella pneumophila vacuole by autophagy. Eur J Immunol 43(5):1333–1344

Kim JJ, Lee HM, Shin DM, Kim W, Yuk JM, Jin HS et al (2012) Host cell autophagy activated by antibiotics is required for their effective antimycobacterial drug action. Cell Host Microbe 11(5):457–468. doi:10.1016/j.chom.2012.03.008

Kim JK, Yuk JM, Kim SY, Kim TS, Jin HS, Yang CS et al (2015) MicroRNA-125a inhibits autophagy activation and antimicrobial responses during mycobacterial infection. J Immunol 194(11):5355–5365. doi:10.4049/jimmunol.1402557.Epub

Kirkin V, Lamark T, Sou YS, Bjørkøy G, Nunn JL, Bruun JA, Shvets E et al (2009) A role for NBR1 in autophagosomal degradation of ubiquitinated substrates. Mol Cell 33(4):505–516

Lam KK, Zheng X, Forestieri R, Balgi AD, Nodwell M, Vollett S et al (2012) Nitazoxanide stimulates autophagy and inhibits mTORC1 signaling and intracellular proliferation of Mycobacterium tuberculosis. PLoSPathog 8(5):e1002691. doi:10.1371/journal.ppat.1002691. Epub

Lam GY, Cemma M, Muise AM, Higgins DE, Brumell JH (2013) Host and bacterial factors that regulate LC3 recruitment to Listeria monocytogenes during the early stages of macrophage infection. Autophagy 9(7):985–995

Li P, Shi J, He Q, Hu Q, Wang YY, Zhang LJ, Chan WT, Chen WX (2015) Streptococcus pneumoniae induces autophagy through the inhibition of the PI3K-I/Akt/mTOR pathway and ROS hypergeneration in A549 cells. PLoS One 10(3):e0122753. doi:10.1371/journal.pone.0122753. eCollection 2015

Lin L, Baehrecke EH (2015) Autophagy, cell death, and cancer. Mol Cell Oncol 2(3):e985913. doi:10.4161/23723556.2014.985913

Lu K, Psakhye I, Jentsch S (2014) A new class of ubiquitin-Atg8 receptors involved in selective autophagy and polyQ protein clearance. Autophagy 10(12):2381–2382

Lu SL, Kuo CF, Chen HW, Yang YS, Liu CC, Anderson R (2015) Insufficient acidification of autophagosomes facilitates group A streptococcus survival and growth in endothelial cells. MBio 6(5):e01435–e01415. doi:10.1128/mBio.01435-15

Matsuda F, Fujii J, Yoshida S (2009) Autophagy induced by 2-deoxy-D-glucose suppresses intracellular multiplication of Legionella pneumophila in A/J mouse macrophages. Autophagy 5(4):484–493

Mizushima N, Yoshimori T, Ohsumi Y (2011) The role of Atg proteins in autophagosome formation. Annu Rev Cell Dev Biol 27:107–132

Mostowy S, Bonazzi M, Hamon MA, Tham TN, Mallet A, Lelek M et al (2010) Entrapment of intracytosolic bacteria by septin cage-like structures. Cell Host Microbe 8(5):433–444. doi:10.1016/j.chom.2010.10.009

Mostowy S, Sancho-Shimizu V, Hamon MA, Simeone R, Brosch R, Johansen T, Cossart P (2011) p62 and NDP52 proteins target intracytosolic Shigella and Listeria to different autophagy pathways. J BiolChem 286(30):26987–26995

Moy RH, Cherry S (2013) Antimicrobial autophagy: a conserved innate immune response in Drosophila. J Innate Immun 5(5):444–455

Nakagawa I, Amano A, Mizushima N, Yamamoto A, Yamaguchi H, Kamimoto T et al (2004) Autophagy defends cells against invading group A Streptococcus. Science 306(5698):1037–1040

Newman AC, Scholefield CL, Kemp AJ, Newman M, McIver EG, Kamal A, Wilkinson S (2012) TBK1 kinase addiction in lung cancer cells is mediated via autophagy of Tax1bp1/Ndp52 and non-canonical NF-κB signaling. PLoS One 7(11):e50672

Nozawa T, Aikawa C, Goda A, Maruyama F, Hamada S, Nakagawa I (2012) The small GTPases Rab9A and Rab23 function at distinct steps in autophagy during Group A Streptococcus infection. Cell Microbiol1 4(8):1149–1165. doi:10.1111/j.1462-5822.2012.01792.x.Epub

O'Seaghdha M, Wessels MR (2013) Streptolysin O and its co-toxin NAD-glycohydrolase protect group A Streptococcus from xenophagic killing. PLoS Pathog 9(6):e1003394. doi:10.1371/journal.ppat.1003394.Epub

Ogawa M, Yoshimori T, Suzuki T, Sagara H, Mizushima N, Sasakawa C (2005) Escape of intracellular Shigella from autophagy. Science 307(5710):727–731. Epub 2004 Dec 2

Ogawa M, Yoshikawa Y, Mimuro H, Hain T, Chakraborty T, Sasakawa C (2011) Autophagy targeting of Listeria monocytogenes and the bacterial countermeasure. Autophagy 7(3):310–314

Osawa T, Mizuno Y, Fujita Y, Takatama M, Nakazato Y, Okamoto K (2011) Optineurin in neurodegenerative diseases. Neuropathology 31(6):569–574

Otto GP, Wu MY, Clarke M, Lu H, Anderson OR, Hilbi H et al (2004) Macroautophagy is dispensable for intracellular replication of Legionella pneumophila in Dictyostelium discoideum. Mol Microbiol 51(1):63–72

Owen KA, Meyer CB, Bouton AH, Casanova JE (2014) Activation of focal adhesion kinase by Salmonella suppresses autophagy via an Akt/mTOR signaling pathway and promotes bacterial survival in macrophages. PLoS Pathog 10(6):e1004159. doi:10.1371/journal.ppat.1004159. eCollection 2014

Pilli M, Arko-Mensah J, Ponpuak M, Roberts E, Master S, Mandell MA et al (2012) TBK-1 promotes autophagy-mediated antimicrobial defense by controlling autophagosome maturation. Immunity 37(2):223–234

Proikas-Cezanne T, Takacs Z, Dönnes P et al (2015) WIPI proteins: essential PtdIns3P effectors at the nascent autophagosome. J Cell Sci 128(2):207–217

Py BF, Lipinski MM, Yuan J (2007) Autophagy limits Listeria monocytogenes intracellular growth in the early phase of primary infection. Autophagy 3(2):117–125

Rich KA, Burkett C, Webster P (2003) Cytoplasmic bacteria can be targets for autophagy. Cell Microbiol 5(7):455–468

Rolando M, Escoll P, Nora T, Botti J, Boitez V, Bedia C et al (2016) Legionella pneumophila S1P-lyase targets host sphingolipid metabolism and restrains autophagy. Proc Natl Acad Sci U S A 113(7):1901–1906. doi:10.1073/pnas.1522067113.Epub

Russell RC, Tian Y, Yuan H et al (2013) ULK1 induces autophagy by phosphorylating Beclin-1 and activating VPS34 lipid kinase. Nat Cell Biol 15(7):741–750

Sakurai A, Maruyama F, Funao J, Nozawa T, Aikawa C, Okahashi N et al (2010) Specific behavior of intracellular Streptococcus pyogenes that has undergone autophagic degradation is associated with bacterial streptolysin O and host small G proteins Rab5 and Rab7. J Biol Chem 285(29):22666–22675. doi:10.1074/jbc.M109.100131.Epub

Sanjurjo L, Amézaga N, Vilaplana C, Cáceres N, Marzo E, Valeri M et al (2013) The scavenger protein apoptosis inhibitor of macrophages (AIM) potentiates the antimicrobial response against Mycobacterium tuberculosis by enhancing autophagy. PLoS one 8(11):e79670. doi:10.1371/journal.pone.0079670. eCollection 2013

Seto S, Tsujimura K, Koide Y (2012) Coronin-1a inhibits autophagosome formation around Mycobacterium tuberculosis-containing phagosomes and assists mycobacterial survival in macrophages. Cell Microbiol 14(5):710–727

Shahnazari S, Yen WL, Birmingham CS, Shiu J, Namolovan A, Zheng YT et al (2010) A diacylglycerol-dependent signaling pathway contributes to regulation of antibacterial autophagy. Cell Host Microbe 8(2):137–146

Sharma G, Dutta RK, Khan MA, Ishaq M, Sharma K, Malhotra H (2014) IL-27 inhibits IFN-γ induced autophagy by concomitant induction of JAK/PI3K/Akt/mTOR cascade and up-regulation of mcl-1 in mycobacterium tuberculosis H37Rv infected macrophages.Int. J Biochem Cell Biol 55(October):335–347. doi:10.1016/j.biocel.2014.08.022. Epub 2014 Sep 4

Shoji-Kawata S, Sumpter R, Leveno M, Campbell GR, Zou Z, Kinch L et al (2013) Identification of a candidate therapeutic autophagy-inducing peptide. Nature 494(7436):201–206. doi:10.1038/nature11866.Epub

Sirianni A, Krokowski S, Lobato-Márquez D, Buranyi S, Pfanzelter J, Galea D et al (2016) Mitochondria mediate septin cage assembly to promote autophagy of Shigella. EMBO Rep 17(7):1029–1043. pii: e201541832

Suzuki T, Franchi L, Toma C, Ashida H, Ogawa M, Yoshikawa Y et al (2007) Differential regulation of caspase-1 activation, pyroptosis, and autophagy via Ipaf and ASC in Shigella-infected macrophages. PLoS Pathog 3(8):e111

Swanson MS, Isberg RR (1995) Association of Legionella pneumophila with the macrophage endoplasmic reticulum. Infect Immun 63(9):3609–3620

Tai SS (2016) Streptococcus pneumoniae serotype distribution and pneumococcal conjugate vaccine serotype coverage among pediatric patients in East and Southeast Asia, 2000-2014: a pooled data analysis. Vaccines (Basel) 4(1):pii: E4. doi:10.3390/vaccines4010004

Tattoli I, Sorbara MT, Vuckovic D, Ling A, Soares F, Carneiro LA, Yang C, Emili A, Philpott DJ, Girardin SE (2012) Amino acid starvation induced by invasive bacterial pathogens triggers an innate host defense program. Cell Host Microbe 11(6):563–575

Tattoli I, Sorbara MT, Yang C, Tooze SA, Philpott DJ, Girardin SE (2013) Listeria phospholipases subvert host autophagic defenses by stalling pre-autophagosomal structures. EMBO J 32(23):3066–3078

Thomas M, Mesquita FS, Holden DW (2012) The DUB-ious lack of ALIS in Salmonella infection: a Salmonella deubiquitinase regulates the autophagy of protein aggregates. Autophagy 8(12):1824–1826

Thurston TL, Ryzhakov G, Bloor S, von Muhlinen N, Randow F (2009) The TBK1 adaptor and autophagy receptor NDP52 restricts the proliferation of ubiquitin-coated bacteria. Nat Immunol 10(11):1215–1221

Thurston TL, Wandel MP, von Muhlinen N, Foeglein A, Randow F (2012) Galectin 8 targets damaged vesicles for autophagy to defend cells against bacterial invasion. Nature 482(7385):414–418

Tsukada M, Ohsumi Y (1993) Isolation and characterization of autophagy-defective mutants of Saccharomyces cerevisiae. FEBS Lett 333(1–2):169–174

Tumbarello DA, Manna PT, Allen M, Bycroft M, Arden SD, Kendrick-Jones J, Buss F (2015) The autophagy receptor TAX1BP1 and the molecular motor myosin VI are required for clearance of Salmonella Typhimurium by Autophagy. PLoS Pathog 11(10):e1005174. doi:10.1371/journal.ppat.1005174. eCollection 2015

Vázquez-Boland JA, Kuhn M, Berche P, Chakraborty T, Domínguez-Bernal G, Goebel W, González-Zorn B, Wehland J, Kreft J (2001) Listeria pathogenesis and molecular virulence determinants. Clin Microbiol Rev 14(3):584–640

Verlhac P, Viret C, Faure M (2015) Dual function of CALCOCO2/NDP52 during xenophagy. Autophagy 11(6):965–966. doi:10.1080/15548627.2015.1046672

von Muhlinen N, Thurston T, Rhyzhakov G, Bloor S, Randow F (2010) NDP52, a novel autophagy receptor for ubiquitin-decorated cytosolic bacteria. Autophagy 6(2):288–289. Epub 2010 Feb 4

Wang J, Yang K, Zhou L, Minhaowu L, Wu Y, Zhu M et al (2013) MicroRNA-155 promotes autophagy to eliminate intracellular mycobacteria by targeting Rheb. PLoS Pathogens 9(10):e1003697. doi:10.1371/journal.ppat.1003697. Epub 2013 Oct 10

Watson RO, Manzanillo PS, Cox JS (2012) Extracellular M. tuberculosis DNA targets bacteria for autophagy by activating the host DNA-sensing pathway. Cell 150(4):803–815. doi:10.1016/j.cell.2012.06.040

Watson RO, Bell SL, MacDuff DA, Kimmey JM, Diner EJ, Olivas J (2015) The cytosolic sensor cGAS detects Mycobacterium tuberculosis DNA to induce type I interferons and activate autophagy. Cell Host Microbe 17(6):811–819. doi:10.1016/j.chom.2015.05.004. Epub 2015 Jun 2

Wild P, Farhan H, McEwan DG, Wagner S, Rogov VV, Brady NR, Richter B et al (2011) Phosphorylation of the autophagy receptor optineurin restricts Salmonella growth. Science 333(6039):228–233

Yamaguchi H, Nakagawa I, Yamamoto A, Amano A, Noda T, Yoshimori T (2009) An initial step of GAS-containing autophagosome-like vacuoles formation requires Rab7. PLoS Pathog 5(11):e1000670. doi:10.1371/journal.ppat.1000670.Epub

Yano T, Mita S, Ohmori H, Oshima Y, Fujimoto Y, Ueda R, Takada H et al (2008) Autophagic control of Listeria through intracellular innate immune recognition in Drosophila. Nat Immunol 9(8):908–916

Yoshikawa Y, Ogawa M, Hain T, Yoshida M, Fukumatsu M, Kim M, Mimuro H et al (2009) Listeria monocytogenes ActA-mediated escape from autophagic recognition. Nat Cell Biol 11(10):1233–1240

Yu HB, Croxen MA, Marchiando AM, Ferreira RB, Cadwell K, Foster LJ et al (2014) Autophagy facilitates Salmonella replication in HeLa cells. MBio 5(2):e00865–e00814. doi:10.1128/mBio.00865-14

Zheng YT, Shahnazari S, Brech A, Lamark T, Johansen T, Brumell JH (2009) The adaptor protein p62/SQSTM1 targets invading bacteria to the autophagy pathway. J Immunol (Baltimore:1950) 183(9):5909–5916

Madhu Puri has completed her M.Sc. (Human Genomics) from Panjab University, Chandigarh, India and Ph.D. (Medical Microbiology) from Justus Liebig University, Giessen, Germany. Her current research interests lie in the fields of host-pathogen interactions, autophagy and molecular biology.

Trinad Chakraborty is Professor of Medical Microbiology and Infectious Diseases at the Justus Liebig University, Giessen, Germany. His interests are in the field of pathogenomics of host-pathogen interactions.

Helena Pillich completed her diploma (Dipl. Biology) and Ph.D. (Doctor of Natural Science) at the Justus Liebig University, Giessen, Germany. Her current research interests are host-pathogen interactions.

5 Emergence of Multidrug-Resistant Bacteria in Freshwater Ecosystems (River) and Screening of Natural Therapeutics Against the Probable Drug Targets of Drug-Resistant Pathogens by Computational Biology Approaches

Sinosh Skariyachan

Abstract

Due to enormous accumulation of industrial effluents and wastes from various sources, the natural status of many freshwater bodies has distorted and created an ideal atmosphere for survival and swift proliferation of many bacterial pathogens. Major populations of these bacteria are coliforms, and they obtain additional characters due to the rapid changes in their chromosomal DNA; a major concern is the acquisition of drug-resistant genes. The emergence of multidrug resistance in these bacteria has become critical issue in many freshwater ecosystems globally. The conventional drugs against these bacteria are not effective therapeutic strategies which necessitates alternative therapeutic approaches. Computer-assisted drug discovery plays a profound role in screening novel herbal-based leads and similar natural leads against the probable drug targets such as drug-resistant gene products and toxins of the pathogenic bacteria. This chapter illustrates the recent concerns and issues related with water pollution in freshwater ecosystems and emphasizes major causes of the emergence of bacterial drug resistance in such water bodies. Further, it investigates the utility of many natural compounds especially herbal-based ligands as lead molecules against the probable drug targets of multidrug-resistant bacterial pathogens by computer-aided virtual screening.

S. Skariyachan (✉)
Department of Biotechnology Engineering, Dayananda Sagar Institutions, Bangalore 560 078, India

Visvesvaraya Technological University, Belagavi, Karnataka, India
e-mail: sinosh-bt@dayanandasagar.edu

P. Shukla (ed.), *Recent Advances in Applied Microbiology*,
DOI 10.1007/978-981-10-5275-0_5

Keywords
Freshwater bodies • Multidrug resistance • Probable lead molecules • Novel phytotherapeutics • Computer-aided virtual screening • Drug-resistant gene products

5.1 Environmental Pollution in Freshwater Ecosystems (Rivers)

Rivers are the major freshwater ecosystems that are crucial for sustenance of all kinds of life. Rivers played a vital role in the lives, history, and culture of the people and their civilization for thousands of years. Water from the river is a fundamental natural resource, essential for human and animal life. The river systems are mainly used for source of drinking water, hydroelectric power generation, irrigation, fishing, route of transportation, and many more developmental activities related to human civilization. These are streams of global strategic importance, contributing major water resources for agricultural, domestic, and industrial sectors. Many of the ancient civilization had thrived, and most developmental activities are dependent upon the rivers. Most of the cities and urban centers are located in basins of various rivers. The urbanizations and industrialization of these places are primarily depending upon the river or similar types of water bodies (Kumar et al. 2016; Zhao et al. 2014). Rivers account for major freshwater systems that constitute 0.01% of earth's water and roughly 0.8% of the surface water which favors at least 100,000 varieties out of approximately 1.8 million described species. Together with inland waters, river water constitutes a precious natural resource in cultural, scientific, educational, and economic purpose (Dudgeon et al. 2006). In addition, the river ecosystems played a major role for maintaining the biodiversity of animal, plants, and microbial life in the earth (Tendall et al. 2014; Baker 2014). Rivers have profound role in maintaining biogeochemical cycling, ecological balance, and fertility of soils, which is a prime concern in agricultural farms and physiochemical features of the soil (Zhao et al. 2014). Rivers also have relevance in mythological perspectives. Many rivers in the world are believed to be sacred by various religions (Belay and Sahile 2013).

Water pollution is a serious problem involving the discharge of dissolved or suspended substances into groundwater, streams, rivers, and oceans (Toroglu and Toroglu 2009). The aquatic ecosystem comprises of several machinery which are indirectly or directly affected by the pollution. The pollution of a water body can always be associated with an industry, domestic, and industrial sewage or agricultural runoff. Before working the water refinement facilities in some developing and developed countries, various rivers are being used as a place to discharge domestic, industrial sewage, and urban wastewater. Wastewater is a major cause of grave public health concern because it contains hazardous chemicals, various xenobiotic compounds, aliphatic, aromatics, alicyclic hydrocarbons, plastics, and other nonbiodegradable substances, microbial load especially bacteria, algae, protozoa, and viruses (Zeng and Jinglu 2013). Water pollution represents a significant and substantial burden on human health that disproportionately affects those living in the developing world with fast economic growth and population increase; the water conditions have been deteriorating and ruined ecological balance (Rather et al. 2016; Tissera and Lee 2013).

The resources from surface water have contributed a significant part throughout history in the expansion of human civilization. The urban surface waters constitute streams, rivers, wetlands, and lake that are directly exposed to elevated population densities of towns and cities. Hence, the rivers are exposed to physicochemical and biological contaminants from their natural surroundings (Sikder et al. 2013). Majority of the rivers in urban regions of developing countries are the destination of effluents discharged from various industrial sectors. The industries such as electroplating, thermal power plants, tanneries, mining operations, battery producers, steel plants, which release toxic metal containing effluents, fertilizers, and pesticides, cause a severe threat to the ecosystem and worsening the quality of water. The toxic substances and heavy metals are frequently disposed of rivers in both developing and developed countries. Urbanization and development of industrial activities proceed with the production of bulk quantities of liquid and solid wastes. The waste materials are discharged into the ecosystem without adequate treatments. Industrial waste materials released in the surroundings are occasionally 10–100 times more accumulated in contaminants than the standards proposed by WHO 2010.

The main risk factors associated with the pollution in water can be classified into two categories: chemical and biological pollutants. The chemicals are mainly heavy metals. The metals such as As, Cd, Cr, Cu, Fe, Co, Mn, Hg, Mb, Pb, Ni, Se, Zn, and V are main heavy metals which contribute to water pollution. These metals have a high atomic mass; hence, they are heavy in that sense (Wu et al. 2016). Many industries such as electroplating industry can produce more volumes of heavy metal-rich effluents than other industries such food processing industry. Both groups are derived from human activity which tends to alter the composition of water in terms of its original state in the nature. The biological contaminants are mainly microorganisms such as bacteria, yeast, protozoa, algae, and viruses especially bacteriophages (Nigam and Shukla 2015). They can survive in the presence of chemicals by utilizing these chemicals as their nutritional requirements. Studies revealed that many water bodies being a normal dumping site for the industries located along its bank have converted into a usual breeding land for many harmful bacteria especially fecal coliforms as well as a hub for large quantities of heavy metals and anthropogenic substances (Umamaheswari and Saravanan 2009; Gazzaz et al. 2012). The discharge of industrial effluents in the polluted water leads to reduced transparency, high turbidity, and augmented suspended solids. This contamination contributes high amount of fecal coliforms that are responsible for many health hazards (Sharma et al. 2012). The high deposition of domestic and industrial wastes together with pesticide remains contributed to enormous multiplication of pathogenic bacteria. These resulted in many waterborne diseases (Ndjama et al. 2008) such as malaria, typhoid, bilharzias, diarrhea, cholera, dysentery, gastroenteritis, and hepatitis A (Ali and Osman 2010). Such kinds of waterborne diseases are mainly due to the microbial load especially fecal coliforms in most of the aquatic ecosystem (Kuitcha et al.

2008). The existence of independent sanitation systems amplifies the risk of pollution in underground and surface water resources (Rashid 2012). Gas and oil and spills are also drastically contributing water pollution. The expansion of small-scale industries in rural areas and accidental urbanization are also severe threat to water. Even though the public alertness about water pollution is much advanced than in the past, there is a scope to expand more advanced strategies and find novel approach to reduce the pollution. Further, existing technologies for contaminated water treatment are costly and inadequate in their use. Hence, there is a high necessity to devise the techniques and practices that are reasonable, easily accessible, and more useful in detecting and dropping water pollution (WHO 2010).

5.2 Water Pollution in Indian Rivers

Water pollution is one of the major ecological issues in all rivers in India (Subrahmanyam et al. 2016; Ndjama et al. 2008). Though three-fourth part of land is being bordered by water, little part of it can be used for drinking purposes. During the preceding decades, the water quality of many Indian rivers has been deteriorating due to constant discharge of domestic and industrial wastes (Raja et al. 2008). There are reports revealing the total and viable microbial count exceeding maximum permissible limits in some of the North Indian Rivers, and these are responsible for many waterborne diseases (Ndjama et al. 2008).

Ganga River is the biggest river of Indian subcontinent. Studies revealed the occurrence of bacterial indicators of fecal origin at several places in most stretch of Gangetic river system (Sood et al. 2008). Similarly, recent studies revealed that the rural areas situated on both the sides of Kosi River are engaged mainly in the agriculture and cattle farming. These cattle while entering in the river transfer fecal matter and other types of pathogens in the river which deteriorates the quality of the river water to a considerable extent (Jalal and Kumar 2013). Similarly, River Hooghly receives industrial wastes of 150 big factories situated along the bank of the river including paper and pulp factories, textile and jute mills, tanneries, and distilleries. This created a massive pollution in the river. A survey focused on tanneries effluents polluted in Palar River basin in Vellore District, Tamil Nadu, demonstrated that it has been polluted with heavy metals especially chromium and associated salts. This has a direct influence of the microbial proliferation in the river (Sundar et al. 2010). Similar reports revealed that elevated levels of biological oxygen demand (BOD), dissolved oxygen (DO), and total phosphorus were found to be adversely affecting the quality of water in Brahmaputra River (Girija et al. 2007). Further, studies conducted in Kerala showed that River Pamba is drastically polluted because of Sabarimala pilgrimage, open flow of wastes, domestic sewage, and fecal contaminants into the river and interference of sea water (Joseph and Jacob 2010). Studies conducted on Pennar River, Kerala, reported that abandoned utilization of chemical fertilizers and discharging of municipal wastes are major reasons for the worsening of water and public health, as it promotes the growth of fecal coliforms (Joseph and Jacob 2010). Varunprasath and Daniel (2010) reported that many rivers in Tamil Nadu such as Aliyar, Amaravathi, Bhavani, Shanmuga, and Siruvani largely undergone pollution by industrial effluents, municipal sewage, and religious credence and subject to seasonable variations, climate change, and influx of waters from various tributaries. These were also resulted in the massive multiplication of various pathogenic microorganisms (Varunprasath and Daniel 2010).

According to the annual report of Central Pollution Control Board, Govt. of India (2012), most of the rivers in India which include Godavari, Yamuna, Narmada, Tapi, Sabarmati, Satluj, Brahmani, Mahanadi, Krishna, Brahmaputra, and Mahi are enormously polluted due to various organic and bacterial pollution (CPCB 2012).

5.2.1 Water Pollution in River Cauvery: A Case Study from South India

River Cauvery is one of the largest rivers of South India and is considered as the "Ganga of South." This river originated from Talacauvery, Kodugu District, Karnataka, India, and covers the southern states of Tamil Nadu, Kerala, and Puducherry covering an area of 81,155 sq. km which is approximately 2.7% of the total geographical locations. The basin extends over 32, 27 km^2 (42.2%) of Karnataka, 2886 km^2 (3.5 km) of Kerala, 43,868 km^2 (54.3%) of Tamil Nadu, and 148 km^2 of Pondicherry (Aatish 2012).

There are many urban centers and industries, such as cotton, textile, cement, mineral, and metals, many other industries, and hospitals situated on the Cauvery basin. Moreover, Bangalore, India, is located near this basin. The river is the chief source for irrigation, hydroelectric power generation, and supply of drinking water to the neighboring rural and urban residents in Karnataka (Venkatesha Raju et al. 2012). Due to rapid industrialization and high population rate, the natural environment of Cauvery River suffers from the detrimental effects of pollution (Vanham et al. 2011). Over the years, the river has been subjected to the interference of human population and water system completely deteriorated.

The water quality assessment of River Cauvery is being carried out in the basin by the State Pollution Control Board, Tamil Nadu, Karnataka, and Kerala at 36 locations. The assessment sites were on mainstream of Cauvery River and on tributaries are Arkavati, Amravati, Laxmantirtha, Hemavati, Shimsa, Bhawani, Yagichi, and Kabini. The water quality observed in Cauvery basin with respect to conductivity, pH, DO, chemical oxygen demand (COD), BOD, and total and fecal coliforms is estimated as high value throughout the year. River Cauvery at 20 regions indicated that dissolved oxygen was in the range of 0.3–9.8 mg/l which was very high. The total coliform count was in the range of 2–9500 MPN/ 100 ml, whereas the fecal coliform count varied from 1 to 3000 MPN/100 ml. High estimation of total and fecal coliforms was observed at KRS dam, Balamuritshetra in Karnataka, and Chirapalayam, Erode in Tamil Nadu, respectively. The BOD level ranges from 1.0 to 12.0 mg/l in Cauvery River with a maximum value of 12 mg/l observed in Karnataka. Previous studies reported the prevalence of multiple drug-resistant bacteria in Byramangala reservoir, one of the main drinking water sources for the residents in urban areas of Bidadi industrial sectors, Bangalore, India. This reservoir is fed by the River Vrishabhavathi, a tributary of River Suvarnamukhi which is sub-tributary of Cauvery River (Skariyachan et al. 2013a, b).

The water qualities of the tributaries of River Cauvery were also analyzed by pollution control board (2010) which revealed that Hemavati, Lakshmantirtha, Shimsa, Yagachi, Arkavati, Bhavani, Kabbani, and Amravati were indicated that dissolved oxygen is not satisfying the water quality standard at most locations except in Bhavani River at Bhavani Sagar, Tamil Nadu, and Arkavati in Kanakapura, Karnataka. River Bhavani showed a BOD range of 0.2–3.4 mg/l with a maximum value of 3.4 mg/l. The fecal coliforms varied from 9 to 5000 MPN/ 100 ml, while the total coliforms ranged from 70 to 30,000 MPN/100 ml (KSPCB 2010).

5.2.2 Global Status of River Pollution

There are worldwide concerns on the massive contamination of valuable freshwater ecosystems in which major organisms are dependent upon these ecosystems (Xiong et al. 2016). Lakes in the lower and middle stretches of River Yangtze, China, form a shallow lake which set unique in the world and gradually more polluted by heavy metals. This was attributed to high level of municipal and untreated industrial sewage produced inside the lake catchments (Kotze 2012).

The drastic fecal microbial pollution and their implications in urban and rural areas of Grand River and its tributaries in Canada were recently reported (Lee et al. 2014). The seasonal changes and resilience of bacterial population in highly polluted urban river in Zenne in Belgium was recently reported. The microbial pollution caused due to *E. coil* and their genetic diversity from household water supply in Dhaka, Bangladesh, was recently reported (García-Armisen et al. 2014). The prevalence of toxigenic *Vibrio cholerae O1* in water systems in New Bell-Douala, Cameroon, was recently reported (Akoachere et al. 2013). A study on the microbial pollution of Canadian drinking water resources with multidrug-resistant *E. coli* was recently published (Coleman et al. 2013). Similarly, the evaluations of chemical and microbial pollutions related to the deterioration of household water supply system were studied. Further, the microbial quality of drinking water in municipal distribution water system in various regions across the world was reviewed by Chowdhury (2012). Similarly, the microbial pollution associated with drinking water facilities and water security issues in Russia was reported by Dudarev et al. (2013). The heavy microbial pollution in water sources and their protection in rural areas of Beijing, China (Ye et al. 2013), and Myanmar (Sakai et al. 2013) were also reported.

There are many reports available on health implications caused due to bacterial pollution in rivers such as Golbasi in Turkey (Toroglu and Toroglu 2009), Vjosa in Albania (Hysko et al. 2010), and Mfoundi River watershed in Cameroon (Dorice et al. 2010). In addition, massive industrialization has also led to heavy water pollution and health hazards in many rivers such as Niagara (Philbert 1991). The genotoxic effect of microbial contamination in Paraguay and Pitimbu Rivers in Brazil was also reported (Pimenta et al. 2008).

There are reports that revealed the heavy metal concentration in Warri River, in Delta state of Nigeria, increased and rendered the water unfit for human consumption (Wogul and Okaka 2011). Similarly, Rosetta Branch of Nile River is subjected to rigorous pollutants of industrial, domestic, and agricultural pollution (Bouraie et al. 2011). The River Olifants, Mpumalanga, is one of the worst polluted river systems located in South Africa. Studies suggested that various anthropogenic actions, especially mining in the higher catchment and agricultural and industrial deeds in the lower catchments, are commonly observed. The levels of various metal content and other micro-constituents were detected at many sites, and they surpassed the permissible limits proposed for water bodies (Kotze 2012).

A study conducted by Rashid (2012) revealed that the domestic water used in the villages west of Lake Nasser, Egypt, contained many pathogenic organisms. Reports also revealed that Buffalo River in South Africa showed higher concentrations of fecal coliform counts at the sampling places situated at lower stretch of the river in comparison with upper stretches. These studies showed that the microbiological qualities of River Buffalo and dams were deprived and suggested that waste was constantly discharged into the Buffalo River (Chigor et al. 2013).

Similar studies revealed that the lower stretch of River Tyume, South Africa, showed most anthropogenic activities where the treated and semi-treated effluents from wastewater treatment facilities are found. Apart from bacteria, some viruses (adenovirus) were also detected from most the sampling sites (Chigor et al. 2013). Another study conducted by Sanchez et al. (2012) reported that the deposition of nitrates, total coliform count, and *Vibrio* sp. for Jamapa River, Mexico, were found to be very high (Sanchez et al. 2012). Their studies revealed that values of environmental parameters estimated in different aquatic systems were beyond the permissible limits proposed by Mexican official standards (Truman et al. 2014).

5.3 Emergence of Multidrug-Resistant Bacteria in Freshwater Ecosystem: A Major Global Health Concern

Due to enormous accumulation of industrial and organic wastes from pharmaceutical industries and hospitals, the normal status of the river water distorted in terms of dissolved oxygen, nutritional contents, pH, temperature, and other physiochemical features (Osińska et al. 2016). These make a suitable setting for the growth and rapid multiplication of various pathogenic microorganism especially coliforms. Along with the quick proliferation, bacteria attain many additional characteristics due to the rapid changes in their chromosomal DNA; a major worry is the gaining of new genes for drug resistance. These coliforms are capable of transferring drug resistance to other bacteria especially enteric pathogens. Reports suggested that the permanent influx of contaminants such as disinfectants, detergents, heavy metals, antibacterial agents, watershed, and livestock wastes may supply to the appearance of antibiotic-resistant pathogens and spread antimicrobial-resistant genes in water bodies (Berglund 2015; Ramírez et al. 2013; Xu et al. 2014). Furthermore, reports depicted that several bacteria eventually mutated and acquired gene code for multidrug resistance (MDR) to current generation drugs and materialized as "superbugs." At present, many freshwater systems have turned as cesspool for antibiotic-resistant bacteria (ARB) (West et al. 2011).

The bacteria became multidrug resistance by natural (intrinsic resistance) or acquired (extrinsic resistance) means. The intrinsic resistance is due to few genes responsible for resistance to its own antibacterial. The extrinsic resistance is due to chromosomal mutation or by the gaining of mobile genetic elements like transposons or plasmids, which carry the antibiotic resistance genes (West et al. 2011). Further, the transfer of resistance between the bacteria might occur by recombination via transformation, transduction, and conjugation. Many antibiotic resistance genes are situated on plasmids, aiding their transfer, and expand multidrug-resistant strains. Thus, a gene responsible for antibiotic resistance may be transferred among different strains (Lupo et al. 2012; Zhang et al. 2011). The detailed mechanism of antibiotic resistance acquired by the bacteria is illustrated in Fig. 5.1.

At present, antibiotic-resistant bacteria are evident in various ecological niches. Selective pressure in favor of bacteria harboring these genes has emerged from the offensive use of antibiotics mainly in hospitals, pharmaceutical sectors, animal farming, and agricultural sectors. Ultimately, they contaminate the drinking water sources which in many parts of the world are consumed without proper treatment. The occurrence of *E. coli* which was resistant to Carbapenem and possessed *NDM-1* gene in drinking water samples in New Delhi, India, was reported (Walsh et al. 2012). The bacteria carried various resistant gene products in spring and tap waters in Turkey (Ozgumus et al. 2007), and drinking water in Mainz, Germany, was established (Schwartz et al. 2003). Further, the occurrence of many harmful bacteria and their gene codes for multidrug resistance toward amoxicillin/ampicillin (bla_{TEM}), β-lactam, tetracycline (*tet*), chloramphenicol (*cmlA*), streptomycin/spectinomycin (*aadA*), methicillin (*mec*), and vancomycin (*van*) in several water bodies was also studied (Thevenon et al. 2012). The occurrence of genes resistant for sulfonamides in many water bodies in Tianjin, China (Gao et al. 2012), and ciprofloxacin and cefotaxime resistance genes in hospital wastewater in Madhya Pradesh, India, was also studied.

Although certain multiple drug-resistant organisms depict resistance to only one drug (e.g., vancomycin-resistant enterococci, methicillin-resistant *Staphylococcus aureus*), these bacteria are often resistant to most available antibacterial agents (Thevenon et al. 2012). The overuse of drugs that are excreted by patients finds their way into the community, and hospital wastewater systems supply an environmental selection pressure for the emergence and perseverance of multidrug-resistant bacteria (Gao et al. 2012). Controlling multidrug-resistant bacteria is imperative because they are resistant to conventional antimicrobial therapy, increase patient morbidity and mortality, add to the treatment cost, and have the potential to spread and operate as a reservoir of drug-resistant genes for the transmission to other bacteria. Antimicrobial resistances threaten health safety and damage economy and trade; the recent occurrence of *NDM-1*-associated infections in India might cause unjustified economic consequences on medical tourism. Multidrug-resistant tuberculosis has

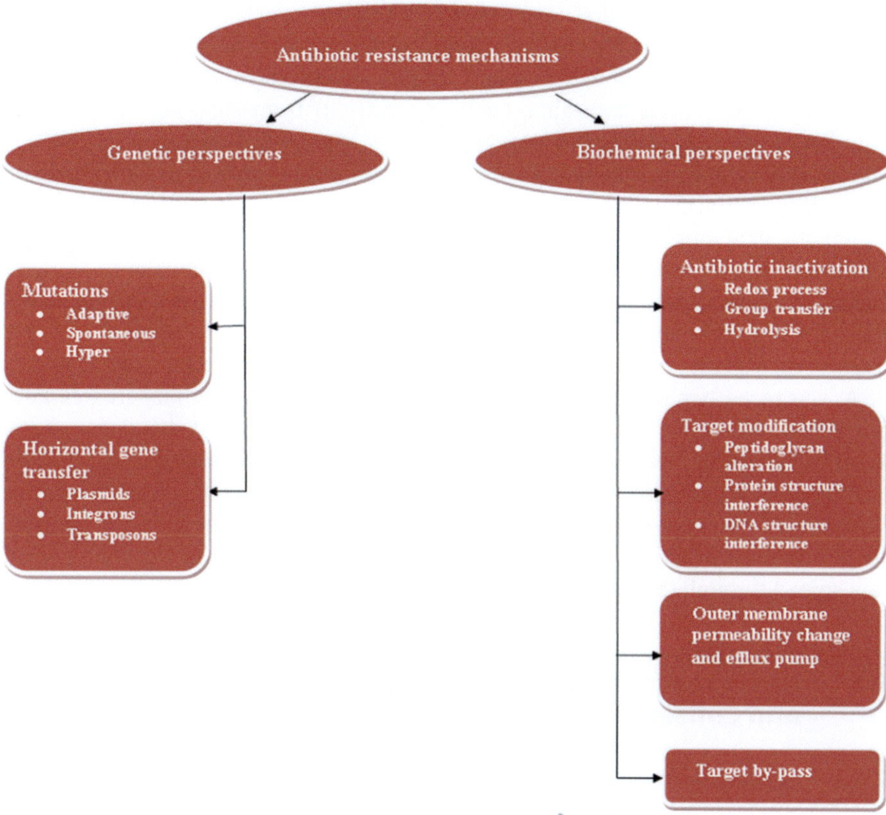

Fig. 5.1 Basic mechanisms of the evolution of antibacterial resistance among bacterial pathogens

reached alarming magnitude in South Africa, draining precious resources that are desirable to fight drug susceptibility. It is estimated that 9.6% of all tuberculosis cases have multidrug-resistant tuberculosis, thereby grading South Africa as one of the utmost multiple drug-resistant tuberculosis burden countries in the world. MDR strains of *Salmonella typhimurium*-type DT104 (resistant to sulfamethoxazole, streptomycin, chloramphenicol tetracycline, and ampicillin) appeared across the USA during the 1990s (Glynn et al. 1998). In 2000, CDS and many state health departments in the USA have identified a rush in the frequency of *Salmonella* serovar Newport-MDRAmpC, particularly multiple drug-resistant strains. These strains were also resistant to tetracycline, chloramphenicol, sulfamethoxazole, streptomycin, and ampicillin. Moreover, many isolates of Newport-MDRAmpC were resistant to amoxicillin/clavulanic acid, cefoxitin, ceftiofur, and cephalothin and showed demonstrated sensitivity to ceftriaxone (Gupta et al. 2003). In addition to bacteria, the occurrence of drug-resistant *Plasmodium falciparum* (malarial parasite) in Southeast Asia, which present belated response to artemisinin-based therapies, was also estimated. The spreading of drug-resistant viruses resulted in stoppage to curb HIV in the case of antiretroviral agents used to treat AIDS, and a more expensive drug combination was required (Volberding and Deeks 2010).

Similarly, multidrug resistance encoding plasmid from *Aeromonas* sp. *P2G1* from River Ter in Ripoll, Spain, was reported by Marti and Balcázar (2012). They have reported the drug-resistant genes for ciproflocaxin, norfloxacin, levofloxacin, amoxicillin, ofloxacin, ceftazidime, nalidixic acid, enrofloxacin, sulfamethoxazole, rifampicin, gentamicin, kanamycin, and erythromycin.

Recent reports revealed that multidrug-resistant bacteria and prevalence of genes responsible for drug resistance are very common in many municipal water treatment plants (WWTPs) in addition to the river water. Occurrence of vancomycin-resistant enterococci (VRE), a leading cause of hospital-acquired infections, at WWTPs, USA, was recently reported (Rosenberg Goldstein et al. 2014). Similar studies were conducted in WWTP in Patancheru; Hyderabad, India, showed that most of the enterobacteriaceae isolated from the tanks were emerged as resistance to β-lactam antibiotics and cephalosporins (Marathe et al. 2013). Similarly, *NDM-1*-positive *Acinetobacter baumannii* in many WWTPs in northern China was recently reported. Such multidrug-resistant bacteria in WWTPs could cause risks to public health if they spread through the environment (Luo et al. 2014).

5.4 Importance of Water Quality Monitoring and Conservation of Freshwater Ecosystem

5.4.1 Conventional Treatment Methods

Wastewater treatments are the most important method used for the treatment of polluted water. There are many wastewater treatment plants (WWTPs) available for effective treatment of contaminated water from rivers, streams, and other sources. The major methods are primary (physical), secondary (biological), and tertiary (chemical) treatments. Physical methods include screening and sedimentation. Biological methods include trickling filter, activated sludge systems, rotating biological contractor, and oxidation ponds. The foremost chemical approach includes halogenation. The objectives of all these methods are the removal of hazardous wastes and particulate materials and reduction of biological oxygen demand (BD). Carey and Migliaccio (2009) reviewed that reduction of nutrient content in wastewater is a prime concept of water quality monitoring as excessive nutritional content prevent water bodies from meeting designated uses (Carey and Migliaccio 2009). The WWTPs employ primary, secondary, and tertiary approaches to improve quality of effluents, but removal of nutrient requires additional treatment and sophisticated infrastructures. Margot et al. (2014) reviewed that many organic micro-pollutants such as pesticides and pharmaceuticals present in wastewater are poorly removed by conventional WWTPs. Advanced wastewater treatment approaches are necessary to remove these substances from the aquatic environment. The most widely used approaches are (1) oxidation subsequently sand filtration and

(2) adsorption of powdered activated carbon (PAC) followed by either ultrafiltration or sand filtration. It is revealed that these methods remove 75% of total pollutants which include pesticides, pharmaceuticals, drug metabolites endocrine disruptors, and common chemicals (Margot et al. 2014). Similarly, various types of community-based wastewater treatment and water quality monitoring approaches available for wastewater treatment systems were recently reported (Lim et al. 2013). Furthermore, centralized municipal wastewater treatment systems and a decentralized aerobic and septic wastewater treatment systems were also reported to be successful for the management of wastewater (Garcia et al. 2013). Recent reports also suggested that Taguchi method is one of the modern approaches for pollution remediation in river water and this method was effectively used in many Rivers such as Danshui River, Taiwan (Yang et al. 2014).

5.4.2 Need for Physiochemical Characterization and Microbial Count Estimation

There are reports suggesting that the physiochemical parameters such as dissolved oxygen (DO), biological oxygen demand (BOD), chemical oxygen demand (COD), temperature, heavy metal content, pH, and many other factors have a direct influence on microbial proliferation (Salem et al. 2011). Reports suggested that the discharge of raw wastewater from municipal and agrochemical environments leads to the accumulation of high organic contents, thereby increasing the DO level in water (Wogul and Okaka 2011). Similarly, studies revealed that anoxygenic conditions in DO permit the growth of common aerobic and anaerobic microorganisms. Water conductivity is also an important parameter. This conductivity level shows high accumulation of dissolved inorganic content signifying that increased concentrations of nutrients are presented for bacterial multiplication (Ramírez Castillo et al. 2013). The consequence of BOD has a direct relationship with microbial population in water bodies especially the multiplication of fecal coliforms (Downing and Nerenberg (2008). The BOD values surpassed the recommended maximum concentration (RMC) proposed by the European Union for water quality which is 3.0–6.0 mg/L. The BOD value of clean waters was characteristically 2 mg/L or less, while polluted water was up to 10 mg/L or more (Jorgensen et al. 1979; Chapman 1996). It was reported that the variation in BOD values was responsible for odor and taste (Lee 2013).

Reports also suggested that an elevated level of inorganic and organic nutrients and the favorable physicochemical condition accelerates massive multiplications of various microorganisms in water bodies. The urban runoff sites and industrial wastes tended to have higher viable counts in comparison with agricultural farms and effluents from wastewater treatment plant (Ramírez Castillo 2013). The bank of river harbors many industries and urban centers which contributed tons of industrial sewages and urban wastes.

From the literature, it is evident that the standard plate count is the measure of viable bacterial count on a sample. The investigation is based on an assumption that each isolate will form a visible colony in an agar medium with appropriate nutrients. These counts indicate the total bacterial count in any environmental samples. These tests were performed by using serial dilutions of the water samples followed by preparation of pour plates, using standard plate count agar in replicates of three (Jorgensen et al. 1979). Similarly, the selective detection of fecal coli forms, which are Gram-negative, non-sporulating, facultative gastroenteric pathogens, often known as indicator microorganism, is possible by most probable number (MPN) test. This technique is evaluated for detecting and enumerating coli forms in water. The test comprises presumptive test, confirmed test, and completed test (Highsmith and Abshire 1975).

Microbiological characterization is the first level characterization of unknown bacteria from any samples. The characterization is primarily based on Bergey's manual of determinative bacteriology (Holt et al. 1993). The steps include the study of morphological, physiological, and biochemical characteristics of bacteria. Morphological studies comprise the identification of size, shape, and arrangement of microbial cells. Physiological characteristics included the nutritional requirements and growth conditions, colony characteristics, and environmental conditions for the optimum growth. Biochemical characteristics are the ability of particular bacteria to metabolize certain substances. These tests include IMViC, catalase, urease, hydrogen sulfide production, oxidase, nitrate reduction, coagulase, starch hydrolysis, triple sugar iron, lysine decarboxylase, arginine decarboxylase and ornithine decarboxylase, and many more (Sulkin and Willett 1940; Moeller 1955; Sperber and Tatini 1975; Mac Faddin 2000; Colonna et al. 1992; Murray et al. 2003; Roser et al. 2005).

5.4.3 Need for Antibiotic Susceptibility Profiling

From the literature, it is evident that there is increasing number of antibacterial available for the treatment of various infections; the designing of simple and accurate methods to estimate the antibiotic sensitivity of microorganisms is of rising importance. The most commonly used method is the filter paper disk diffusion assay (Bauer et al. 1959). The major classes of antibiotics commonly used in clinical microbiology are aminoglycosides, ansamycins, carbacephem, cephalosporins (first to fifth generation), glycopeptides, lipopeptide, macrolides, monobactams, nitrofurans, oxazolidonones, penicillins, polypeptides, quinolones/fluoroquinolone, sulfonamides, tetracyclines, etc. (Butler and Cooper 2011; O'Neill 2008). However, carbapenems are the broad spectrum antibiotics with high bactericidal activities (Murakami et al. 1982). At present, carbapenems are considered as the strongest antibiotics which possess the broadest spectrum of activity with high potency. These are considered as "last-line drugs" when patients with disease become seriously sick or are assumed of prevailing drug-resistant bacteria especially Gram-negative pathogens. Resistances to recent and potent antibacterial are making the therapeutic alternatives inadequate to certain antimicrobial agents such as fosfomycin, carbapenem, and colistin (Tanwar et al. 2014). Sadly, the current occurrence of MDR pathogens gravely threatens most class of conventional drugs. Various studies conducted worldwide demonstrated that resistance to carbapenems and colistins is raising concern worldwide. Infections with MDR Gram-negative bacteria are difficult to treat which result high morbidity and mortality. Pathogenic MDR bacteria have emerged as a severe hazard to human health. It is anticipated that in the USA, approximately two million people develop hospital-acquired infections every year; the major cause of these infections are MDR pathogens (Kadouri et al. 2013). Studies depicted that efficient awareness and situation of prevailing MDR bacterial pathogens are of main importance for the appropriate use of antibacterial drugs and the policy making to battle multidrug resistance (Tanwar et al. 2014).

5.5 Need for an Alternative Therapy

The drug resistance is the foremost healthcare concern in the treatment of bacterial pathogens by various antimicrobials (van Duin and Paterson 2016). Understanding the genetic distinction among plasmids from various species of bacteria is a main action toward studying the mechanism of evolution and virulence. Understanding their pathogenicity helps in developing more effective antibacterial against the drug-resistant microorganisms. The current accessibility of high-throughput sequencing strategies contributes the capacity for fast and effective understanding of genomes (Threlfall et al. 2000). Drug design is a laborious, complex, and interdisciplinary approach. For the pharmaceutical company, the period of time required to initiate a new drug into the market is approximately 12–14 years which costs an estimate of $1.2–$1.4 billion. For example, if 10,000 molecules are initially undergone testing in various animal models, around 10 would identify for further clinical studies, and one may reach into the market finally (Zhao et al. 2010).

By all concerns associated with antimicrobial resistance, there is a terrible scope for identifying novel therapeutic agents (Daniels 2011). The modern drug discovery process has been transformed with the advent of bioinformatics, combinatorial chemistry, genomics, proteomics, and high-throughput screening and molecular modeling. The major concepts of computation in drug designing are virtual screening, de novo design, computational ADMET prediction, and study of ligand-receptor interactions. Computational ADMET prediction is carried out alongside of the in vitro data generated, for understanding the structure of targets for probable binding orientation, generating best possible conformation, predicting the druggability of ligands, molecular docking study, ranking the best docked conformations based on minimum energies, and optimizing the lead molecules for improving the binding potential. Bioinformatics and system biology tools offer the benefit of delivering novel therapeutic molecules with suitable druggish and pharmacokinetic features (Karumuri et al. 2015, Singh and Shukla 2015).

5.5.1 Need for Computational Modeling and Virtual Screening

Recent reports suggested that antibacterial losing their grade as the "miracle drug" and "failure in treatment" is a common condition. The drug resistances became prevalent in the ecosystem as existing resistant bacteria appeared as dominant populations and emerged as the superbugs (Wallace et al. 2000; Kadouri et al. 2013). As the bacteria demonstrated resistance to most of the routinely used antibacterial, there is a need to screen prioritized drug targets and search for additional therapeutic molecules. One promising approach is to inhibit the major virulent factors of drug-resistant pathogens by novel lead molecules which are not based on accessible synthetic drugs (Gowrishankar et al. 2013). Antimicrobial action of *Sarcostemma intermedium*, an extraordinary and effective medicinal plant, was also determined against few pathogenic microorganisms (Dahiya et al. 2014). The novel strategies which have to be implemented include identification of probable drug targets, selection and screening of new lead molecules, detection of ideal treatment approaches, and characterization of drug-resistant bacterial pathogens and its susceptibility to the routinely used antibacterial. Further, it requires deeper understanding of the machinery by which the bacteria acquire resistance to conventional antibacterial which will assist in selecting prioritized targets for various drugs (Coleman et al. 2013).

Computational virtual screening became gradually more vital in various disciplines such as gene or protein prediction, homology or comparative modeling, detection of functional sites, prediction of active site for binding, molecular docking of lead molecules, study of protein-ligand interactions, and molecular simulations. The outcome of computational prediction yields inputs ahead of the present in vitro and in vivo promises and can be used to direct and improve an immense group of studies (Schjørring and Krogfelt 2011; Chah et al. 2006; Gago 2004). Studies emphasize that the identification of the homologous relationship between the proteins is the key concept of the structural and functional explanation of newly screened drug resistance genes (Muller et al. 1999).

The fundamental requirement of computational drug discovery is the 3D structure of major drug targets such as resistant gene products/toxins. The 3D structures of most the targets are not yet elucidated experimentally. Thus, there is a scope for high-quality 3D atomic model of the targets. This is possible by computational structure prediction (three dimensional) approaches such as ab initio (de novo) prediction, fold recognition (threading), and homology modeling or comparative modeling (Marti-Renom et al. 2000). Homology modeling is an approach to create high-quality atomic model of a protein from its basic amino acid sequence based on the structure of homologous proteins that have been elucidated experimentally. Modeler is a homology modeling tool, and it implements structure modeling by satisfaction of spatial restraints and generates a good quality model. The protein to be modeled will be provided with best homologous experimentally solved structures known as template. The prediction process includes fold assignment, structural alignment, model building, model refinement, and evaluation (Marti-Renom et al. 2000).

There are studies suggesting that herbal-based bioactive compounds are effective therapeutic substances (Nair et al. 2005). Most of the herbal-based medications are well studied, and their bioactive mechanisms have been established. Moreover, mode of action, the bioactivity assay, and inhibitory action of many herbal-derived compounds against various drug targets are well understood (Briskin 2000). Computer-assisted drug design (CADD) is an initial platform to select and screen various lead molecules with drug likeliness and pharmacokinetic and bioavailability features (Bharath et al. 2011). Currently CADD is used for the identification of active drug molecules and selection and optimization of lead candidates which transforms bioactive molecules into appropriate drugs by improving their drug likeliness and pharmacokinetic features. Computational virtual screening is used to select novel lead candidates from chemical scaffolds by database searching and other approaches (Kapetanovic 2008). CADD is the primary aspect of structure-based drug discovery that uses a variety of computational predictions and models to select a new lead candidate with efficacy, safety, and selectivity. The study of ligand-receptor interaction is the fundamental spotlight of structure-based drug screening, and the prediction of these bindings by computational models has scope and applications (Lyskov and Gray 2008). Macromolecular docking or computational docking is a useful approach to study the binding of receptor and ligand which is a basic aspect behind computer-assisted drug discovery. Molecular docking is an approach of predicting the binding affinity of the lead molecule or inhibitor in the binding site of the receptor. Most of the docking approaches depend on search algorithms which determine the binding of ligand in the active site of the receptor and a scoring function which estimate the binding potential and how accurately the ligand binds with the receptor (Dhanik and Kavraki 2012). The major interactions that stabilized the receptor-ligand complexes are weak interactions such as hydrophobic interactions, electrostatic interaction van der Waals forces, and hydrogen bonding. Thus, the main factors essential to scrutinize stable-docked complexes are the binding energy (kcal/mol), extent of electrostatic interactions, and number of hydrogen bonding that stabiles the docked complex. An outline of macromolecular docking process is shown in Fig. 5.2.

There are various studies revealing the application of computational virtual screening toward the identification of best therapeutic candidates. A report suggested that the inhibitory potential of bioactive substances was screened from essential oils toward probable drug targets of *Streptococcus mutans* (Galvão et al. 2012) by computational modeling. Similarly, reports suggested that various phytochemicals screened from medicinal plants demonstrated significant binding potential toward prioritized targets of various MDR clinical isolates (Dahiya and Purkayastha 2012). Further, the inhibitory properties of Kurarinone, a flavonoid characterized from *Sophora flavescens*, against selected drug targets of vancomycin-resistant *Streptococcus* sp., methicillin-resistant *Staphylococcus aureus*, and *Streptococcus mutans* were also proposed (Chen et al. 2010). It has been suggested that novel herbal-based molecules screened by virtual screening depicted inhibitory potential toward streptolysin O of MDR *Streptococcus pyogenes* (Skariyachan et al. 2014). Similarly, previous study suggested that the herbal-derived leads screened by

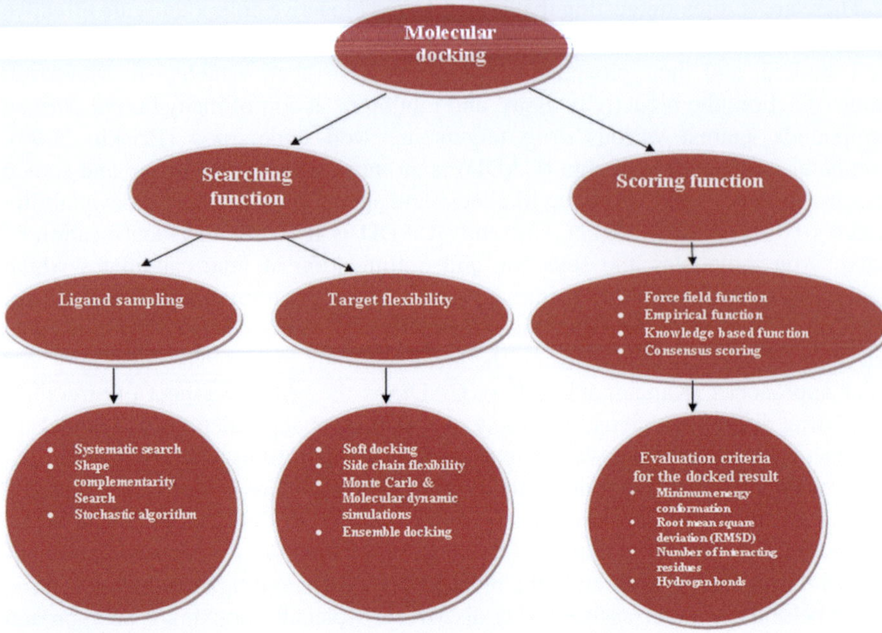

Fig. 5.2 The basic components involved in molecular docking process

computational modeling showed good binding potential toward the drug-resistant gene products of *Salmonella typhi*, *Staphylococcus aureus*, and *Vibrio cholerae* (Skariyachan et al. 2013a, b). These herbal leads showed profound binding properties with the probable drug targets (the drug-resistant genes) with minimum binding energies and various stabilizing interactions (Fig. 5.3). Furthermore, previous study suggested various herbal-derived molecules against virulent toxins of bacterial superbugs (Skariyachan et al. 2012).

The virtual screening of inhibitors for bacterial biotin carboxylase using structure-based approaches was reported (Brylinski and Waldrop 2014). Further, the screening of new inhibitors of the glyoxylate pathways in Gram-negative bacteria was also proposed (Fahnoe et al. 2012). Computational screening of new inhibitors for various drug targets in *Mycobacterium tuberculosis* was also reported (Chung et al. 2013; Speck-Planche et al. 2012). Furthermore, studies suggested the application of polyphosphate kinase (PPK) of *E. coli* as probable drug target and computational virtual screening of novel small molecules toward these targets (Saha and Verma 2013). The inhibitory activities of new lead molecules toward serine protease from various MDR bacteria by computational screening were also reported (Mandal et al. 2014).

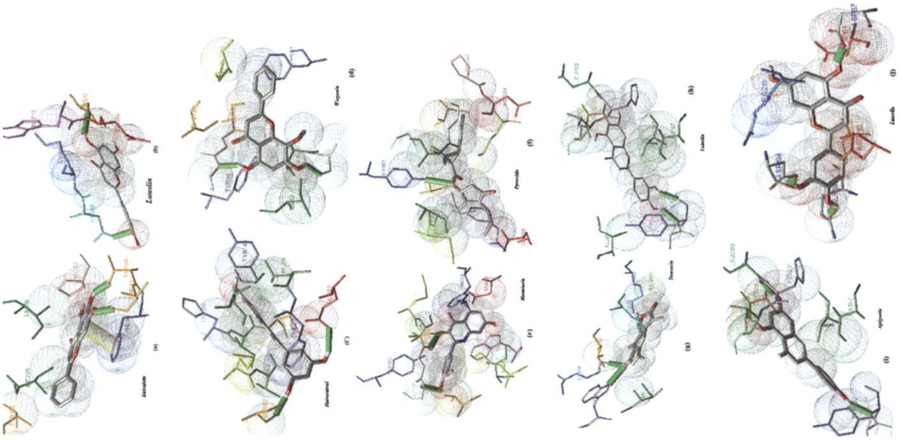

Fig. 5.3 The binding efficiency of herbal-based ligands toward the probable drug target-multidrug-resistant gene product of various Gram-negative bacteria predicted by molecular docking. The ligands and interacting residues are shown in stick figures. Hydrogen bonds are shown in thick stick figures. (**a**) Phytoligand Baicalien interacted with modeled aph protein of *Salmonella typhi* by the formation of two hydrogen bonds (binding energy −6.39 kcal/mol). (**b**) Interaction of luteolin with aph protein *of Salmonella typhi* is stabilized by two hydrogen bonds (binding energy −6.42 kcal/ mol). (**c**) Herbal ligand resveratrol binding with dihydrofolate reductase (dfrA) of *Salmonella typhi* by two hydrogen bonds (binding energy −7.58 kcal/ mol). (**d**) Interaction between Wogonin and dfrA1 is stabilized by two hydrogen bonds (binding energy be -7.28 Kcal/ mol. (**e**) Herbal ligand Herniarin binding with dihydrofolate reductase type 1(dfrA1) of *Vibrio cholerae* by two hydrogen bonds (binding energy −8.06 kcal/mol) (**f**) Interaction between Pyrocide and dfrA1 is stabilized by a hydrogen bond (binding energy −8.93 kcal/mol). (**g**) Herbal ligand Taraxacin of methicillin resistance protein (mecI) of *Staphylococcus aureus* interacted via two hydrogen bonds (binding energy −7.28 kcal/mol). (**h**) Herbal lead luteolin interacted with mecI by two hydrogen bonds (binding energy −7.58 kcal/mol). (**i**) The interaction between herbal apigenin and vancomycin resistance protein (vanH) of *Staphylococcus aureus* stabilized by two hydrogen bonds (binding energy −6.07 kcal/mol). (**j**) Herbal ligand luteolin binds with vanH by three hydrogen bonds (6.32 kcal/mol)

5.6 Conclusion

The major conclusion from the chapter is highlighted below:

- The discharging of domestic and industrial waste materials facilitates the emergence of multidrug-resistant bacteria in freshwater ecosystems which are one of the global threats. There should be high priority to address these issues and preserving the natural status of the freshwater ecosystems from the detrimental impact of pollution.
- Most bacteria emerged as extreme drug-resistant strains against currently prescribed chemotherapeutics, and the present antibacterial agents have limited applications in healthcare sectors.

- Thus, there is high priority to screen novel lead molecules by understanding of the major concepts by which bacteria acquire resistance to present generation antibacterial.
- Computational biology and high-throughput virtual screening serve as promising platforms to select novel therapeutic leads, especially from herbal origin, against drug targets and provide remarkable insights for further experimental validation and industrial scale up.

References

Aatish P (2012) Cauvery water dispute essay (www.publishyourarticles.org/eng). Accessed 25 Mar 2014

Akoachere JF, Masalla TN, Njom HA (2013) Multi-drug resistant toxigenic *Vibrio cholerae* O1 is persistent in water sources in New Bell-Douala, Cameroon. BMC Infect Dis 13:366. doi:10.1186/1471-2334-13-366

Ali MS, Osman GA (2010) Microbial load as pollution indicator in water of El-Khadra Lake at Wadi El-Natroun, Egypt. J Am Sci 6:489–496

Baker GJ (2014) Christianity and eugenics: the place of religion in the British eugenics education society and the American eugenics society, c.1907-1940. Soc Hist Med 27:281–302

Bauer AW, Perry DM, Kirby WMM (1959) Single-disk antibiotic-sensitivity testing of staphylococci: an analysis of technique and results. AMA Arch Intern Med 104:208–216

Belay A, Sahile S (2013) The effects of Dashen brewery wastewater treatment effluent on the bacteriological and physicochemical quality of Shinta River in Gondar, North West Ethiopia. World Environ 3:29–36

Berglund B (2015) Environmental dissemination of antibiotic resistance genes and correlation to anthropogenic contamination with antibiotics. Infect Ecol Epidemiol 5:28564

Bharath EN, Manjula SN, Vijaychand A (2011) In silico drug design tool for overcoming the innovation deficit in the drug discovery process. Int J Pharm Pharm Sci 3:8–12

Bouraie MME, Barbary AAE, Yehia MM, Motawea EA (2011) Water quality of Rosetta branch in Nile delta, Egypt. Suo 61:1–12

Briskin DP (2000) Medicinal plants and phytomedicines. Linking plant biochemistry and physiology to human health. Plant Physiol 124:507–514

Brylinski M, Waldrop GL (2014) Computational redesign of bacterial biotin carboxylase inhibitors using structure-based virtual screening of combinatorial libraries. Molecules 19:4021–4045

Butler MS, Cooper MA (2011) Antibiotics in the clinical pipeline in 2011. J Antibiot (Tokyo) 64:413–425

Carey RO, Migliaccio KW (2009) Contribution of wastewater treatment plant effluents to nutrient dynamics in aquatic systems: a review. Environ Manag 44:205–217

Chah KF, Eze CA, Emuelosi CE, Esimone CO (2006) Antibacterial and wound healing properties of methanolic extracts of some Nigerian medicinal plants. J Ethnopharmacol 104:164–167

Chapman D (1996) Water quality assessments: a guide to the use of biota, sediments and water in environmental monitoring, 2nd edn. Chapman & Hall, London

Chen CY, Lindsey RL, Strobaugh TP Jr, Frye JG, Meinersmann RJ, Chen CY, Lindsey RL, Strobaugh TP Jr, Frye JG, Meinersmann RJ (2010) Prevalence of ColE1-like plasmids and kanamycin resistance genes in *Salmonella enterica* serovars. Appl Environ Microbiol 76:6707–6714

Chigor VN, Sibanda T, Okoh AI (2013) Studies on the bacteriological qualities of the Buffalo River and three source water dams along its course in the Eastern Cape Province of South Africa. Environ Sci Pollut Res Int 20:4125–4136

Chowdhury S (2012) Heterotrophic bacteria in drinking water distribution system: a review. Environ Monit Assess 184:6087–6137

Chung BK, Dick T, Lee DY (2013) In silico analyses for the discovery of tuberculosis drug targets. J Antimicrob Chemother 68:2701–2709

Coleman BL, Louie M, Salvadori MI, McEwen SA, Neumann N, Sibley K, Irwin RJ, Jamieson FB, Daignault D, Majury A, Braithwaite S, Crago B, McGeer AJ (2013) Contamination of Canadian private drinking water sources with antimicrobial resistant Escherichia coli. Water Res 47:3026–3036

Colonna P, Leloup V, Buléon A (1992) Limiting factors of starch hydrolysis. Eur J Clin Nutr 46:S17–S32

Central Pollution Control Board (CPCB), Annual Report-2012, Ministry of Environment & Forests (http://www.cpcb.nic.in/upload/AnnualReports). Accessed 1 Apr 2015

Dahiya P, Purkayastha S (2012) Phytochemical screening and antimicrobial activity of some medicinal plants against multi-drug resistant bacteria from clinical isolates. Indian J Pharm Sci 74:443–450

Dahiya T, Baweja M, Shukla P (2014) A swift description on antimicrobial action of *Sarcostemma intermedium*, an extraordinary scarce medicinal plant against few pathogenic microorganisms. Res J Pharm Biol Chem Sci 5(2):1064–1069

Daniels R (2011) Surviving the first hours in sepsis: getting the basics right (an intensivist's perspective). J Antimicrob Chemother 66:11–23

Dhanik A, Kavraki LE (2012) Protein–ligand interactions: computational docking. In: eLS. Wiley, Chichester

Dorice K, Joséphine N, Margaret TA, Gaston L, Beyala KK, Bessa A, Henriette, Emmanuel EG (2010) Bacterial contamination of water points of the upper Mfoundi watershed, Yaounde, Cameroon. Afr J Microbio Res 4:568–574

Downing LS, Nerenberg R (2008) Effect of bulk liquid BOD concentration on activity and microbial community structure of a nitrifying, membrane-aerated biofilm. Appl Microbiol Biotechnol 81:153–162

Dudarev AA, Dushkina EV, Sladkova YN, Alloyarov PR, Chupakhin VS, Dorofeyev VM, Kolesnikova TA, Fridman KB, Evengard B, Nilsson LM (2013) Food and water security issues in Russia II: water security in general population of Russian Arctic, Siberia and Far East, 2000-2011. Int J Circumpolar Health 72:22646. doi:10.3402/ijch.v72i0.22646

Dudgeon D, Arthington AH, Gessner MO, Kawabata Z, Knowler DJ, Lévêque C, Naiman RJ, Prieur-Richard AH, Soto D, Stiassny ML, Sullivan CA (2006) Freshwater biodiversity: importance, threats, status and conservation challenges. Biol Rev Camb Philos Soc 81:163–182

Fahnoe KC, Flanagan ME, Gibson G, Shanmugasundaram V, Che Y, Tomaras AP (2012) Nontraditional antibacterial screening approaches for the identification of novel inhibitors of the glyoxylate shunt in gram-negative pathogens. PLoS One 7:e51732

Gago F (2004) Modelling and simulation: a computational perspective in anticancer drug discovery. Curr Med Chem Anticancer Agents 4:401–403

Galvão LC, Furletti VF, Bersan SM, da Cunha MG, Ruiz AL, de Carvalho JE, Sartoratto A, Rehder VL, Figueira GM, Teixeira Duart MC, Ikegaki M, de Alencar SM, Rosale PL (2012) Antimicrobial activity of essential oils against *Streptococcus mutans* and their antiproliferative effects. Evid Based Complement Alternat Med 2012:751435

Gao P, Mao D, Luo Y, Wanga L, Xu B, Xu L (2012) Occurrence of sulfonamide and tetracycline-resistant bacteria and resistance genes in aquaculture environment. Water Res 46:2355–2364

Garcia SN, Clubbs RL, Stanley JK, Scheffe B, Yelderman JC Jr, Brooks BW (2013) Comparative analysis of effluent water quality from a municipal treatment plant and two on-site wastewater treatment systems. Chemosphere 92:38–44

García-Armisen T, Inceoğlu O, Ouattara NK, Anzil A, Verbanck MA, Brion N, Servais P (2014) Seasonal variations and resilience of bacterial communities in a sewage polluted urban river. PLoS One 9:e92579

Gazzaz MN, Yusoff KM, Ramli MF, Aris AZ, Juahir HA (2012) Characterization of special patterns in river water quality using chemo metric pattern recognition techniques. Mar Pollut Bull 64:688–698

Girija TR, Mahanta C, Chandramouli V (2007) Water quality assessment of an untreated effluent impacted urban stream: the Bharalu tributary of the Brahmaputra River, India. Environ Monit Assess 130:221–236

Glynn MK, Bopp C, Dewitt W, Dabney P, Mokhtar M, Angulo FJ (1998) Emergence of multidrug-resistant Salmonella enterica serotype typhimurium DT104 infections in the United States. N Engl J Med 338:1333–1338

Gowrishankar S, Thenmozhi R, Balaji K, Pandian SK (2013) Emergence of methicillin-resistant, vancomycin-intermediate *Staphylococcus aureus* among patients associated with group A Streptococcal pharyngitis infection in southern India. Infect. Genet Evol 14:83–389

Gupta A, Fontana J, Crowe C, Bolstorff B, Stout A, Duyne SV, Hoekstra MP, Whichard JM, Barrett TJ, Angulo FJ (2003) Emergence of multidrug-resistant *Salmonella enterica* serotype Newport infections resistant to expanded-spectrum cephalosporins in the United States. J Infect Dis 188:1707–1716

Highsmith AK, Abshire RL (1975) Evaluation of a most probable- number technique for the enumeration of Pseudomonas Aeruginosa. Appl Microbiol 30:596–601

Holt JG, Bergey DH, Breed RS (eds) (1993) Bergey's manual of determinative bacteriology, 9th edn. Lippincott Williams and Wilkins, Baltimore

Hysko M, Gace B, Puto K (2010) Microbial water pollution of Vjosa River and the vegetable contamination by irrigation. In: BALWOIS-international conference on water, climate and environment, water observation and information system for decision support, Ohrid, Republic of Macedonia, 25–29 May

Jalal FN, Kumar S (2013) Water quality assessment of Pamba River of Kerala, India in relation to pilgrimage season. Int J Res Chem Environ 3:341–347

Jorgensen JH, Lee JC, Alexander GA, Wolf HW (1979) Comparison of Limulus assay, standard plate count, and total coliform count for microbiological assessment of renovated wastewater. Appl Environ Microbiol 37:928–931

Joseph PV, Jacob C (2010) Physicochemical characteristics of Pennar River, a fresh water wetland in Kerala, India. E-Journal of Chemistry 7:1266–1273

Kadouri DE, To K, Shanks RM, Doi Y (2013) Predatory bacteria: a potential ally against multidrug-resistant gram-negative pathogens. PLoS One 8:e63397. doi:10.1371/journal.pone.0063397

Kapetanovic IM (2008) Computer-aided drug discovery and development (CADDD): in silico-chemico-biological approach. Chem Biol Interact 171:165–176

Karumuri S, Singh PK, Shukla P (2015) *In Silico* analog design for terbinafine against *Trichophyton rubrum*: a preliminary study. Ind J Microbiol 55(3):333–340. doi:10.1007/s12088-015-0524-x

Kotze PJ (2012) Aspects of water quality, metal contamination of sediment and fish in the Olifants River, Mpumalanga. Thesis (M. Sc.), Rand Afrikaans University, 1997

Karnataka State Pollution Control Board (KSPCB) (2010) Annual report-August 2010 (http://kspcb.kar.nic.in/notifications.htm). Accessed on Jan 2014

Kuitcha D, Kamgang KBV, Sigha NL, Lienou G, Ekodeck GE (2008) Water supply, sanitation and health risks in Yaoundé, Cameroon. Afr J Envir Sc Technol 2:379–386

Kumar V, Sharma A, Chawla A, Bhardwaj R, Thukral AK (2016) Water quality assessment of river Beas, India, using multivariate and remote sensing techniques. Environ Monit Assess 188(3):137. doi:10.1007/s10661-016-5141-6

Lee Y (2013) An evaluation of microbial and chemical contamination sources related to the deterioration of tap water quality in the household water supply system. Int J Environ Res Public Health 10:4143–4160

Lee DY, Lee H, Trevors JT, Weir SC, Thomas JL, Habash M (2014) Characterization of sources and loadings of fecal pollutants using microbial source tracking assays in urban and rural areas of the Grand River watershed, Southwestern Ontario. Water Res 53:123–1231

Lim HS, Lee LY, Bramono SE (2013) Community-based wastewater treatment systems and water quality of an Indonesian village. J Water Health 12:196–209

Luo Y, Yang F, Mathieu J, Mao D, Wang Q, Alvarez PJJ (2014) Proliferation of multidrug-resistant New Delhi Metallo-β-lactamase genes in municipal wastewater treatment plants in Northern China. Environ Sci Technol Lett 1:26–30

Lupo A, Coyne S, Berendonk TU (2012) Origin and evolution of antibiotic resistance: the common mechanisms of emergence and spread in water bodies. Front Microbiol 3:1–13

Lyskov S, Gray JJ (2008) The Rosetta Dock server for local protein-protein docking. Nucleic Acids Res 36:W233–W238

Mac Faddin JF (2000) Biochemical tests for identification of medical bacteria, 3rd edn. Williams and Wilkins, Baltimore, Lippincott

Mandal SM, Porto WF, De D, Phule A, Korpole S, Ghosh AK, Roy SK, Franco OL (2014) Screening of serine protease inhibitors with antimicrobial activity using iron oxide nanoparticles functionalized with dextran conjugated trypsin and *in silico* analyses of bacterial serine protease inhibition. Analyst 139:464–472

Marathe NP, Regina VR, Walujkar SA, Charan SS, Moore ER, Larsson DG, Shouche YS (2013) A treatment plant receiving waste water from multiple bulk drug manufacturers is a reservoir for highly multi-drug resistant integron-bearing bacteria. PLoS One 8:e77310

Margot J, Kienle C, Magnet A, Weil M, Rossi L, de Alencastro LF, Abegglen C, Thonney D, Chèvre N, Schärer M, Barry DA (2014) Treatment of micropollutants in municipal wastewater: ozone or powdered activated carbon? Sci Total Environ 461-462:480–498

Marti E, Balcázar JL (2012) Multidrug resistance-encoding plasmid from *Aeromonas* sp. strain P2G1. Clin Microbiol Infect 18:E366–E368

Marti-Renom MA, Stuart A, Fiser A, Sánchez R, Melo F, Sali A (2000) Comparative protein structure modeling of genes and genomes. Annu Rev Biophys Biomol Struct 29:291–325

Moeller V (1955) Simplified tests for some amino acid decarboxylases and for the arginine dihydrolase system. Acta Pathol Microbiol Scand 36:158–172

Muller A, Mac Callum RM, Sternberg MJ (1999) Benchmarking PSI- BLAST in genome annotation. J Mol Biol 293:1257–1271

Murakami K, Doi M, Yoshida T (1982) Asparenomycins A, B and C, new carbapenem antibiotics. J Antibiot (Tokyo) 35:32–38

Murray PR, Baron JH, Pfaller MA, Jorgensen JH, Yolken RH (eds) (2003) Manual of clinical microbiology, 8th edn. American Society for Microbiology, Washington, DC

Nair R, Kalariya T, Chanda S (2005) Antibacterial activity of some selected Indian medicinal flora. Turk J Biol 29:41–47

Ndjama J, Kamgang KBV, Sigha NL, Ekodeck GE, Tita MA (2008) Water supply, sanitation and health risks in Douala, Cameroon. Afr J Environ Sc Technol 2:422–429

Nigam VK, Shukla P (2015) Enzyme based biosensors for detection of environmental pollutants-a review. Microb Biotechnol 25(11):1773–1781

O'Neill AJ (2008) New antibacterial agents for treating infections caused by multi-drug resistant gram-negative bacteria. Expert Opin Investig Drugs 17:297–302

Osińska A, Harnisz M, Korzeniewska E (2016) Prevalence of plasmid-mediated multidrug resistance determinants in fluoroquinolone-resistant bacteria isolated from sewage and surface water. Environ Sci Pollut Res Int 23(11):10818–10831

Ozgumus OB, Celik-Sevim E, Alpay-Karaoglu S, Sandalli C, Sevim A (2007) Molecular characterization of antibiotic resistant *Escherichia coli* strains isolated from tap and spring waters in a coastal region in Turkey. J Microbiol 45:379–387

Philbert FJ (1991) The Niagara River: a water quality management overview. Environ Monit Assess 17:157–166

Pimenta VM, Nepomuceno JC, Pavanin LA (2008) Genotoxicity of water from the Paraguay River near Caceres-MT, Brazil in the Drosophila wing-spot test. Environ Mol Mutagen 49:458–466

Raja P, Amarnath AM, Elangovan R, Palanivel M (2008) Evaluation of physical and chemical parameters of river Kaveri, Tiruchirappalli, Tamil Nadu, India. J Environ Biol 29:765–768

Ramírez Castillo FY, Avelar González FJ, Garneau P, Márquez Díaz F, Guerrero Barrera AL, Harel J (2013) Presence of multi-drug resistant pathogenic *Escherichia coli* in the San Pedro river located in the state of Aguascalientes, Mexico. Front Microbiol 4:147. doi:10.3389/fmicb.2013.00147. eCollection 2013

Rashid A (2012). Water pollution: a rising life-threatening problem. Articles and reports, News Central Asia

Rather MI, Rashid I, Shahi N, Murtaza KO, Hassan K, Yousuf AR, Romshoo SA, Shah IY (2016) Massive land system changes impact water quality of the Jhelum River in Kashmir Himalaya. Environ Monit Assess 188(3):185. doi:10.1007/s10661-016-5190-x

Rosenberg Goldstein RE, Micallef SA, Gibbs SG, George A, Claye E, Sapkota A, Joseph SW, Sapkota AR (2014) Detection of vancomycin-resistant enterococci (VRE) at four U.S. wastewater treatment plants that provide effluent for reuse. Sci Total Environ 466-467:404–411

Roser DJ, Ashbolt NHG, Mathew K, Nair J, Ryken-Rapp D, Toze S (2005) Hydrogen sulphide production tests and the detection of groundwater fecal contamination by septic seepage. Water Sci Technol 51:291–300

Saha SB, Verma V (2013) *In silico* analysis of Escherichia coli polyphosphate kinase (PPK) as a novel antimicrobial drug target and its high throughput virtual screening against PubChem library. Bioinformation 9:518–523

Sakai H, Kataoka Y, Fukushi K (2013) Quality of source water and drinking water in urban areas of Myanmar. Sci World J 2013:854261. doi:10.1155/2013/854261

Salem IB, Ouardani I, Hassine M, Aouni M (2011) Bacteriological and physicochemical assessment of wastewater in different region of Tunisia: impact on human health. BMC Res Notes 4:144. doi:10.1186/1756-0500-4-144

Sanchez CL, Reynoso FL, Chavez MR, Villa IG, Gavrilov IN, Garcia MP, Velazquez CR, Rodriguez MMM (2012) Assessment of water pollution in different aquatic systems: aquifers, aquatic farms on the Jamapa River, and coastal lagoons of Mexico. J Agr Sci 4:186–193

Schjørring S, Krogfelt KA (2011) Assessment of bacterial antibiotic resistance transfer in the gut. Int J Microbiol 2011:312956. doi:10.1155/2011/312956

Schwartz T, Kohnen W, Jansen B, Obst U (2003) Detection of antibiotic-resistant bacteria and their resistance genes in wastewater, surface water, and drinking water biofilms. FEMS Microbiol Ecol 43:325–335

Sharma A, Bhattacharya A, Bora CR, Shukla V, Parihar P (2012) Diversity of enteropathogens in river Narmada and their environmental and health implications. Microorg Environ Manag Microbes and Environ 1:35–60

Sikder MT, Kihara Y, Yasuda M, Yustiawati M, Mihara Y, Tanaka S, Odgerel D, Mijiddorj B, Syawal SM, Hosokawa T, Saito T, Kurasaki M (2013) River water pollution in developed and developing countries: judge and assessment of physicochemical characteristics and selected dissolved metal concentration. CLEAN Soil, Air, Water 41:60–68

Singh PK, Shukla P (2015) Systems biology as an approach for deciphering microbial interactions. Brief Funct Genomics 14(2):166–168. doi:10.1093/bfgp/elu023

Skariyachan S, Mahajanakatti AB, Sharma N, Karanth S, Rao S, Rajeswari N (2012) Structure based virtual screening of novel inhibitors against multidrug resistant superbugs. Bioinformation 8:420–425

Skariyachan S, Jayaprakash N, Bharadwaj N, Narayanappa R (2013a) Exploring insights for virulent gene inhibition of multidrug resistant *Salmonella typhi, Vibrio cholerae*, and *Staphylococcus areus* by potential phytoligands via in silico screening. J Biomol Struct Dyn 32:1379–1395

Skariyachan S, Lokesh P, Rao R, Kumar AU, Vasist K, Narayanappa R (2013b) A pilot study on water pollution and characterization of multidrug resistant superbugs from Byramangala tank, Ramanagara district, Karnataka, India. Environ Monit Assess 185:5483–5495

Skariyachan S, Narayan NS, Aggimath TS, Nagaraj S, Reddy MS, Narayanappa R (2014) Molecular modeling on streptolysin-O of multidrug resistant *Streptococcus pyogenes* and computer aided screening and *in vitro* assay for novel herbal inhibitors. Curr Comput Aided Drug Des 10(1):59–74

Sood A, Singh KD, Pandey P, Sharma S (2008) Assessment of bacterial indicators and physicochemical parameters to investigate pollution status of Gangetic river system of Uttarakhand (India). Ecol Indic 8:709–717

Speck-Planche A, Kleandrova VV, Luan F, Cordeiro MN (2012) *In silico* discovery and virtual screening of multi-target inhibitors for proteins in Mycobacterium tuberculosis. Comb Chem High Throughput Screen 15:666–673

Sperber WH, Tatini SR (1975) Interpretation of the tube coagulase test for identification of *Staphylococcus aureus*. Applied Microbiol 29:502–505

Subrahmanyam G, Shen JP, Liu YR, Archana G, Zhang LM (2016) Effect of long-term industrial waste effluent pollution on soil enzyme activities and bacterial community composition. Environ Monit Assess 188(2):112. doi:10.1007/s10661-016-5099-4

Sulkin SE, Willett JC (1940) A triple sugar-ferrous sulfate medium for use in identification of enteric organisms. J Lab Clin Med 25:649–653

Sundar K, Vidya R, Mukherjee A, Chandrasekaran N (2010) High chromium tolerant bacterial strains from Palar River basin: impact of tannery pollution. Res J Environ Earth Sci 2:112–117

Tanwar J, Das S, Fatima Z, Hameed S (2014) Multidrug resistance: an emerging crisis. Interdiscip Perspect Infect Dis 2014:541340. doi:10.1155/2014/541340

Tendall DM, Hellweg S, Pfister S, Huijbregts MA, Gaillard G (2014) Impacts of river water consumption on aquatic biodiversity in life cycle assessment--a proposed method, and a case study for Europe. Environ Sci Technol 48:3236–3244

Thevenon F, Adatte T, Wildi W, Poté J (2012) Antibiotic resistant bacteria/genes dissemination in lacustrine sediments highly increased following cultural eutrophication of Lake Geneva (Switzerland). Chemosphere 86:468–476

Threlfall EJ, Ward LR, Frost JA, Willshaw GA (2000) The emergence and spread of antibiotic resistance in food-borne bacteria. Int J Food Microbiol 62:1–5

Tissera S, Lee SM (2013) Isolation of extended spectrum ß-lactamase (ESBL) producing bacteria from urban surface waters in Malaysia, Malays. J Med Sci 20:14–22

Toroglu E, Toroglu S (2009) Microbial pollution of water in Golbasi Lake in Adiyaman, Turkey. J Environ Biol 30:33–38

Truman AW, Kwun MJ, Cheng J, Yang SH, Suh JW, Hong HJ (2014) Antibiotic resistance mechanisms inform discovery: identification and characterization of a novel amycolatopsis strain producing ristocetin. Antimicrob Agents Chemother 58:5687–5695

Umamaheswari S, Saravanan AN (2009) Water quality of Cauvery River basin in Trichirappalli, India. Int J Lakes and Rivers 2:1–20

van Duin D, Paterson DL (2016) Multidrug-resistant bacteria in the community: trends and lessons learned. Infect Dis Clin N Am 30(2):377–390. doi:10.1016/j.idc.2016.02.004

Vanham D, Weingartner R, Rauch W (2011) The Cauvery river basin in southern India: major challenges and possible solutions in the 21st century. Water Sci Technol 64:122–131

Varunprasath K, Daniel AN (2010) Comparison studies of three freshwater rivers (Cauvery, Bhavani and Noyyal) in Tamil Nadu, India. Iran J Energy Environ 1:315–320

Venkatesha Raju K, Somashekar RK, Prakash KL (2012) Heavy metal status of sediment in river Cauvery, Karnataka. Environ Monit Assess 184:361–373

Volberding PA, Deeks SG (2010) Antiretroviral therapy and management of HIV infection. Lancet 376:49–62

Wallace AJ, Stillman TJ, Atkins A, Jamieson SJ, Bullough PA, Green J, Artymiuk PJ (2000) E. coli hemolysin E (HlyE, ClyA, She A): X-ray crystal structure of the toxin and observation of membrane pores by electron microscopy. Cell 100:265–276

Walsh TR, Weeks J, Livermore DM, Toleman MA (2012) Dissemination of NDM-1 positive bacteria in the New Delhi environment and its implications for human health: an environmental point prevalence study. Lancet Infect Dis 1:355–362

West BM, Liggit P, Clemans DL, Francoeur SN (2011) Antibiotic resistance, gene transfer, and water quality patterns observed in waterways near CAFO farms and wastewater treatment facilities. Water Air Soil Pollut 217:473–489

World Health Organisation (WHO) (2010) Global report on antimalarial drug efficacy and drug resistance: 2000–2010. World Health Organization, Geneva

Wogul MD, Okaka CE (2011) Pollution studies on Nigerian rivers: heavy metals in surface water of warri river, Delta State. J Biodivers Environ Sci 1:7–12

Wu Q, Zhou H, Tam NF, Tian Y, Tan Y, Zhou S, Li Q, Chen Y, Leung JY (2016) Contamination, toxicity and speciation of heavy metals in an industrialized urban river: implications for the dispersal of heavy metals. Mar Pollut Bull 104(1–2):153–161

Xiong W, Li J, Chen Y, Shan B, Wang W, Zhan A (2016) Determinants of community structure of zooplankton in heavily polluted river ecosystems. Sci Rep 6:22043. doi:10.1038/srep22043

Xu ZQ, Flavin MT, Flavin J (2014) Combating multidrug-resistant gram-negative bacterial infections. Expert Opin Investig Drugs 23:163–182

Yang TM, Hsu NS, Chiu CC, Wang HJ (2014) Applying the Taguchi method to river water pollution remediation strategy optimization. Int J Environ Res Public Health 11:4108–4124

Ye B, Yang L, Li Y, Wang W, Li H (2013) Water sources and their protection from the impact of microbial contamination in rural areas of Beijing, China. Int J Environ Res Public Health 10:879–891

Zeng H, Jinglu WJ (2013) Heavy metal pollution of lakes along the mid-lower reaches of the Yangtze River in China: intensity, sources and spatial patterns. Int J Environ Res Public Health 10:793–807

Zhang Q, Lambert G, Liao D, Kim H, Robin K, Tung CK, Pourmand N, Austin RH (2011) Acceleration of emergence of bacterial antibiotic resistance in connected microenvironments. Science 333:1764–1767

Zhao F, Bai J, Wu J, Liu J, Zhou M (2010) Sequencing and genetic variation of multidrug resistance plasmids in *Klebsiella pneumoniae*. PLoS One 5(4):e10141. doi:10.1371/journal.pone.0010141

Zhao G, Mu X, Strehmel A, Tian P (2014) Temporal variation of stream flow, sediment load and their relationship in the yellow river basin, china. PLoS One 9:e91048. doi:10.1371/journal.pone. 0091048

Sinosh Skariyachan has obtained his M.Sc. (Microbiology) and M.Sc. (Bioinformatics) from Bharathiar University, Coimbatore, Tamil Nadu, India and Ph.D. (Biotechnology) from Visvesvaraya Technological University, Belgaum, Karnataka, India. He has currently working as an Associaate Professor in teh department of Biotechnology Engg., Dayananda Sagar Instiutions, Bangalore, India. His current research interests are Microbial and Environmental Biotechnology, Food Biotechnology, Antimicrobial resistance & public health Microbiology, Genomics, Proteomics and Computational biology, Molecular modeling and Computer aided drug designing. He has more than 35 international publications in reputed journals with high impact. He aauthored a text book and many book chapters for reputed publisers. He has received several awards for his reasearch and scientific contributions.

Probiotics for Human Health: Current Progress and Applications

6

Ruby Yadav and Pratyoosh Shukla

Abstract

Probiotics are live microbial cultures which enhance the beneficial gut microflora to improve the overall health of the host. It is a rising field in dairy food industry with significant growth potential. Probiotic food supplements have attracted a lot of attention and revealed a remarkable growth in this field. Various bacteria, yeast, and molds can be used as probiotics, but most commonly used microorganisms are lactic acid bacteria (LAB). LAB is involved in the fermentation of dairy products, foods, and beverages and produces lactic acid as the end product of fermentation. Among LAB, most commonly used bacteria which exhibit excellent probiotic properties belong to *Lactobacillus* and *Bifidobacterium* genus. These bacteria produce a variety of compounds such as organic acids (lactic acid and acetic acid), antimicrobial compounds (bacteriocins), nutraceuticals, vitamins, enzymes, etc. Probiotics are also consumed in combination with prebiotics known as synbiotics. Prebiotics are nondigestible carbohydrates, which pass through the small intestine in unmetabolized form and undergo fermentation in the large intestine. The fermentation products act as an energy source for indigenous gut microflora. The food products containing probiotics and prebiotics result in the enhancement of the microflora which promotes the overall gut health. This chapter enclosed a brief knowledge of different probiotic strains, probiotic foods, and their health applications.

Keywords

Probiotics • Lactic acid bacteria • Prebiotics • Health application

---

R. Yadav • P. Shukla (✉)
Enzyme Technology and Protein Bioinformatics Laboratory, Department of Microbiology, Maharshi Dayanand University, Rohtak 124001, Haryana, India
e-mail: rubyyadav67@gmail.com; pratyoosh.shukla@gmail.com

P. Shukla (ed.), *Recent Advances in Applied Microbiology*,
DOI 10.1007/978-981-10-5275-0_6

6.1 Introduction

Probiotics are microorganisms that are involved in providing health benefits to the consumer, when administrated in adequate amounts (Hill et al. 2014). As probiotics exhibit various health benefits, these are used for the production of food products like yogurt, fermented milk, etc. Lilly and Stillwell described the term probiotics and defined as the substances which are secreted by one microbe and in turn affect the other microbes. Fermented milk was the first recorded probiotic food. Elie Metchnikoff published his book *The Prolongation of Life* which boosted the research in probiotic area. After experimenting in the Bulgarian peasants, he concluded that longevity in them was due to consumption of large amounts of sour milk. This milk contains LAB which removes the pathological reactions like autotoxication by the gut's normal flora (Metchnikoff 1907). After this, the researchers started working on LAB importance in human health and diseases. Probiotics include bacteria, non-pathogenic yeast, and molds (Ouwehand et al. 2002). LAB produce lactic acid as their end product of fermentation and are ordinarily used for the production of yogurt, fermented milk, or other fermented foods. These are gram-positive, oxidase- and catalase-negative, sporulating rods and cocci. The genera comprise of *Lactobacillus*, *Leuconostoc*, *Lactococcus*, *Streptococcus*, *Pediococcus*, *Enterococcus*, *Carnobacterium*, *Sporolactobacillus*, *Bifidobacterium*, *Oenococcus*, *Sporolactobacillus*, *Vagococcus*, *Tetragenococcus*, *Weissella*, etc. These are widely spread in nature and also found in many food products as well as in oral, genital, and intestinal cavity of animals and human. *Lactobacillus* is the most important genus of LAB which include more than 80 species, and these are present in milk and dairy products like cheese and yogurt. Most microorganisms which are used as probiotics belong to genera *Lactobacillus*, *Enterococcus*, and *Bifidobacterium* (Holzapfel and Wood 2014). Some of strains which are showing good probiotic properties are *Lactobacillus acidophilus*, *L. lactis*, *L. casei*, *L. plantarum*, *L. helveticus*, *L. salivarius*, *L. johnsonii*, *L. bulgaricus*, *L. reuteri*, *L. rhamnosus*, *L. delbrueckii*, *L. fermentum*, *Bifidobacterium bifidum*, *B. longum*, *B. breve*, *Streptococcus thermophilus*, *Enterococcus faecium*, and *Saccharomyces boulardii*.

Various beneficial effects produced by LAB consumption involve improvement of the health of the intestinal tract, increasing the capability of the immune system, synthesis and increase of availability of the nutrients, reduction of the symptoms of lactose intolerance, decreasing the frequency of the allergy in the individuals who are susceptible to allergy, and reducing the risk of cancers in the consumer (Goyal et al. 2013, Savadago et al. 2006). The mechanisms of the action of probiotic by which they employ their effects are largely unidentified but may involve in the modification of the pH of the gut. They decrease the concentration of the pathogens by producing antimicrobial compounds and by competing for the binding sites and receptor sites for the growth factors and for the nutrients available to the pathogen. The food products which contain probiotics and prebiotics affect the functionality of the foods, which results in the enhancement of the microflora that promotes the gut health. The probiotic bacterial species increases the growth of the beneficial microorganisms, removes of the harmful bacteria, and supports the innate immunity of the body. A probiotic strain should fulfill the following selection criteria: safety, viability during storage, acid tolerant, bile tolerant, other epithelial cells of the gut, and able to colonize the intestinal tract, stimulate immune responses, and modulate normal microflora (Saad et al. 2013; Lee et al. 2014). The viability and functionality cannot be lost during their technological integration into food products, and it must not involve the creation of unpleasant textures or flavors. The technological, operational, and safety characteristics are accepted for the selection of the probiotics.

Processes like food digestion and nutrient integration occur in the small intestine, whereas the indigestible carbohydrates known as prebiotics can cross the small intestine, and, thus, they pass into the large intestine for the stimulation of the growth of probiotic bacteria. Prebiotics can be defined as nondigestible food products which stimulate growth and activity of bacteria in the host's intestine. So, these are dietary substances which support the growth of wanted bacteria over the unwanted one present in the gut. Prebiotics mainly include breast milk oligosaccharides, short-chain carbohydrates, inulin, galacto-oligosaccharides, oligofructose, and lactulose. Food containing prebiotics are garlic, wheat, bananas, honey, leeks, onion, etc. Synbiotics are the combination of probiotics and prebiotics. These contain probiotics which are beneficial bacteria and prebiotics which are indigestible products for the enhancement of growth of good bacteria. Fermented dairy products, kefir, and yogurt are the examples of the synbiotic food products. The most common synbiotics include fructooligosaccharides (FOS) and bifidobacteria, inulins and *Lactobacillus* GG, and bifidobacteria and lactobacilli with fructooligosaccharides. Nowadays metabolic engineering techniques for designing new probiotics in food industry have gained a lot of attention (Singh and Shukla 2014). Genetic modification of microbes involves the introduction of desired genes that may have a positive impact on the food industry (Gupta and Shukla 2015).

6.2 Probiotic Foods

Most commonly used probiotic product is yogurt. In spite of this, cheese, milk either fermented or unfermented, smoothies, juice, nutrition bars, cereals, and infant formula feed all are examples of probiotic food (Ranadheera et al. 2010). Along with food, probiotics can also be available in the form of liquid, powder, gel, granules, paste, capsules, sachets, drugs, and as dietary supplements. All these forms contain a large number of bacteria which remain in a stable condition. These forms of products are more convenient as we can deliver a large number of bacteria from manufacturer to customer in stable condition. There are various products made by different companies using different strains of probiotics. These products have various clinical or therapeutic applications. The use of probiotics in food or other products depends on many factors like stability of product, humidity, pH, age of customer, quantity or number of bacteria used, etc. These products are helpful for all age groups like children, infants, and old age. The aim of using these microbes as probiotics is mainly to increase the beneficial flora of the host.

In the United States, dairy products (e.g., fermented milk and yogurt) are the food products that contain the probiotic exclusions. LAB are associated with the fermented milk (Shah 2015). The most commonly used bacteria in dairy products (containing probiotics) include *Lactobacillus* and *Bifidobacterium* (Backhed 2012). Probiotics commonly are not the colonizers of the GI tract for long term, but they can stick provisionally to the epithelial layer. They remain metabolically active, although they divide very slowly in the intestine. Milk and milk products contain probiotic bacteria, which improve the beneficial microbiota in the intestine (Isolauri et al. 2001). Probiotics are a group of microorganisms that are involved directly in increasing the resistance of bacteria against intestinal pathogens and thus are involved in the prevention of the diseases. Probiotic bacteria involved in the production of a variety of compounds, which shows an inhibitory effect to the growth of pathogenic microorganisms, include bacteriocins, reuterin, and organic acids such as acetic acids and lactic acids. For the delivery of probiotic microorganisms in the body, food is the common medium. Probiotic microorganisms that are given by food systems have to fulfill some conditions like they have to first survive during the transfer through the upper gastrointestinal tract and then survive in the gut to produce beneficial effects to the host. Fermented foods can have probiotics, prebiotics, or both, and they are associated with good health. Yogurt is a well-known probiotic used today. It contains a very good nutritional value and provides health benefits to the host. The various factors that affect the growth of probiotic bacteria are pH, the presence of hydrogen peroxide and dissolved oxygen, buffering capacity, and concentration of metabolites (lactic acid, acetic acid, etc.).

6.3 Probiotic Microorganisms

Various bacteria, yeast, and molds can be used as probiotics, but most commonly used microorganisms are bacteria, and among bacteria, LAB are more popular. An overview of applications of LAB is given in Fig. 6.1.

Probiotics may be formed by a group of different strains or single bacterium. Probiotic preparations consist of specific strains of *Lactobacillus*, *Streptococcus*, and *Bifidobacterium* either alone or in combination. These three genera are considered to be safe and might be capable of preventing the overgrowth of pathogenic organisms as these are important components of the gastrointestinal flora. The commonly used probiotic bacteria are summarized in Table 6.1.

6.4 Probiotics: Mechanism of Action

Probiotics exhibit numerous and various effects to the host. The probiotic bacteria decrease the luminal pH resulting in inhibition of the establishment of pathogenic bacteria, inhibit the bacterial attack and the attachment of pathogenic bacteria to epithelial cells, and produce the antimicrobial compounds, e.g., defensins and

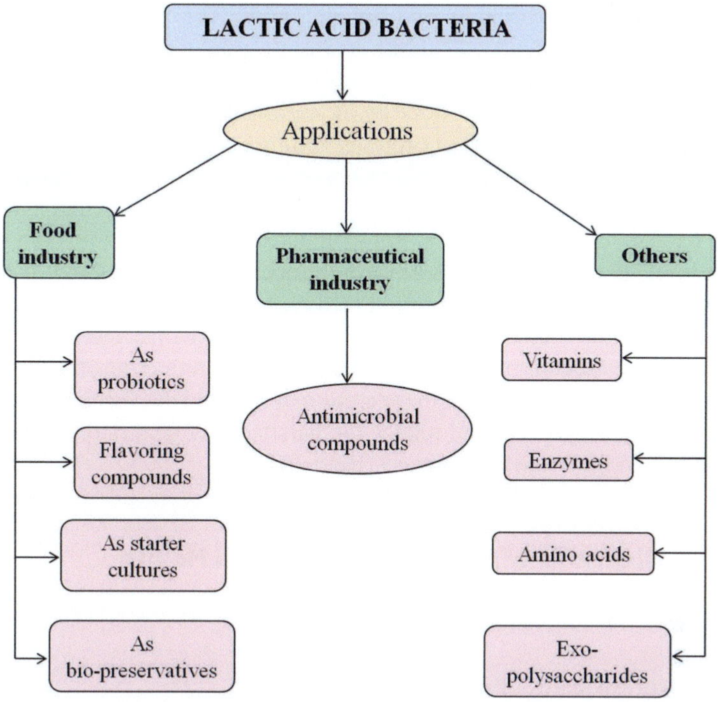

Fig. 6.1 Applications of LAB

Table 6.1 Bacterial species used as probiotics

Lactobacillus species	*Bifidobacterium* species	*Enterococcus* species	*Streptococcus* species
L. acidophilus	*B. bifidum*	*E. faecium*	*S. diacetylactis*
L. casei	*B. animalis*	*E. faecalis*	*S. cremoris*
L. paracasei	*B. infantis*		*S. salivarius*
L. rhamnosus	*B. thermophilum*		*S. intermedius*
L. johnsonii	*B. longum*		
L. delbrueckii ssp. *(bulgaricus)*	*B. adolescentis*		
	B. lactis		
L. brevis			
L. curvatus			
L. fermentum			
L. lactis			
L. plantarum			
L. reuteri			
L. cellobiosus			

bacteriocins, hydrogen peroxide, and organic acids. The action of LAB with the lymphoid cells in the gut and cells of the gastrointestinal tract increases the immune response of the gut against pathogens. Development of the function of mucosal barrier against ingested pathogens is done by increasing the production of mucus by the variation in the phosphorylation of proteins in tight junctions and cytoskeleton. The probiotic bacteria strive for the binding sites of epithelial layer with the pathogenic bacteria and inhibit the multiplication of strains like *E. coli* and *Salmonella*. Probiotic bacteria cooperate with the gut epithelial cells, directly (via compounds of cells like lipoteichoic acids, cell-surface polysaccharides, and DNA) or indirectly (by bioactive metabolite production). Probiotics influence the acquired and the innate immunity, thus having an important role in the human diseases. Probiotics remove the neoplastic cells of the host. Furthermore, the production of B-lymphocytes and antibody results in increase of the secretion of IgA and the vaccination response. Currently, probiotics prevent and reduce the harshness of respiratory infections, by increasing IgA in the bronchial mucosa (Reid 2016).

6.5 Applications of Probiotics in Human Health

There are several health benefits of probiotics:

Increase nutritional status of the individual
Increase vitamins, minerals, and trace elements availability to the body
Help in secretion of digestive enzyme, e.g., secretion of β-galactosidase
Prevention and treatment of diarrhea due to infection, traveler's diarrhea, acute viral diarrhea in children, diarrhea associated with overdose of antibiotics, and irradiation exposure diarrhea
Lowers down body cholesterol
Improvement of the immune system
Increase large intestinal motility that helps to relief constipation
Maintain mucosal integrity
Maintain intestinal microbes by antimicrobial activity
Decrease symptoms of lactose intolerance
Prevent food-borne allergies
Exhibit anticarcinogenic activities

An overview of health benefits of probiotic microorganisms has been given in Table 6.2.

6.5.1 Treatment of Diarrhea

Probiotics have a beneficial role in the prevention of diarrhea. It has been established that various probiotic strains like *L. rhamnosus* GG, *L. casei*, *L. reuteri*, *Bifidobacterium* spp., *S. boulardii*, etc., are useful in decreasing the duration and

Table 6.2 Probiotic strains and their health benefits

Probiotic strains	Health benefits	References
L. acidophilus NCFO1748	Constipation treatment, lowers fecal enzymes, prevents radiotherapy-related diarrhea	Giralt et al. (2008)
L. acidophilus LC1	Immune booster, maintains intestinal microflora balance, adjuvant for vaccines	Bernet et al. (1994)
L. acidophilus L1	Decreases total cholesterol level	Anderson and Gilliland (1999)
L. acidophilus LA5	Anti-infection, antidiarrhea, immune booster	Sadana et al. (2015)
L. rhamnosus GG	Anti-allergy, improves oral health	Schulz et al. (2015)
L. casei Shirota	Improves digestion, intestinal microflora reposition	Yadav and Shukla (2015)
L. gasseri	Reduction of fecal enzymes, intestinal tract survival	Pedrosa et al. (1995)
B. lactis	Enhances immune response, increases nonspecific immune functions	Mohan et al. (2008)
B. longum	Lowers serum concentration of total cholesterol, improvement of LDL/HDL cholesterol ratio, reduces lactose intolerance	Xiao et al. (2003)
B. adolescentis	Stimulates immune system	He et al. (2008)
B. animalis	Prevention of acute diarrhea	Shah (2007)
B. infantis	Increases anti-mutagenic activity	Hsieh and Chou (2006)
Streptococcus salivarius K12	Improvement of oral health, immune booster	Patel et al. (2015)
Enterococcus faecium SF68	Treatment and prevention of intestinal disorders	Hajela et al. (2015)
Saccharomyces boulardii	Immune booster	Patel et al. (2015) and Yadav and Shukla (2015)

severity of diarrhea (Isolauri 2004). The duration of acute diarrhea can be decreased by the oral administration of probiotics in children who are approximately of 1 day. The timing of administration is also very important for the action of probiotics. In antibiotic-associated diarrhea, *L. rhamnosus* GG and *S. boulardii* are effective in children or in adults who receive antibiotic therapy. The causative agent of antibiotic-associated diarrhea is *C. difficile*, which occurs after antibiotic treatment. It is an indigenous resident of a healthy intestine in low numbers. The antibiotic treatment may lead to disruption of healthy gut microflora, which results in increasing the number of *C. difficile*, which leads to diarrhea symptoms (Vasiljevic and Shah 2008; Ollech et al. 2016). Probiotic administration is very effective in the restoration of gut microflora. It is observed that *L. casei*, *B. longum*, *L. plantarum*, *B. breve*, *L. acidophilus*, *B. infantis*, *L. delbrueckii*, and *S. thermophilus* are very effective in the treatment of diarrhea caused by radiations. *L. casei* DN-114001 is very effective in the prevention of the radiation-induced diarrhea (Giralt et al. 2008). Probiotics are also able to prevent traveler's diarrhea and diarrhea caused by rotavirus (Vanderhoof 2000). It inhibits rotavirus adherence by modifying glycosylation state of epithelial cell receptors by excreting soluble factors (Freitas et al. 2003). Probiotic microorganisms prevent the diarrhea caused by microbes either by competitive binding to gut epithelial cells or by producing antimicrobial bacteriocins such as nisin. Also the probiotic strains can modulate the innate immune response (Braat et al. 2004).

6.5.2 Eradication of *Helicobacter pylori* Infections

Probiotic strains produce a variety of antimicrobial substances such as organic acids (acetic acid and lactic acid), bacteriocins, hydrogen peroxide, fatty acids, and antifungal peptides. The organic acids lower the pH of the gastrointestinal tract, which has an inhibitory effect on pathogens. *H. pylori* is an intestinal pathogen which is associated with peptic ulcers, chronic gastritis, and gastric cancer (Plummer et al. 2004; Fallone et al. 2016). The lactic acid produced by *L. salivarius* was found to inhibit *H. pylori* growth in vitro (Aiba et al. 1998). Modulating the diet by adding the probiotics may reduce the bacterial load and inflammation (Khulusi et al. 1995). *L. casei* Shirota, *L. gasseri* OLL2716, and *L. johnsonii* La1 were found to reduce the colonization and inflammation caused by *H. pylori* (Felley et al. 2001, Sgouras et al. 2004). A study reported by a group of researchers concluded that regular intake of probiotic yogurt containing mixture of *L. acidophilus* La5 and *B. animalis* Bb12 may suppress the infection by decreasing the bacterial load (Wang et al. 2004).

6.5.3 Cardiovascular Diseases

Cardiovascular diseases include coronary artery diseases, stroke, hypertensive heart diseases, etc.

Cholesterol-rich diet increases the risk of coronary heart diseases by increasing the serum cholesterol level. Mann and Spoerry (1974) first time reported the decreased serum cholesterol level due to consumption of fermented milk. Regular intake of probiotics may decrease the concentration of serum cholesterol level. A group of researchers demonstrated the role of *L. plantarum*, *B. longum*, *E. faecium*, and *Propionibacterium freudenreichii* in hypercholesterolemia and cardiovascular diseases (Kiatpapan et al. 2001; Xiao et al. 2003; Nguyen et al. 2007).

6.5.4 Cancer Prevention

Some fermented foods such as yogurt, dahi, sauerkraut, kefir, kimchi, and fermented milk comprise of anticarcinogenic activities (Mohania et al. 2013, Kwak et al. 2014). Regular intake of probiotic yogurt containing *Lactobacillus* and *Bifidobacterium* could reduce the risk of cervical, bladder, and colon cancer

(Chandan and Kilara 2013). A study has been reported for the treatment of cancer by using kefir (Yanping et al. 2009). Fermented milk contains *L. acidophilus* which activates the immune system of the host and removes procarcinogens (Macouzet et al. 2009). A possible mechanism of cancer control by probiotic could be in the following ways: probiotics can cause inhibition of tumor cells and suppression of bacteria that produces such enzymes, which catalyze procarcinogen conversion to carcinogens. Also, they can destroy the carcinogens.

6.5.5 Prevention of Allergy

The prevention of atopic dermatitis (allergy) is the strongest evidence when some probiotics are supplemented to the newborns (up to the age of 6) and to pregnant mothers. Some specific probiotic strains are very operative for the treatment of the patients suffering from atopic eczema. *B. bifidum*, *B. lactis*, *E. coli*, and *L. lactis* have been reported for their beneficial roles in the treatment of eczema and food allergies (Niers, et al. 2009). Hong et al. (2010) reported anti-allergic effect of *L. kefiranofaciens* M1, isolated from kefir grains. *Lactobacillus* strain isolated from fermented kimchi has ability to modulate the balance of Th1/Th2 by producing interferons IL-12 and IFN-γ which reduce symptoms of food allergies and atopic dermatitis (Won et al. 2011; Koletzko 2016). Omega-3 fatty acid-rich fermented fish oil alleviates the allergic sensitization (Han et al. 2012).

6.5.6 Irritable Bowel Syndrome

An intestinal disorder with symptoms of belly pain, gas, diarrhea, and constipation is known as irritable bowel syndrome. *L. reuteri* may be involved in the improvement of symptoms after 1-week treatment. In summary, there is a data which suggests that some probiotics such as *B. infantis* 35624, *E. coli* DSM17252, and *B. breve* may be involved in the improvement of symptoms (Enck et al. 2009).

6.5.7 Inflammatory Bowel Disease

The major symptoms of inflammatory bowel disease (IBD) are diarrhea, abdominal pain, and gastrointestinal bleeding (Hanauer 2006). Ulcerative colitis and Crohn's disease are two categories of IBD, which are relapsing, remitting, and chronic diseases. Ulcerative colitis is a Th2-driven immune response characterized by production of interleukin (IL)-5, a pro-inflammatory cytokine. Further, Crohn's disease is Th1 immune response with a predominant increase in IL-12, interferon (IFN)-g, and tumor necrosis factor (TNF). Probiotic *L. acidophilus*, *E. coli* Nissle 1917, and *Bifidobacterium* have shown positive effects on ulcerative colitis (Imaoka et al. 2008). In vitro studies of IBD have shown that probiotic *L. rhamnosus* GG can modulate the host immune system by carrying out downregulation of TNF (Zhang et al. 2005). Similarly, in vivo studies on animals indicated the beneficial role of *B. lactis* Bb12 in immune modulation and prevention of intestinal inflammation (Ruiz et al. 2005). A study revealed that fermented milk consumption could prevent the effect of ulcerative colitis (Ishikawa et al. 2003). In a similar study, *B. animalis* strains showed the reduction of IBD symptoms (Guyonnet et al. 2007).

6.5.8 Lactose Malabsorption

The condition of incomplete digestion of lactose (principle carbohydrate of milk) is known as lactose malabsorption. Due to deficiency of β-galactosidase enzyme, lactose does not completely break into glucose and galactose in the small intestine and passes to the large intestine (Shah 2015). The undigested lactose is fermented by the indigenous microflora of the large intestine, which results in production of short-chain fatty acids and gases (CO_2, CH_4, H_2). The gas production causes gastrointestinal disturbances such as abdominal pain, diarrhea, and flatulence (Granato et al. 2010). Probiotic microorganisms are extensively known to alleviate the symptoms of lactose malabsorption. Studies have been reported that *Bifidobacterium* may increase the production of β-galactosidase enzyme which can improve lactose digestion in the small intestine (Parracho et al. 2007). Furthermore, consumption of yogurt containing *B. animalis*, *L. delbrueckii* subsp. *bulgaricus*, and *S. thermophilus* could improve the gut microflora and reduce the symptoms of lactose malabsorption (Shah et al. 2013). Kefir is also an excellent source of β-galactosidase enzyme for persons suffering from lactose intolerance (Hertzler and Clancy 2003).

6.5.9 Immune System Modulation

Human immune system is a complex system which includes two types of immunity: innate (natural) and adaptive (acquired). The innate immunity of an individual is by birth and acts as the first line of defense to external stimulus. Natural killer cells are the key components of innate immunity and are involved in recognition and lysis of tumor cells, virus-infected cells, etc. On the other hand, the adaptive immunity is acquired through the lifetime of an individual. Both the immunities are key aspects of understanding the mechanisms of autoimmunity, allergy, vaccination, and carcinogenicity. The epithelial cells of the intestine remain in direct contact with gut microflora and interface with the immune system (Vasiljevic and Shah 2008). The epithelial cell-surface receptors recognize probiotics, and they beneficially affect the immune system (Isolauri et al. 2001). Dairy products containing probiotic bacteria could stimulate the mucosal immune system and increase the IgA+ cell count, which acts as a first line of defense (Lollo et al. 2013). Numerous studies reported immune system modulation by bifidobacteria. In a study, infants suffering with necrotizing enterocolitis were fed with breast milk containing *B. infantis* and *L. acidophilus*. As a result of probiotic intake, the severity of disease is reduced (Lin et al. 2005). A similar study revealed that fermented milk and yogurt contain some nonbacterial components such as peptides and fatty acids which are produced during fermentation. These components were shown to modulate the immune system. Probiotic *L. plantarum* DSMZ 12028 and *Bacillus circulans* PB7 were also reported for immune modulatory effect (Cammarota et al. 2009).

6.6 Application in Animal and Plant Health

The microflora of the gastrointestinal tract of the animals is very important for the normal digestion. Probiotics attach to the mucus wall and adjust with the immune response of the host and then compete with the pathogenic microorganisms. Probiotics enhance the immune responses by the uptake of necessary nutrients for the body and are involved in the removal of pathogenic bacteria and increase of growth of nonpathogenic strains. The soil becomes more fertile by the presence of beneficial bacteria and fungi in the soil. Plant probiotic products could be used as biopesticides, biofetilizers, and plant stimulators. Probiotic bacteria also influence the hormonal equilibrium in plants (Berg 2009). Some commercial plant products that use probiotic cultures are Kodiak (*Bacillus subtilis* GB03), YiedShield (*B. pumilis* GB34), Rotex (*Phlebiopsis gigantea*), Cedomon (*Pseudomonas chlororaphis*), etc. (Song et al. 2012).

6.7 Conclusion and Future Perspectives

In this chapter, there is a brief knowledge of applications of different probiotic strains in human health. Effectiveness of probiotics could be improved by using the mixture of probiotics and prebiotics. For this purpose, *Bifidobacteria* can act as a valuable adjuvant for improvement of probiotic functionality. A person acquires his/her microflora at the time of birth from his/her mother's health as well as from surroundings. The presence of good microflora leads to good health conditions. External factors such as foreign microorganisms, diseases, and excess use of antibiotics may lead to disturbance of gut microflora. Probiotic microorganisms are able to restore the microflora which imparts various health benefits to the consumer. Fermented products such as yogurt, fermented milk, curd, kimchi, kefir, etc., are a good source of probiotics. In addition, researchers have reported various studies of probiotic applications in the treatment, prevention, and management of diseases. So, there is a requirement of designing new and improved form of probiotics for their applications in the field of food and health.

Effects of probiotics could be improved by the development of nano-encapsulated probiotics (the shelf life of the product can be enhanced by the encapsulation) by using nanotechnology applications. The WHO also suggested that probiotics will be the most important tool to fight against many infectious and noninfectious diseases in place of antibiotics which show many adverse effects like antibiotic resistance.

Antibiotic resistance cases can be treated by probiotics that is termed as microbial interference therapy. Thus, that time is not so far when probiotics will become the most commonly used therapeutic tool by medical personnels. Recent advancement in technology helps to isolate and colonize microorganisms to determine their specific therapeutic properties and uses. In countries like Japan, Europe, and Australia, probiotics and their related products currently occupy the largest sector in the food market. The European Commission has sponsored research projects for the safety and efficacy of the products.

References

Aiba Y, Suzuki N, Kabir AM, Takagi A, Koga Y (1998) Lactic acid-mediated suppression of *Helicobacter pylori* by the oral administration of *Lactobacillus salivarius* as a probiotic in a gnotobiotic murine model. Am J Gastroenterol 93:2097–2101

Anderson JW, Gilliland SE (1999) Effect of fermented milk (yogurt) containing *Lactobacillus acidophilus* L1 on serum cholesterol in hypercholesterolemic humans. J Am Coll Nutr 18:43–50

Backhed F, Fraser CM, Ringel Y, Sanders ME, Sartor RB, Sherman PM et al (2012) Defining a healthy human gut microbiome: current concepts, future directions, and clinical applications. Cell Host Microbe 12:611–622

Berg G (2009) Plant-microbe interactions promoting plant growth and health: perspectives for controlled use of microorganisms in agriculture. Appl Microbiol Biot 84:11–18

Bernet MF, Brassart D, Neeser JR, Servin AL (1994) *Lactobacillus acidophilus* LA-1 binds to cultured human intestinal cell-lines and inhibits cell attachment and cell invasion by enterovirulent bacteria. Gut 35:483–489

Braat H, de Jong EC, van den Brande JM, Kapsenberg ML, Peppelenbosch MP, van Tol EA, van Deventer SJ (2004) Dichotomy between *Lactobacillus rhamnosus* and *Klebsiella pneumoniae* on dendritic cell phenotype and function. J Mol Med 82:197–205

Cammarota M, De Rosa M, Stellavato A, Lamberti M, Marzaioli I, Giuliano M (2009) In vitro evaluation of *Lactobacillus plantarum* DSMZ 12028 as a probiotic: emphasis on innate immunity. Int J Food Microbiol 135:90–98

Chandan RC, Kilara A (2013) Manufacturing yogurt and fermented milks, 2nd edn. Wiley, Chichester, p 477

Enck P, Zimmermann K, Menke G, Klosterhalfen S (2009) Randomized controlled treatment trial of irritable bowel syndrome with a probiotic *E. coli* preparation (DSM17252) compared to placebo. Z. Gastroenterol 47:209–214

Fallone CA, Chiba N, van Zanten SV, Fischbach L, Gisbert JP, Hunt RH, Jones NL, Render C, Leontiadis GI, Moayyedi P, Marshall JK (2016) The Toronto consensus for the treatment of *Helicobacter pylori* infection in adults. Gasteroenterol 151:51–69

Felley CP, Corthesy-Theulaz I, Rivero JL, Sipponen P, Kaufmann BP, Wiesel PH, Brassart D, Pfeifer A, Blum AL, Michetti P (2001) Favourable effect of an acidified milk (LC-1) on *Helicobacter pylori* gastritis in man. Eur J Gastroenterol Hepatol 13:25–29

Freitas M, Tavan E, Cayuela C, Diop L, Sapin C, Trugnan G (2003) Host-pathogens cross-talk. Indigenous bacteria and probiotics also play the game. Biol Cell 95:503–506

Giralt J, Regadera JP, Verges R, Romero J, de la Fuente BA, Villoria J, Cobo JM, Guarner F (2008) Effects of probiotic Lactobacillus casei DN-114 001 in prevention of radiation-induced diarrhea: results frommulticenter, randomized, placebo-controlled nutritional trial. Int J Radiat Oncol Biol Phys 71:1213–1219

Goyal S, Raj T, Banerjee C, Imam J, Shukla P (2013) Isolation and ecological screening of indigenous probiotic microorganisms from curd and chili sauce samples. Int J Probiotics Prebiotic 8:91

Granato D, Branco GF, Nazzaro F, Cruz AG, Jose AF (2010) Functional foods and nondairy probiotic food development: trends, concepts, and products. Compr Rev Food Sci Food Saf 9:292–302

Gupta SK, Shukla P (2015) Advanced technologies for improved expression of recombinant proteins in bacteria: perspectives and applications. Crit Rev Biotechnol 18:1–10

Guyonnet D, Chassany O, Ducrotte P, Picard C, Mouret M, Mercier CH, Matuchansky C (2007) Effect of a fermented milk containing *Bifidobacterium animalis* DN-173 010 on the health-related quality of life and symptoms in irritable bowel syndrome in adults in primary care: a multicentre, randomized, double-blind, controlled trial. Aliment Pharmacol Ther 26:475–486

Hajela N, Ramakrishna BS, Nair BG, Abraham P, Gopalan S, Ganguly NK (2015) Gut microbiome, gut function, and probiotics: implications for health. Ind J Gastroenterol 34:93–107

Han S, Kang G, Ko Y, Kang H, Moon S, Ann Y, Yoo ES (2012) Fermented fish oil suppresses T helper 1/2 cell response in a mouse model of atopic dermatitis via generation of CD4+CD25+Foxp3+ T cells. BMC Immunol 13:44

Hanauer SB (2006) Inflammatory bowel disease: epidemiology, pathogenesis, and therapeutic opportunities. Inflamm Bowel Dis 12:3–9

He T, Priebe MG, Zhong Y, Huang CM, Harmsen HJ, Raangs GC, Antoine JM, Welling GW, Vonk RJ (2008) Effects of yogurt and bifidobacteria supplementation on the colonic microbiota in lactose-intolerant subjects. J Appl Microbiol 104:595–604

Hertzler SR, Clancy SM (2003) Kefir improves lactose digestion and tolerance in adults with lactose maldigestion. J Am Diet Assoc 103:582–587

Hill C, Guarner F, Reid G, Gibson GR, Merenstein DJ, Pot B et al (2014) Expert consensus document: the international scientific association for probiotics and prebiotics consensus statement on the scope and appropriate use of the term probiotic. Nat Rev Gastroenterol Hepatol 11:506–514

Hong W, Chen Y, Chen M (2010) The antiallergic effect of kefir Lactobacilli. J Food Sci 75:244–253

Holzapfel WH, Wood BJB (2014) Lactic acid bacteria: biodiversity and taxonomy. Wiley-Blackwell, New York, p 632

Hsieh ML, Chou CC (2006) Mutagenicity and antimutagenic effect of soymilk fermented with lactic acid bacteria and bifidobacteria. Int J Food Microbiol 111:43–47

Imaoka A, Shima T, Kato K, Mizuno S, Uehara T, Matsumoto S, Setoyama H, Hara T, Umesaki Y (2008) Anti-inflammatory activity of probiotic *Bifidobacterium:* enhancement of IL-10 production in peripheral blood mononuclear cells from ulcerative colitis patients and inhibition of IL-8 secretion in HT-29 cells. World J Gastroenterol 14:2511–2516

Ishikawa H, Akedo I, Umesaki Y, Tanaka R, Imaok AI, Otani T (2003) Randomized controlled trial of the effect of bifidobacteria-fermented milk on ulcerative colitis. J Am Coll Nutr 22:56–63

Isolauri E (2004) Dietary modification of atopic disease: use of probiotics in the prevention of atopic dermatitis. Curr Allergy Asthma Rep 4:270–275

Isolauri E, Sutas Y, Kankaanpa P, Arvilommi H, Salminen S (2001) Probiotics: effects on immunity. Am J Clin Nutr 73:444–450

Khulusi S, Mendall MA, Patel P, Levy J, Badve S, Northfield TC (1995) *Helicobacter pylori* infection density and gastric inflammation in duodenal ulcer and non-ulcer subjects. Gut 37:319–324

Kiatpapan P, Yamashita M, Kawaraichi N, Yasuda T, Murooka Y (2001) Heterologous expression of a gene encoding cholesterol oxidase in probiotic strains of *Lactobacillus plantarum* and *Propionibacterium freudenreichii* under the control of native promoters. J Biosci Bioeng 92:459–465

Koletzko S (2016) Probiotics and prebiotics for prevention of food allergy: indications and recommendations by societies and institutions. J Pediatr Gastroenterol Nutr 63:9–10

Kwak SH, Cho YM, Noh GM, Om AS (2014) Cancer preventive potential of kimchi lactic acid bacteria (*Weissella cibaria, Lactobacillus plantarum*). J Cancer Prevent 19:253–258

Lee NK, Kim SY, Han KJ, Eom SJ, Paik HD (2014) Probiotic potential of *Lactobacillus* strains with anti-allergic effects from kimchi for yogurt starters. LWT Food Sci Tech 58:130–134

Lin HC, Su BH, Chen AC, Lin TW, Tsai CH, Yeh TF, Oh W (2005) Oral probiotics reduce the incidence and severity of necrotizing enterocolitis in very low birth weight infants. Pediatrics 115:1–4

Lollo PCB, de Moura CS, Morato PN, Cruz AG, de Freitas CW, Betim CB, Nisishima L, Faria JAF, Junior MM, Fernandes CO, Amaya-Farfan J (2013) Probiotic yogurt offers higher immune-protection than probiotic whey beverage. Food Res Int 54:118–124

Macouzet M, Lee BH, Robert N (2009) Production of conjugate linoleic acid by probiotic *Lactobacillus acidophilus* La-5. J Appl Microbiol 106:1886–1891

Metchnikoff E (1907) The prolongation of life. Optimistic studies London, pp 16–183

Mohan R, Koebnick C, Schildt J, Mueller M, Radke M, Blaut M (2008) Effects of *Bifidobacterium lactis* Bb12 supplementation on body weight, fecal pH, acetate, lactate, calprotectin, and IgA in preterm infants. Pediatric Res 64:418–422

Mohania D, Kansal VK, Sagwal R, Shah D (2013) Anticarcinogenic effect of probiotic dahi and piroxicam on DMH-induced colorectal carcinogenesis in Wister rats. Am J Cancer Ther Pharmacol 1:8–24

Nguyen TD, Kang JH, Lee MS (2007) Characterization of *Lactobacillus plantarum* PH04, a potential probiotic bacterium with cholesterol-lowering effects. Int J Food Microbiol 113:358–361

Niers L, Martin R, Rijkers G, Sengers F, Timmerman H, van Uden N, Smidt H, Kimpen J, Hoekstr M (2009) The effects of selected probiotic strains on the development of eczema (the P and a study). Allergy 64:1349–1358

Ollech JE, Shen NT, Crawford CV, Ringel Y (2016) Use of probiotics in prevention and treatment of patients with *Clostridium difficile* infection. 30:111–118

Ouwehand AC, Salminen S, Isolauri E (2002) Probiotics: an overview of beneficial effects. Antonie Van Leeuwen 82:279–289

Parracho H, McCartney AL, Gibson GR (2007) Probiotics and prebiotics in infant nutrition. Proc Nutr Soc 66:405–411

Patel S, Shukla R, Goyal A (2015) Probiotics in valorization of innate immunity across various animal models. J Func Foods 14:549–561

Pedrosa MC, Golner BB, Goldin BR, Barakat S, Dallal GE, Russell RM (1995) Survival of yogurt-containing organisms and *Lactobacillus gasseri* (ADH) and their effect on bacterial enzyme activity in the gastrointestinal tract of healthy and hypochlorhydric elderly subjects. Am J Clin Nutr 61:353–359

Plummer M, Franceschi S, Munoz N (2004) Epidemiology of gastric cancer. IARC Sci Publ 157:311–326

Ranadheera R, Baines S, Adams M (2010) Importance of food in probiotic efficacy. Food Res Int 43:1–7

Reid G (2016) Probiotics: definition, scope and mechanisms of action. Best Pract Res Clin Gasteroenterol 30:17–25

Ruiz PA, Hoffmann M, Szcesny S, Blaut M, Haller D (2005) Innate mechanisms for *Bifidobacterium lactis* to activate transient pro-inflammatory host responses in intestinal epithelial cells after the colonization of germ-free rats. Immunol 115:441–450

Saad N, Delattre C, Urdaci M, Schmitter JM, Bressollier P (2013) An overview of the last advances in probiotic and prebiotic field. LWT Food Sci Technol 50:1–16

Sadana G, Sikri C, Sadana S (2015) Role of probiotics in general and oral health: a review. Ind J Compr Dent Care 5:571–577

Savadago A, Ouattara CAT, Bassol IHN, Traore SA (2006) Bacteriocins and lactic acid bacteria–a mini review. Afr J Biotec 5:678–683

Schulz C, Koch N, Schutte K, Pieper DH, Malfertheiner P (2015) *H. pylori* and its modulation of gastrointestinal microbiota. J Dig Dis 16:109–117

Sgouras D, Maragkoudakis P, Petraki K, Martinez-Gonzalez B, Eriotou E, Michopoulos S, Kalantzopoulos G, Tsakalidou E, Mentis A (2004) In vitro and in vivo inhibition of *Helicobacter pylori* by *Lactobacillus casei* strain Shirota. Appl Env Microbiol 70:518–526

Shah NP (2007) Functional cultures and health benefits. Int Dairy J 17:1262–1277

Shah NP (2015) Functional properties of fermented milks. In: Tamang JP (ed) Health benefits of fermented foods. CRC Press, New York, pp 261–274

Shah NP, da Cruz AG, Faria JDAF (2013) Probiotics and probiotic foods: technology, stability and benefits to human health. Nova Science Publishers, New York

Singh PK, Shukla P (2014) Systems biology as an approach for deciphering microbial interactions. Brief Func Genomics 14:166–168

Song X, Hu X, Ji P, Li Y, Chi G, Song Y (2012) Phytoremediation of cadmium-contaminated farmland soil by the hyperaccumulator beta vulgaris L. Var. Cicla. Bull Environ Contam Toxicol 88:623–626

Vanderhoof JA (2000) Probiotics and intestinal inflammatory disorders in infants and children. J Pediatr Gastroenterol Nutr 30:34–38

Vasiljevic T, Shah NP (2008) Probiotics-from Metchnikoff to bioactives. Int Dairy J 18:714–728

Wang KY, Li SN, Liu CS, Perng DS, Su YC, Wu DC, Jan CM, Lai CH, Wang TN, Wang WM (2004) Effects of ingesting Lactobacillus- and Bifidobacterium-containing yogurt in subjects with colonized Helicobacter pylori. Am J Clin Nutr 80:737–741

Won TJ, Kim B, Song DS, Lim YT, Oh ES, Lee DI, Park ES, Min H, Park SY, Hwang KW (2011) Modulation of Th1/Th2 balance by *Lactobacillus* strains isolated from kimchi via stimulation of macrophage cell line J774A.1 in vitro. J Food Sci 76:55–61

Xiao J, Kondo S, Takahashi N, Miyaji K, Oshida K, Hiramatsu A, Iwatsuki K, Kokubo S, Hosono A (2003) Effects of milk products fermented by *Bifidobacterium longum* on blood lipids in rats and healthy adult male volunteers. J Dairy Sci 86:2452–2461

Yadav R, Shukla P (2015) An overview of advanced technologies for selection of probiotics and their expediency: a review. Crit Rev Food Sci Nutr. doi:10.1080/10408398.2015.1108957

Yanping W, Nv X, Aodeng X, Zaheer A, Bin Z, Xiaojia B (2009) Effect of *Lactobacillus plantarum* MA2 isolated from Tibet kefir on lipid metabolism and intestinal microflora of rats fed on high-cholesterol diet. App Microbiol Biotechnol 84:341–347

Zhang L, Li N, Caicedo R, Neu J (2005) Alive and dead *Lactobacillus rhamnosus* GG decrease tumor necrosis factor-alpha-induced interleukin-8 production in Caco-2 cells. J Nutr 135:1752–1756

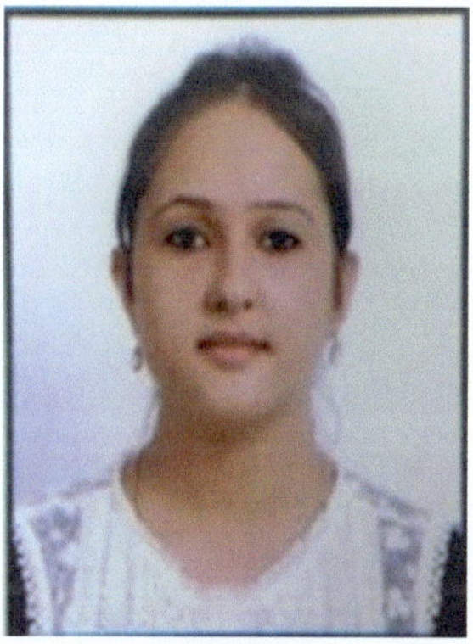

Ruby Yadav graduated in Biotechnology in 2010 from Kurukshetra University, Kurukshetra. She completed her M.Sc. (Microbiology) from Maharshi Dayanand University, Rohtak, India. She is currently pursuing Ph.D in Microbiology, Maharshi Dayanand University, Rohtak, India. Her current research interests are in Probiotics and Food Microbiology.

Pratyoosh Shukla is M.Sc. (Applied Microbiology Biotechnology) from Dr. H.S. Gour University, Sagar, India and Ph.D. in Microbiology and presently working as Professor and Head at Department of Microbiology, Maharshi Dayanand University, Rohtak, India. He was awarded with Indo-US Visiting research Professorship by American Society of Microbiology (ASM) in 2014. His current interests are Bioenergy, Enzyme technology and Protein bioinformatics.

Part III

Microbial Interactions

Functionalities of Phosphate-Solubilizing Bacteria of Rice Rhizosphere: Techniques and Perspectives

7

Nilima Dash, Avishek Pahari, and Tushar Kanti Dangar

Abstract

Phosphorus (P) is the second essential macronutrient next to nitrogen. It is a constituent of the essential macromolecules like DNA, RNA, ATP, phospholipids, etc. and indispensable at all growth stages of all plants. With rice being the staple cereal diet of about 50% in the world and 85% of the Indian population and P being an essential plant nutrient, understanding P metabolism in rice rhizosphere is important to assess fertility status of rice soils. Oxidation-reduction reactions in rice soil open up challenges, opportunities and potentials for manifold microbial activities which are significant for maintenance of fertility and sustainability in rice production systems. Rice production should be enhanced to meet the world food need. Most agricultural soils are P deficient and, therefore, require application of phosphatic fertilizers to sustain crop production. But excess chemical P fertilizer application can cause environmental hazards like pollution, eutrophication, etc. This situation warrants for eco-friendly and economical alternate strategy like biofertilizer application for improving crop production in P-deficient soils. Phosphate-solubilizing microorganisms, viz. *Bacillus*, *Pseudomonas*, *Azotobacter*, *Aspergillus* spp. etc. can mineralize insoluble P thereby increasing P availability to plants. Therefore, efficient P-metabolizing biofertilizer application would promote plant growth, improve soil health and protect plants from different pathogens without disturbing the environment.

Keywords

Phosphate solubilization • Plant growth promotion • Rice • Phosphate-solubilizing microbe

N. Dash • A. Pahari • T.K. Dangar (✉)
Microbiology Laboratory, Crop Production Division, National Rice Research Institute, Cuttack 753006, Odisha, India
e-mail: dashnilima@gmail.com; avishekpahari@gmail.com; dangartk@gmail.com

P. Shukla (ed.), *Recent Advances in Applied Microbiology*,
DOI 10.1007/978-981-10-5275-0_7

7.1 Introduction

Soil is a heterogeneous and discontinuous structure, generally nutrient and energy resource poor and inhabited with microorganisms specific for the microhabitats (Stotzky 1997). Influence of the plant communities on microbes is significant. A large number of diverse microbes inhabit on and around the plant root system (rhizosphere) and are influenced mainly by secretions of the roots (Tilak et al. 2005). Thus, they initiate complex interactions (viz. positive, negative and neutral) among themselves and with the plants. Numerous rhizospheric microbes produce plant growth-promoting substances or nutrients which directly influence growth of the plants, or otherwise, they have negative effect on the pathogens and protect the plants (Stotzky 1997, Tilak et al. 2005). Co-evolution of the interacting biological components effected association between plants and microbes of the rhizospheric soil (Brimecombe et al. 2001), and microbial diversity analysis reveals the microbial composition and succession of the microbes of different ecologies, which opens up its functionality. A large number of microbial species, viz. *Rhizobium*, *Bradyrhizobium*, *Frankia*, *Azotobacter*, *Azospirillum*, *Klebsiella*, *Bacillus*, *Nitrosomonas*, *Pseudomonas*, *Mycobacterium*, *Micrococcus*, *Flavobacterium*, *Penicillium*, *Fusarium*, *Aspergillus*, *Thiobacillus*, *Arthrobacter*, *Acinetobacter*, *Burkholderia*, *Enterobacter*, *Erwinia* spp. etc. are associated with a wide array of plant rhizosphere (Anand et al. 2016).

Next to nitrogen, P is the important macroelement required for growth and development of plants. It constitutes about 0.2% plant dry weight and is a repository of chemical energy required for metabolism and promotion of N_2 fixation in leguminous plants (Saber et al. 2005). P is available in water, soil and sediments and accounts for about 0.05% soil content, out of which only 0.1% is available to the plants. Depending on the soil properties, the phosphate anions react with cations viz. Ca^{2+}, Mg^{2+}, Al^{3+} and Fe^{3+} and precipitate. As the cation bound forms of P are highly insoluble, it becomes unavailable to plants which reduces overall P use efficiency after phosphatic fertilizer application also (Vassilev and Vassileva 2003). Hence, to maintain crop production, repeated application of soluble forms of inorganic P is required which leaches to the groundwater causing eutrophication of aquatic environments (Smyth et al. 2011). The phosphate-solubilizing microbes (PSM) mineralize phosphate which reduces P deficiency in soil, increases soluble phosphate availability and enhances plant growth, biological nitrogen fixation (BNF) efficiency, plant growth promotion regulator (PGR) metabolism and availability of trace elements like iron, zinc, boron, copper, molybdenum, manganese etc. (Saber et al. 2005, Ponmurugan and Gopi 2006).

Development of root, tillering, early flowering and ripening of rice essentially requires P. About 1.8–4.2 kg of P is required per ton of rice grain yield. For upland crops, P availability is optimum at a pH range of 6.0–6.5 of the soil. In acid soils ($pH < 6.0$), P mobilization to the plants is reduced as it is coupled with iron (Fe^{3+}) and aluminium (Al^{3+}) compounds ($FePO_4$, $AlPO_4$). At >6.5 soil pH, P could not be easily mineralized as it forms calcium and magnesium complexes ($Ca_3(PO_4)_2$, $Mg_3(PO_4)_2$) and therefore, P uptake is limited in alkaline soils. P availability increases in submerged soils as flooding moderates the pH towards neutrality which enhances P availability for plants grown in flooded conditions, viz. rice (Morales et al. 2014). Besides, P is generally deficient in sandy soil containing low organic matter and very acidic and alkaline soils.

7.2 Types of Phosphorous in Soil

In soil, two types of phosphorous i.e. organic and inorganic are present. Organic phosphorus constitutes about 50% (4–90% in most soils) of the total P in soil. Organic P compounds occur mostly as esters of orthophosphoric acid like (1) phospholipids, (2) inositol phosphates and (3) nucleic acids. Mono- to hexaphosphate esters would be the constituents of inositol phosphates. The most abundant (10–50%) organophosphate in soil is phytin (a Ca-Mg salt of phytic acid). Out of total organic P in soil, phospholipid content is 1–5% (insoluble in water) and nucleic acids content is 0.2–2.5% which is mineralized by microorganisms in soil. Inorganic phosphates of soil are mostly calcium, iron and aluminium compounds. The important P compounds are variscite ($AlPO_4.2H_2O$) and strengite ($FePO_4.2H_2O$) in acid soils and tricalcium phosphate ($Ca_3(PO_4)_2$), dicalcium phosphate dehydrate ($CaHPO_4.2H_2O$), dicalcium phosphate ($CaHPO_4$), hydroxyapatite ($Ca_5(PO_4)_3OH$), fluorapatite ($Ca_5(PO_4)_3F$) and octacalcium phosphate ($Ca_4H(PO_4)_3.2\text{-}5H_2O$) in neutral and alkaline soils (Yadav and Verma 2012).

7.3 Plant Growth-Promoting Bacteria

The diverse microbial communities maintain important functional network of vital processes of the habitats and are essential to sustain the functioning of ecosystem (Bardgett and Shine 1999). A number of bacterial species such as *Acinetobacter*, *Alcaligenes*, *Azotobacter*, *Azospirillum*, *Burkholderia*, *Bacillus*, *Enterobacter*, *Flavobacterium*, *Pantoea*, *Pseudomonas*, *Rhizobium*, *Serratia* spp. etc. can benefit growth of plants and are designated as plant growth-promoting bacteria (PGPB). Similarly, the naturally occurring plant growth-promoting rhizobacteria (PGPR) support growth and development of plants by production/secretion of different metabolites in the rhizosphere (Ahemad and Kibret 2013). They increase nitrogen fixation in legumes, enhance nutrient (P, S, Fe, Cu, etc.) availability, metabolize plant growth regulators (PGR), favour other beneficial microbes and suppress diseases and pests (Saharan and Nehra 2011; Sharma et al. 2011).

7.4 Bacteria Involved in Phosphate Mineralization

Both organic and inorganic types of phosphorus are available in soil (Khan et al. 2009). Majority of P is in insoluble forms and cannot be used by plants. As phosphatic fertilizers are expensive with inherent anti-environmental effects, there is a need to find out eco-friendly and cost-effective alternatives to improve crop production in P-poor soils. The phosphate-solubilizing microorganisms (PSM) participate in biogeochemical P cycling in natural and agricultural ecosystems which grants its access to the plants; hence, they might substitute (at least partially) chemical phosphatic fertilizers (Ahemad and Kibret 2013). The PSM transform the insoluble P to soluble forms viz. HPO_4^- and $H_2PO_4^-$ by acidification, chelation, ion exchange reactions etc. Application of these microbes around the plants, in soil and with insoluble rock fertilizers releases phosphorus, promotes plant growth, improves soil quality and prevents pathogens to cause harm to plants Therefore, application of phosphate-solubilizing microbes with P rock fertilizers would be a cost-effective and environmentally healthy and promising approach (Chang and Yang 2009). The *Azotobacter*, *Burkholderia*, *Bacillus*, *Enterobacter*, *Erwinia*, *Microbacterium*, *Rhizobium*, *Pseudomonas* spp. are a few major phosphate-solubilizing bacteria in soil (Bhattacharyya and Jha 2012). Furthermore, *Bacillus*, *Pseudomonas*, *Rhizobium*, *Enterobacter* spp. of bacteria, *Aspergillus* and *Penicillium*, *Trichoderma* spp., *Rhizoctonia solani* of fungi, *Streptomyces* and *Micromonospora* spp. of actinomycetes were identified as potent P-metabolizing microbes of rice rhizosphere (Sharma et al. 2013). Nevertheless, 0.35 kg of P_2O_5 is utilized for production of 50 kg rice/ha. The efficient PSM can mineralize insoluble inorganic phosphate like rock phosphate, tricalcium phosphate, iron and aluminium phosphates etc. and organic phosphate compounds as well. They are known to supplement about 30–35 kg of P_2O_5/ha/year. Besides, microbial biomass itself contributes a significant amount of P in soil i.e. up to about 100 mg P/kg/soil/year is cycled through the microbial biomass.

7.5 Mechanisms of Phosphate Solubilization

Microorganisms release organic acids (OA) which enhance chelation of divalent cations of Ca^{2+} by lowering medium pH and forming soluble complexes of the metal ions of insoluble P and release P from insoluble phosphates. Organic acids are the products of microbial metabolism. These acids, especially gluconic acid, are main effectors of P solubilization. Glucose dehydrogenase (GDH) oxidizes glucose to produce acids and reduces pH at the site of P solubilization i.e. periplasmic space. The rice rhizospheric microbes like *Bacillus*, *Pseudomonas*, etc. were reported to produce succinic acid, propionic acid, oxalic acid and malic acid and mineralize insoluble P to make it available to rice plants (Panhwar et al. 2012; Khan et al. 2014).

Glucose is oxidized to gluconic acid by the quinoprotein GDH in the direct oxidation (DO) pathway by using the redox cofactor 2,7,9-tricarboxyl-1H-pyrrolo [2,3-f] quinoline-4,5-dione (PQQ) and requires the metal ions such as Ca^{2+} (or Mg^{2+} in vitro). Gluconate dehydrogenase oxidizes gluconic acid to 2-keto-gluconic acid, which is again oxidized to 2,5-diketo gluconic acid (Goldstein et al. 2003, Tripura et al. 2005, Sharma et al. 2013, Krishnaraj and Dahale 2014). Exogenous amino acids and other compounds are taken up through transmembrane proton motive

Table 7.1 Cloned genes of DO pathway

Gene/recombinant clone	Source	Assigned function	Remarks
gdh A	*Acinetobacter calcoaceticus*	Glucose oxidation to gluconic acid	Gene encodes membrane-bound apo enzyme
gdh B	*Acinetobacter calcoaceticus*	Glucose oxidation to gluconic acid	Gene encodes soluble form of enzyme
gdh	*Escherichia coli*	Glucose oxidation to gluconic acid	Gene encodes membrane-bound apo enzyme
gdh	*Enterobacter asburiae*	Glucose oxidation to gluconic acid	Gene encodes membrane-bound apo enzyme
gadh	*Erwinia cypripedii* ATCC 29267	Gluconic acid oxidation to 2-keto-gluconic acid	Gene cluster encodes three subunits of membrane-bound GADH
pKG3791 (DNA fragment)	*Serratia marcescens*	Induces PQQGDH in *E. coli* DH5α	DNA fragment involved in MPS
pqq genes	*Acinetobacter calcoaceticus*	PQQ biosynthesis	PQQ is the cofactor for GDH and is involved in the formation of holo GDH
*pqq*A, *pqq*B, *pqq*C, *pqq*D, *pqq*E, *pqq*F	*Klebsiella pneumoniae*	PQQ biosynthesis	PQQ forms holo GDH
pqq synthase	*Erwinia herbicola*	Involved in PQQ biosynthesis	*E. coli* produces gluconic acid and has MPS+ phenotype
*gab*Y	*Pseudomonas cepacia*	Expression and/or regulation of DO pathway	Enhances gluconic acid in *E. coli* JM109
pKKY (DNA fragment)	*Enterobacter agglomerans*	Genes involved in MPS	Enhance phosphate-solubilizing ability in *E. coli* JM109
pqq genes	*Rahnella aquatilis*	PQQ biosynthesis	Expression of these genes conferred gluconic acid production to *E. coli* HB101
pqq genes	*Gluconobacter oxydans*	PQQ biosynthesis	PQQ is the cofactor for GDH

Adapted from Tripura et al. (2005)

force (PMF) generated by the oxidation. The H^+ ions solubilize the calcium phosphate complexes. The *gdh* gene of the DO pathway of *A. calcoaceticus* and *E. coli* has been cloned and characterized. The *gabY* gene may act as a functional *Mps* gene *in vivo* as it regulates the DO pathway in *P. cepacia* (Tripura et al. 2005, Table 7.1). The pyrroloquinoline quinone (PQQ)-synthesizing enzymes of *K. pneumoniae* and *Rahnella aquatilis* have also been cloned which is a cluster of six open reading frames (*pqqA, B, C, D, E, F*) (Kim et al. 2003). It was observed that different organisms like *Pseudomonas cepacia, P. aeruginosa, Rahnella aquatilis, Serratia marcescens* etc. solubilize phosphate by gluconic acid production. The pqq and other genes involved in phosphate solubilization in *Erwinia herbicola, Rahnella*

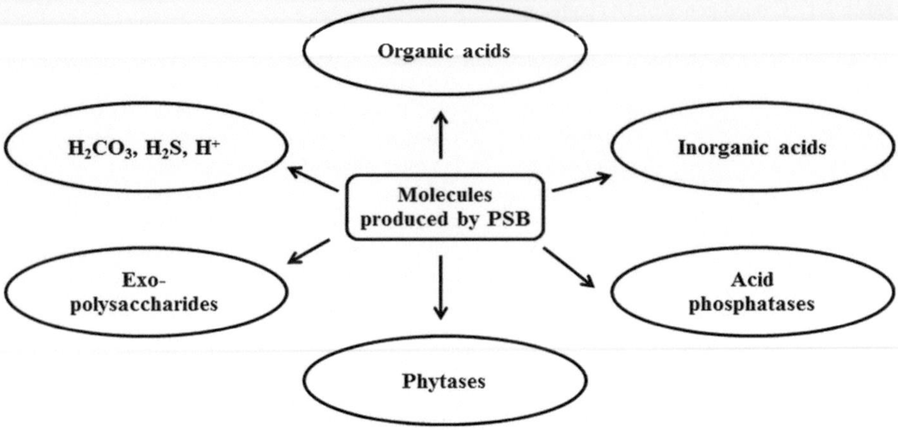

Fig. 7.1 Different molecules produced by phosphate-solubilizing bacteria

aquatilis, *Pseudomonas cepacia* and *Enterobacter agglomerans* have already been cloned in *E. coli* (Ahmed and Shahab 2009). Production of the chelating substances, H_2S, CO_2, mineral acids, siderophores and plant hormones like indoles, gibberellins and cytokinins are also associated with phosphate solubilization. P solubilization is affected by the ring-structured complex formed by coordinate bonds between anionic or polar molecule and cation (Whitelaw 2000). Ferric phosphate reacts with H_2S produced by some bacteria to form ferrous sulphate and releases phosphate. *Nitrosomonas* and *Thiobacillus* species produce nitric and sulphuric acids, respectively, which dissolve insoluble phosphate compounds (Fig. 7.1) (Sharma et al. 2013). Plants rely on NH_4^+ rather than NO_3^- to reduce the pH for phosphate solubilization. Ammonium sulphate is the best N source to support P solubilization. The P-solubilizing genes like *pqq*, *gcd*, *gdh*, *gabY* etc. were well documented from the rice rhizospheric microbes like *Acinetobacter calcoaceticus*, *K. pneumonia* and *Pseudomonas cepacia* (Sashidhar and Podile 2010, Krishnaraj and Dahale 2014).

In soil, organic form of P is about 15–85% of total content. The organic phosphorus compounds like nucleic acids, phospholipids, lecithin, phytin etc. are derived from plants and microorganisms. Phytin is the calcium-magnesium salt of phytic acid. Phospholipids contain 10% of cellular phosphorus. Organic P mineralization in soil is important for phosphorus cycling. Organic phosphorus is mineralized in soil by three groups of enzymes: (1) nonspecific phosphatases dephosphorylate phosphor-ester or phosphor-anhydride bonds in organic matter, (2) phytases release P from phytic acid and (3) phosphonatases and C-P lyases cleave the C-P bond in organophosphonates (Sharma et al. 2013). In aerobic rice, PSB inoculated treatments proved that the phosphatase and phytase enzymes may dissolve the soil organic P (Panhwar et al. 2013). The soil yeast *Candida tropicalis* HY (CtHY) produced phytase which mobilized phosphate from insoluble tricalcium phosphate and stimulated growth of rice seedling (Amprayn et al. 2012).

7.6 Techniques Used to Evaluate Phosphate-Solubilizing Bacteria

Phosphate solubilization is generally tested preliminarily and qualitatively on Pikovskaya's agar (g/l: yeast extract 0.5, dextrose 10, $Ca_3(PO_4)_2$ 5, $(NH4)_2SO_4$ 0.5, KCl 0.2, $MgSO_4.7H_2O$ 0.1, $MnSO_4.7H_2O$ 0.0001, $FeSO_4.7H_2O$ 0.0001, agar 15, pH 7) and NBRIP (National Botanical Research Institute's phosphate medium) agar (g/l: glucose 10, $Ca_3(PO_4)_2$ 5, $MgCl_2.6H_2O$ 5, $MgSO_4.7H_2O$ 0.25, KCl 0.2, $(NH4)_2SO_4$ 0.1, agar 18, pH 7) plates by spot inoculation. P solubilization is indicated by halo zone formation around the bacterial growth. However, plate assay method cannot detect all P-mineralizing microbes as the non-acid-producing P-metabolizing organisms do not produce clear zone, but they can be detected in liquid culture (quantitative assay, mentioned elsewhere) method. The P metabolic efficiency on plate assay is depicted by P solubilization index (PSI), and the P solubilization efficiency (PSE) is calculated from the formula (Premono et al. 1996, Ponmurugan and Gopi 2006):

$$\mathrm{PSI} = \frac{Z}{C}$$

$$\mathrm{PSE} = \frac{(Z - C)}{C} \times 100$$

where Z is the clearing zone diameter and C is the colony diameter.

Inorganic phosphate ($Ca_3(PO_4)_2$) solubilization in broth culture can be estimated quantitatively after growing the organisms in NBRIP (without agar) broth. Free P in culture filtrate after desired growth of the microbes is quantified as μg phosphorus released/ml culture medium recording A660 using sulfo-molybdate reagent (Olsen et al. 1954) in the assay reaction mixture containing 1 ml of culture supernatant, 2 ml each of 2.5% sulfomolybdate solution, 0.25% p-nitrophenol indicator, H_2SO_4 (1N) and 1 ml of stannous chloride solution (40% w/v in 12N HCl). Concomitant pH change of the medium is also checked to ascertain its relation with P release.

Organic phosphate (phytate) mineralization is qualitatively tested on phytate screening media (g/l: D-glucose 10, sodium phytate 4, $CaCl_2.2H_2O$ 2, NH_4NO_3 5, KCl 0.5, $MgSO_4.7H_2O$ 0.5, $MnSO_4.H_2O$ 0.01, $FeSO_4.7H_2O$ 0.01, agar 15, pH 7) (Kerovuo et al. 1998). The bacterial isolates are spotted on phytate-screening plates. Clearing area around growth of the bacteria is considered for phytate mineralization and acidification of the PSM medium is visualized by red colouration of the clear zone with the pH indicator methyl red reagent (0.02%). Qualitative estimation of free inositol in phytate broth culture is used for organic phosphate mineralization efficiency of PSM.

7.7 Phosphate Solubilization in Rhizospheric Soil and Effect on Crop Productivity Including Rice

Soil fertility is important to improve agricultural production. Microbes in the rhizosphere of soya bean, which solubilize P *in vitro*, possess other PGP traits that also increase soya bean growth. Phosphate-solubilizing endophytic *E. coli* isolates of sugarcane (*Saccharum* sp.) and rye grass (*Lolium perenne*) help in growth of those plants (Saharan and Nehra 2011). It was observed that PB-21 efficiently solubilizes and releases P from insoluble rock phosphate and either alone or in combination with VAM improves growth of root and shoot of black pepper cuttings (Ramachandran et al. 2003). Shoot length increased to 21%, shoot fresh weight to 42%, shoot dry weight to 24%, root length to 11%, root fresh weight to 59%, root dry weight to 35% and chlorophyll content to 32% in maize due to application of PSB with rock phosphate (Manzoor et al. 2017). *Pantoea* sp. Pot1 can solubilize tricalcium phosphate at a rate of 956 mg/L. When tomato plants were inoculated with *Pantoea* sp. Pot1, more P was incorporated and much higher biomass weights of plants were recorded (Sharon et al. 2016).

It was recorded that in rice PSB biofertilizer could increase 1–11% grain yield and 6–8% grain phosphorus uptake than control (Vahed et al. 2012) and the PSB population was more in rhizosphere and endosphere than the non-rhizospheric soil. Besides, rhizospheric microbial population (8.78 $\log_{10}$ cfu/g) was recorded more than the non-rhizosphere and endosphere of plants. Application of organic acids along with PSB in aerobic rice seedlings recorded enhancement of soluble P in the soil, improvement of root growth and increase in plant biomass without affecting soil pH (Panhwar et al. 2013). Increase in phosphorus content in roots and grains was recorded in bacteria inoculated rice plants. *B. subtilis* worked as efficient bio-inoculants for rice (Trivedi et al. 2007). Phosphorus content and uptake by rice plants were increased by PSB inoculation. The PSBs *Burkholderia* sp. (MTCC 8369) and *Gluconacetobacter* sp. (MTCC 8368) were recorded as potent microbial formulations for plant growth enhancement (Stephen et al. 2015). Organic phosphorus is mineralized by synthesis of different phosphatases, which catalyse the hydrolysis of phosphoric esters (Glick, 2012). Several PGPR viz. *Bacillus*, *Pseudomonas*, *Arthrobacter*, *Azotobacter*, *Azospirillum*, *Serratia* spp. etc. mineralize insoluble P and support growth of different plants including rice (Ashrafuzzaman et al. 2009).

The reporting investigators have recorded (unpublished data) that the PSB populations in the rice soils of Chandipur, Talapada, Talasari and Cuttack (NRRI) in Odisha, India were diverse in physiological and genetic characters, osmotic stress tolerance, molecular compositions and antibiotic sensitivity. The organisms were halotolerant and therefore, they can be exploited for maintenance of soil health and nutrition in both saline and nonsaline ecologies and control of plant pathogens. The predominant microbes were *Bacillus* spp. The potent organism i.e. *Bacillus* sp. enhanced rice (*Oryza sativa* L. var. Naveen) growth.

7.8 Biofertilizers

PGPR can function as potential bio-inoculants for promotion of growth and development of plants especially rice. Peat is generally used as inoculant carrier. PSB application maintains soil nutrient status and structure. The positive impact of PSB inoculation on P availability to crops has led to development of the inoculum, phosphobacterin (Rathi and Gaur 2016). They can promote root and shoot growth, improve the vigour of seedlings, thereby, increase the grain yield. They increase the nutrient (NPK) uptake of plants, reduce surface run-off and increase the efficiency of applied fertilizers (Duarah et al. 2011). Biofertilizer application generally increases uptake of mineral and water, root development, vegetative growth and nitrogen fixation. Efficiency of microbes in biofertilizer formulation for rice cultivation on acid sulphate soils was proven by the improvement of volume of root and dry weight of seedling through bacterial treatment (Panhwar et al. 2014). *Allorhizobium*, *Azorhizobium*, *Bradyrhizobium*, *Mesorhizobium*, *Rhizobium and Sinorhizobium* spp. were reported as potent PGPR strains as biofertilizers (Bhattacharyya and Jha 2012). When *napA* phosphatase gene was transferred to *Burkholderia cepacia* IS-16 (a biofertilizer) from *Morganella morganii* through the broad-host range vector pRK293, the recombinant strain's extracellular phosphatase activity was increased (Sharma et al. 2013).

Commercial biofertilizers of mixed bacterial cultures having phosphate solubilization efficiency have been developed. Some commercial phosphate biofertilizers are Phylazonit M (permission No. 9961, 1992, Ministry of Agriculture of Hungary), which is a blend of *Azotobacter chrococcum* and *Bacillus megaterium* that can increase P and N supply to the plants, and Kyusei EM (EM Technologies, Inc.), which is a mixture of lactic acid bacteria is also a well-established biofertilizer (Rodriguez and Fraga 1999). Phosphobacteria are one of the biofertilizers developed by Nivshakti Bioenergy (2003) in India. Liquid biofertilizers can replace the traditional chemical fertilizers and carrier-based biofertilizers and play a key role in restoring the soil health (Pindi and Satyanarayana 2012). Other commercially available formulations are BIO-NPK and Bharpur, Biophos and Get-Phos (contain strain of *Bacillus megaterium* var. *phosphaticum*).

Root colonization influences the role of inoculants. Majority of microbes of soil are associated with plant roots with population density up to 10^9–10^{12}/g soil which can supplement 500 kg/ha biomass. In rice, abundance of microbes in the rhizosphere was observed due to secretion of high amount of root exudates like organic acids, amino acids and sugars (Adhya et al. 2015). The physical and chemical properties of soil are modified by the exudation of different chemical compounds (amino acids like α-alanine, β-alanine, aspartate, glycine, leucine, methionine, threonine, serine, etc.; organic acids like citric acid, oxalic acid, malic acid, succinic acid, fumaric acid, butyric acid, etc.; sugars like glucose, galactose, fructose, ribose, etc.; vitamins like thiamine, biotin, etc.; purines/nucleosides; enzymes like acid/alkaline phosphatases, amylase, invertase, etc.; inorganic ions and gaseous molecules) which regulate the microbial community in soil of root zone (Ahemad and Kibret 2013).

7.9 Conclusion

Phosphorus is an important crop nutritional element. Harmful effects on environment have led to the search for an alternative of chemical fertilizers to meet food of the ever-increasing population. Microorganisms can recycle P in different ecosystems as well as control pest too. Efficiency of P-solubilizing microorganisms depends on their ability to compete, colonize, survive and proliferate in the rhizosphere. Screening, selection and identification of efficient P can be achieved by different molecular tools. More attention is to be given to improve the PSB strains. Field trials should be properly undertaken for maximum exploitation of effective strains. The mechanisms of phosphate solubilization and functional variations of the strains should be studied in detail. The manipulation of mineral phosphate-solubilizing genes would have a noteworthy impact on modern-day agriculture. Biofertilizer application would be important for nutrient management, as it is economical and a renewable source of plant nutrients which would reduce the use of chemical fertilizers. It is expected to be more useful in the long run once it reaches the farmers. The use of potent phosphate-solubilizing microbes as biofertilizers will help to boost up crop production and contribute to sustainability in agriculture.

References

Adhya TK, Kumar N, Reddy G, Podile AR, Bee H, Samantaray B (2015) Microbial mobilization of soil phosphorus and sustainable P management in agricultural soils. Curr Sci 108(7):1280–1287

Ahemad M, Kibret M (2013) Mechanisms and applications of plant growth promoting rhizobacteria: current perspective. J King Saud Univ Sci 26:1–20

Ahmed N, Shahab S (2009) Phosphate solubilization: their mechanism genetics and application. Internet J Microbiol 9(1):1–19

Amprayn K, Rose MT, Kecskés M, Pereg L, Nguyen HT, Kennedy IR (2012) Plant growth promoting characteristics of soil yeast (Candida tropicalis HY) and its effectiveness for promoting rice growth. Appl Soil Ecol 61:295–299

Anand K, Kumari B, Mallick MA (2016) Phosphate solubilizing microbes: an effective and alternative approach as biofertilizers. Int J Pharm Pharm Sci 8(2):37–40

Ashrafuzzaman M, Hossen FA, Ismail MR, Hoque MA, Islam MZ, Shahidullah SM, Meon S (2009) Efficiency of plant growth promoting rhizobacteria (PGPR) for the enhancement of rice growth. Afr J Biotechnol 8(Suppl 7):1247–1252

Bardgett RD, Shine A (1999) Linkages between plant litter diversity, soil microbial biomass and ecosystem function in temperate grasslands. Soil Biol Biochem 31:317–321

Bhattacharyya PN, Jha DK (2012) Plant growth-promoting rhizobacteria (PGPR): emergence in agriculture. World J Microbiol Biotechnol 28:1327–1350

Brimecombe MJ, De Leij FA, Lynch JM (2001) The effect of root exudates on rhizosphere microbial populations. In: Pinton R, Varaninin Z, Nannipieri P (eds) The rhizosphere: biochemistry and organic substances at the soil-plant interface. Marcel Dekker, New York, pp 95–140

Chang CH, Yang SS (2009) Thermo-tolerant phosphate-solubilizing microbes for multi-functional biofertilizer preparation. Bioresour Technol 100:1648–1658

Duarah I, Deka M, Saikia N, Deka Boruah HP (2011) Phosphate solubilizers enhance NPK fertilizer use efficiency in rice and legume cultivation. 3 Biotech 1(4):227–238

Glick B (2012) Plant growth-promoting bacteria: mechanisms and applications. Scientifica 2012(5):1–15

Goldstein AH, Lester T, Brown J (2003) Research on the metabolic engineering of the direct oxidation pathway for extraction of phosphate from ore has generated preliminary evidence for PQQ biosynthesis in *Escherichia coli* as well as a possible role for the highly conserved region of quinoprotein dehydrogenases. Biochim Biophys Acta 1647:266–271

Kerovuo J, Lauraeus M, Nurminen P, Kalkinen N, Apajalahti J (1998) Isolation, characterization, molecular gene cloning, and sequencing of a novel phytase from Bacillus Subtilis. Appl Environ Microbiol 64:2079–2085

Khan MS, Zaidi A, Wani PA, Ahemad M, Oves M (2009) Functional diversity among plant growth-promoting rhizobacteria. In: Khan MS, Zaidi A, Musarrat J (eds) Microbial strategies for crop improvement. Springer, Berlin, pp 105–132

Khan MS, Zaidi A, Ahmad E (2014) Mechanism of phosphate solubilization and physiological functions of phosphate-solubilizing microorganisms. In: Khan MS (eds) Phosphate solubilizing microorganisms, doi:10.1007/978-3-319-08216-5_2, © Springer International Publishing, Switzerland

Kim CH, Han SH, Kim KY, Cho BH, Kim YH, Koo BS, Kim YC (2003) Cloning and expression of pyrroloquinoline quinine (PQQ) genes from a phosphate-solubilizing bacterium *Enterobacter intermedium*. Curr Microbiol 47:457–461

Krishnaraj PU, Dahale S (2014) Mineral phosphate Solubilization: concepts and prospects in sustainable agriculture. Proc Indian Natn Sci Acad 80(2):389–405

Manzoor M, Kaleem Abbasi M, Sultan T (2017) Isolation of phosphate solubilizing bacteria from maize Rhizosphere and their potential for rock phosphate Solubilization–mineralization and plant growth promotion. Geomicrobiol J 34(1):81–95

Morales LA, Vázquez EV, Paz-Ferreiro J (2014) Spatial distribution and temporal variability of ammonium-nitrogen, phosphorus, and potassium in a rice field in Corrientes, Argentina. Sci World J 2014:135906, 1–12

Olsen SR, Cole CV, Watanabe FS, Dean LA (1954) Estimation of available phosphorus in soil by extraction with sodium bicarbonate. USDA Circ. 939. U.S. Government Printing Office, Washington, DC

Panhwar QA, Othman R, Rahman ZA, Meon S, Ismail MR (2012) Isolation and characterization of phosphate-solubilizing bacteria from aerobic rice. Afri J Biotechnol 11(11):2711–2719

Panhwar QA, Jusop S, Naher UA, Othman R, Razi MI (2013) Application of potential phosphate-solubilizing bacteria and organic acids on phosphate solubilization from phosphate rock in aerobic rice. Sci World J Article ID 272409, 10 pages

Panhwar QA, Naher UA, Shamshuddin J, Othman R, Latif MA, Ismail MR (2014) Biochemical and molecular characterization of potential phosphate-solubilizing bacteria in acid sulfate soils and their beneficial effects on rice growth. PLoS One 9(10):e97241

Pindi PK, Satyanarayana SDV (2012) Liquid microbial consortium- a potential tool for sustainable soil health. J Biofertil Biopestici 3:124. doi:10.4172/2155-6202.1000124

Ponmurugan P, Gopi C (2006) Distribution pattern and screening of phosphate solubilising bacteria isolated from different food and forage crops. J Agron 5(4):600–604

Premono ME, Moawad AM, Vlek PLG (1996) Effect of phosphate-solubilizing *Pseudomonas putida* on the growth of maize and its survival in the rhizosphere. Indones J Agric Sci 11:13–23

Ramachandran K, Srinivasan V, Hamza S, Anandaraj M (2003) Phosphate solubilizing bacteria isolated from the rhizosphere soil and its growth promotion on black pepper (Piper nigrum L.) cuttings. In: Vela´zquez E, Rodrı´guez-Barrueco C (eds) First international meeting on microbial phosphate solubilization, pp 325–331

Rathi M, Gaur N (2016) Phosphate solubilizing bacteria as biofertilizer and its applications. J Pharm Res 10(3):146–148

Rodríguez H, Fraga R (1999) Phosphate solubilizing bacteria and their role in plant growth promotion. Biotechnol Adv 7(4–5):319–339

Saber K, Nahla LD, Chedly A (2005) Effect of P on nodule formation and N fixation in bean. Agron Sustain Dev 25:389–393

Saharan BS, Nehra V (2011) Plant growth promoting rhizobacteria: a critical review. Life Sci Med Res 2011: LSMR-21, 1–30

Sashidhar B, Podile AR (2010) Mineral phosphate solubilization by rhizosphere bacteria and scope for manipulation of the direct oxidation pathway involving glucose dehydrogenase. J Appl Microbiol 109(1):1–12

Sharma S, Kumar V, Tripathi RB (2011) Isolation of phosphate solubilizing microorganism (PSMs) from soil. J Microbiol Biotech Res 1(2):90–95

Sharma SB, Sayyed RZ, Trivedi MH, Gobi TA (2013) Phosphate solubilizing microbes: sustainable approach for managing phosphorus deficiency in agricultural soils. SpringerPlus 2:587

Sharon JA, Hathwaik LT, Glenn GM, Imam SH, Lee CC (2016) Isolation of efficient phosphate solubilizing bacteria capable of enhancing tomato plant growth. J Soil Sci Plant Nutr 16(2):525–536

Smyth EM, McCarthy J, Nevin R, Khan MR, Dow JM, O'Gara F, Doohan FM (2011) *In vitro* analyses are not reliable predictors of the plant growth promotion capability of bacteria; a *Pseudomonas fluorescens* strain that promotes the growth and yield of wheat. J Appl Microbiol 111:683–692. doi:10.1111/j.1365-2672. 2011.05079.x

Stephen J, Shabanamol S, Rishad KS, Jisha MS (2015) Growth enhancement of rice (*Oryza sativa*) by phosphate solubilizing *Gluconacetobacter* sp. (MTCC 8368) and *Burkholderia* sp. (MTCC 8369) under greenhouse conditions. 3 Biotech 5:831–837

Stotzky G (1997) Soil as an environment for microbial life. In: van Elsas JD, Trevors JT, Wellington EMH (eds) Modern soil microbiology. Marcel Dekker, New York, pp 1–20

Tilak KVBR, Ranganayaki N, Pal KK, De R, Saxena AK, Nautiyal CS, Mittal S, Tripathi AK, Johri BN (2005) Diversity of plant growth and soil health supporting bacteria. Curr Sci 89(1):136–150

Tripura CB, Sashidhar B, Podile AR (2005) Transgenic mineral phosphate solubilizing bacteria for improved agricultural productivity. In: Satyanarayana T, Johri BN (eds) Microbial diversity current perspectives and potential applications. I. K. International Pvt. Ltd, New Delhi, pp 375–392

Trivedi P, Kumar B, Pandey A, Palni LMS (2007) Growth promotion of rice by phosphate solubilizing bioinoculants in a Himalayan location. In: Velazquez E, Rodriguez-Barrueco C (eds) Proceedings books of first international meeting on microbial phosphate solubilization. Kluwer, Dordrecht, pp 291–299

Vahed HS, Shahinrokhsar P, Heydarnezhad F (2012) Performance of phosphate solubilizing bacteria for improving growth and yield of rice (*Oryza Sativa* L.) in the presence of phosphorus fertilizer. Int J Agri Crop Sci 4(17):1228–1232

Vassilev N, Vassileva M (2003) Biotechnological solubilization of rock phosphate on media containing agro-industrial wastes. Appl Microbiol Biotechnol 61(5):435–440

Whitelaw MA (2000) Growth promotion of plants inoculated with phosphate-solubilizing fungi. Adv Agron 69:99–151

Yadav BK, Verma A (2012) Phosphate solubilization and mobilization in soil through soil microorganisms under arid ecosystems, the functioning of ecosystems. In: Ali M (ed) In Tech. ISBN: 978-953-51-0573-2

Nilima Dash completed M.Sc. (Microbiology) from Orissa University of Agriculture and Technology, Bhubaneswar, Odisha, India. Current research interests are genetics and mechanisms of phosphate solubilization by microbes, metagenomics and plant growth promotion (especially rice) by microorganisms.

Avishek Pahari passed M.Sc. (Microbiology) from Orissa University of Agriculture and Technology, Bhubaneswar, Odisha, India. His research interests are siderophore production by bacteria and plant (vegetables) growth promotion.

Tushar Kanti Dangar completed M.Sc. and Ph.D. from Burdwan University, Burdwan, West Bengal, India. Presently, working as a Principal Scientist (Microbiology) in National Rice Research Institute, Cuttack, Odisha. Broad areas of research include microbial diversity of stressed rice ecologies, physiology and genetics of nitrogen fixation, mechanism of abiotic stress tolerance, PGPR, mass production and formulation of PGPR and biocides especially *Bacillus thuringiensis*, *Pseudomonas* spp., *Beauveria bassiana, Metarhizium anisopliae* etc.

Tolerance of Microorganisms in Soil Contaminated with Trace Metals: An Overview

8

Dhritiman Chanda, G.D. Sharma, D.K. Jha, and Mohamed Hijri

Abstract

Trace metal (TM) pollution of soil is a worldwide problem threatening the quality of human life and a proper environment. We investigated fungal and bacterial diversity of trace metal-polluted site contaminated with paper mill effluent in India. Twelve fungal dominant isolates, viz. *Aspergillus*, *Penicillium*, *Fusarium*, *Cunninghamella*, *Simplicillium*, *Trichoderma*, *Rhizomucor*, *Cladosporium* and *Hypocrea*, were identified. Subsequent screening approach to assess their TM tolerance was performed in vitro cultures which revealed that the majority of the isolates were tolerant to Ni-, Cu-, Zn- and Cd-amended medium. The minimum inhibitory concentration (MIC) for Ni, Cu, Zn and Cd was also determined in isolated strains of *Aspergillus*, *Penicillium*, *Rhizomucor*, *Trichoderma* and *Fusarium* to study the concentration of growth against various trace metals. A total of 22 bacterial isolates was also isolated using 16S rRNA, and the dominant genera such as *Bacillus*, *Rhizobium*, *Microbacterium*, *Arthrobacter*, *Kribbella*

D. Chanda (✉)
Department of Botany, G.C. College, Assam University, Silchar, 788004, Assam, India
e-mail: dhriti.delhi@gmail.com

G.D. Sharma
Bilaspur University, Bilaspur, 495009, Chhattisgarh, India
e-mail: gduttasharma@yahoo.co.in

D.K. Jha
Department of Botany, Gauhati University, Guwahati, 781014, Assam, India
e-mail: dhruvakjha@gauhati.ac.in

M. Hijri
Department of Biological Sciences, Institut de recherche en biologie végétale (IRBV), University De Montreal, Montreal, Canada
e-mail: mohamed.hijri@umontreal.ca

P. Shukla (ed.), *Recent Advances in Applied Microbiology*,
DOI 10.1007/978-981-10-5275-0_8

and *Chitinophaga* were identified. The relative growth and LD_{-50} were also estimated against the different trace metals from concentration 0.1 to 4 mM. Thus, these fungal and bacterial isolates showed a high TM tolerance and would be a great interest for their use in bioremediation to clean up TM-polluted soil.

Keywords
Trace metal • Fungi • Bacteria • Biodiversity • Paper mill effluent • Bioremediation

8.1 Introduction

8.1.1 Microbial Diversity in Extreme Habitats

Soil harbours a variety of fungi, bacteria and other soil microorganisms. Soil microorganisms are the living component of soil organic matter and are responsible for mineralization of nutrients, decomposition and degradation or transformation of toxic compounds. Metals and metal compounds are natural constituents of all ecosystems, moving between atmosphere, hydrosphere, lithosphere and biosphere (Bargagli 2000; Wuana and Okieimen 2011). One of the challenges facing the mankind in recent times is the degradation and pollution of soil. Since soil is a vital natural resource, its degradation threatens the basic life support system. The industrial influent's sludge and solid waste are the sources of potentially harmful inorganic as well as organic contaminants. Microorganisms growing in such habitats evolved under conditions that permitted their survival and growth (Thakre and Shanware 2015). They multiplied in accordance with natural selection. For such adapted microorganisms, the conditions of these habitats are not 'extreme' but rather the normal physiological conditions for their growth in their natural habitats (Ali et al.2013; Akponah 2013; Kumar et al. 2014; Smith et al. 2015). In metal-contaminated soils, the siderophores and plant growth hormones are produced by plant-associated microbes (Pattus and Abdallah 2000; Wu et al. 2006; Schalk et al. 2011; Ullah et al. 2015). The secretion of siderophores by fungi and bacteria is dependent on several factors like soil pH, nutrient availability in soils and type and concentration of trace metals (Rajkumar et al. 2010; Sessitsch et al. 2013; Yu et al. 2014).

8.1.2 Heavy Metal Resistance in Fungi

Trace metals like Cu, Ni, Zn, Cd and Mn present in paper mill effluent can be removed by indigenous fungi isolated from effluent itself (Khan 2000, Karn and Reddy 2012). Biosorption of metal is carried out by (1) extracellular accumulation/precipitation, (2) cell surface sorption/precipitation and (3) intracellular accumulation through the cell wall of microorganisms (Volesky and Holan 1995; Valix et al. 2001; Madhaiyan et al. 2007; Ma et al. 2016). *Penicillium, Aspergillus, Trichoderma, Cladosporium*, etc., are found to be very useful for the removal of trace metals (Dursun 2008; Ezzourhi et al. 2009; de Lima et al. 2011). El-Morsy (2004) reported that *Cunninghamella echinulata* biomass could be employed as a biosorbent of metal ions in waste water. De Lima et al. (2013) and Bello and Abdullahi (2016) also studied the cadmium tolerance by *Cunninghamella elegans* by the polyphosphate metabolism. *Trichoderma* sp. produces organic acids like fumaric acid, citric acid and glycolic acid which can decrease the pH in alkaline soil and thus increase the solubility of macro- and micronutrients necessary for plant growth and metabolism (Malgorzata et al. 2014; Song et al. 2015).

8.1.3 Heavy Metal Resistance of Bacteria

Heavy metals can decrease carbon mineralization, nitrogen transformation and soil enzyme activities, microbial numbers (CFU), biomass (Borjesson et al. 2012) and frequency of trace metal-resistant bacteria (Wang et al. 2007, Kanmami et al. 2012). The molecular fingerprinting techniques are also useful to study the changes in the microbial community in trace metal stress conditions (Anyanwu et al. 2011; Andrew et al. 2013). Bacterial populations negatively affected by trace metals. Bacteria are found to develop five important mechanisms to detoxify the trace metals available in contaminated soils: (1) extracellular detoxification, (2) extracellular sequestration, (3) reduced permeability, (4) intracellular sequestration and (5) export. These resistant mechanisms are encoded in bacterial plasmids and transposons due to spontaneous mutation and gene transfer (Osborn et al. 1997; Karelová et al. 2011; Cetin et al. 2012; Zhou et al. 2013). Pal et al. (2004) reported Ni-resistant genes in Gram-positive and Gram-negative bacterial isolates from Ni-rich serpentine soil. In Gram-negative bacteria, the czc-genes encode for a cation-proton antiporter (CzcABC) which is responsible for the resistance against Cd, Zn and Co metals (Nies 1995; Harriso et al. 2007; Abdelatey et al. 2011; Mindlin et al. 2016).

The trace metal tolerance by a particular group of bacteria or isolate in artificial medium supplemented with trace metal showed high tolerance level as reported by Ahmed et al. (2001), Hayat et al. (2002) and Rajbanshi (2008). Olukoya et al. (1997) isolated 228 trace metal-resistant bacteria belonging to 9 genera, and the most common genera were *Staphylococcus*, *Streptococcus* and *Bacillus* found to be resistant to cobalt, zinc, copper, nickel and mercury. Temperature is also a determined factor that affects the growth of bacteria and bioaccumulation of trace metals (Lee et al. 2011a, b). The gene expression study revealed that mercuric ion (*mer*A) and chromate (*chr*B) genes were downregulated in all the strains of bacteria, i.e. *S. aureus*, *Bacillus subtilis*, *B. cereus*, *Pseudomonas* sp. and *Bordetella* sp., when treated with Co and Cd. The expression level of genes *mer*A, *chr*B, *czc* D and *ncc* A in these bacterial strains was measured by real-time PCR method (Abou-Shanab et al. 2007). Nies (1999) and Hirak and Das (2014) compared the metal resistance physiology in 63 species of bacteria and examined the protein-level similarities and suggested that these metal-resistant bacteria can be developed into metal pollution biosensors. Long et al. (2012) described the importance efflux transporters as a metal tolerance lactic by bacteria. Braud et al. (2010) reported a low level of toxicity of trace metals like Ni, Cu, Zn, Cd and Pb in *Pseudomonas aeruginosa. Chitinophaga eiseniae* was also reported as a trace metal tolerant by Yasir et al. (2011) and Gao et al. (2012);

Stan et al. (2011) studied the significant increase of growth, abundance, genetic diversity, nodulation ability and efficacy in the diversity of *Rhizobium* sp. in the soil polluted with copper, zinc and lead. Hemida et al. (2012) and Hao et al. (2014) also discussed the potential role of legume-rhizobia symbiosis in aiding phytoremediation. Hijri et al. (2014) also studied the linkage between fungal and bacterial communities in rhizosphere in hydrocarbon-contaminated soil and their significant effect for plant productivity.

The present study was carried out to understand and evaluate the status of heavy metal-resistant fungi, bacteria and actinomycetes in the Hindustan Paper Corporation (HPC), Assam. Geographically the site is situated at longitude of 24°41′29.9″N and latitude at 92°45′25.9″E.

8.1.4 Characterization of Fungal and Metal-Resistant Bacteria Isolates

The fungal isolates were isolated and were identified to species level using colony diameter and spore measurement following references and monographs adopted by Gilman (1957) and Raper and Fennell (1965). The fungal DNA was isolated with help of nucleic acid and protein purification kit (Macherey-Nagel, USA). The fungal strains have been characterized by PCR with (forward) ITS1 5′-TCCGTAGGTGAACCTGCGG-3′ and (reverse) ITS43′TCCTCCGCTTATTGATATGC-5′ (White et al. 1990).

The isolation and purification of chromosomal DNA as well as the amplification and sequencing of partial 16S rRNA gene of potential metal-resistant bacteria isolate was carried out. Bacterial 16S rDNA sequences were amplified using the 27F Lane (1991) and 1492R Turner et al. (1999) primer sets.

The selected bacterial isolate was tested for their resistance to different trace metals by their growth in nutrient broth tubes containing various concentrations of trace metals (0.1, 0.5, 2.0, 4.0 mM). The metals selected for the present investigation included Ni, Cu, Zn and Cd. The bacterial growth was determined by measuring the optical density using spectrophotometer at 540 nm. Relative growth of the isolate was expressed as the percentage of those obtained in untreated control. Lethal dose (LD-50) was estimated for all the tested bacterial isolates (Essam et al. 2013; Anderson and Hughes 2014). DNA sequencing was performed on an Applied Biosystems 3730xl DNA Analyzer (Applied Biosystems, Carlsbad, CA). The nucleotide sequence of bacterial isolate is thus obtained by the use of database using the 'NCBI BLAST' (Altschul et al. 1990).

8.1.5 Metal-Resistant Fungal Isolates

Twelve fungal strains were isolated from polluted soil contaminated with trace metals in paper mill effluents and tested with different trace metals in different concentrations. The 12 genera like *Aspergillus*, *Penicillium*, *Cladosporium*, *Cunninghamella*, *Trichoderma*, *Fusarium and Hypocrea* showed significant tolerance against various trace metals (Table 8.1). Minimum inhibitory concentration

Table 8.1 Genetic characteristics of isolated fungal strains

Sl no:	Isolated fungal strains	Accession number	Hit in NCBI database	Max indent (%)
1.	*Penicillium* sp.	KC602310	*Penicillium aculeatum*	99
2.	*Trichoderma* sp.	KC602314	*Trichoderma koningiopsis*	94
3.	*Cunninghamella* sp.	KC602315	*Cunninghamella* sp.	90
4.	*Trichoderma* sp.	KC602331	*Trichoderma harzianum*	97
5.	*Penicillium* sp.	KC602344	*Penicillium simplicissimum*	98
6.	*Rhizomucor* sp.	KC602345	*Rhizomucor variabilis*	99
7.	*Fusarium* sp.	KC602349	*Fusarium proliferatum*	99
8.	*Aspergillus* sp.	KC602350	*Aspergillus tamarii*	98
9.	*Penicillium* sp.	KC602359	*Penicillium janthinellum*	99
10.	*Aspergillus* sp.	KC602371	*Aspergillus niger*	99
11.	*Hypocrea* sp.	KC602373	*Hypocrea lixii*	95
12.	*Cladosporium* sp.	KC602374	*Cladosporium tenuissimum*	100

Table 8.2 Minimum inhibitory concentration (MIC) for tested fungal strains

Fungal isolates accession numbers		MIC (Mm)			
		Ni	Cu	Zn	Cd
Penicillium sp.	KC602310	10<mic>15	1<MIC>2.5	10<mic>15	2.5<MIC>5
Trichoderma sp.	KC602314	10<mic>15	1 < MIC>2.5	20<mic>25	15<mic>20
Cunninghamella sp.	KC602315	10<mic>15	5<MIC>10	15<mic>20	5<MIC>2.5
Trichoderma sp.	KC602331	15<mic>20	1<MIC>2.5	20<mic>25	Cont<MIC>1
Penicillium sp.	KC602344	15<mic>20	1<MIC>2.5	20<mic>25	15<mic>20
Rhizomucor sp.	KC602345	15<mic>20	15<mic>25	15<mic>25	15<mic>20
Fusarium sp.	KC602349	5<MIC>10	1<MIC>2.5	15<mic>25	Cont<MIC>1
Aspergillus sp.	KC602350	5<MIC>10	1<MIC>2.5	15<mic>25	5<MIC>10
Penicillium sp.	KC602359	15<mic>20	2.5<MIC>5	20<mic>25	1<MIC>2.5
Aspergillus sp.	KC602371	10<mic>15	1<MIC>2.5	20<mic>25	15<mic>20
Hypocrea sp.	KC602373	15<mic>20	2.5<MIC>5	15<mic>20	1<MIC>2.5
Cladosporium sp.	KC602374	15<mic>20	5<MIC>10	15<mic>20	2.5<MIC>5

(MIC) of the isolated fungal strains against the different concentration of trace metals was estimated and found that, at higher metal ion concentrations, most of the tested fungal strains were found tolerant and showed strong growth (Table 8.2).

In the presence of various concentrations of nickel, the fungal strains which were able to grow in 15–20 mM were *Trichoderma* sp., *Penicillium* sp., *Rhizomucor* sp., *Cladosporium* sp. and *Hypocrea* sp. The other tested strains like *Penicillium*, *Aspergillus* and *Cunninghamella* were also to grow in MIC of 10–15 mM (Plates 8.1 and 8.2).

In the presence of various concentrations of copper, most of the tested strains showed a very low MIC except *Cunninghamella* and *Cladosporium* where MIC range was 5–10 mM. Their mycelia became diffused compared with the control. All strains studied could not grow in higher concentrations except *Rhizomucor* sp. (KC602345) which showed the highest MIC of 15–25 mM. The white colour of the

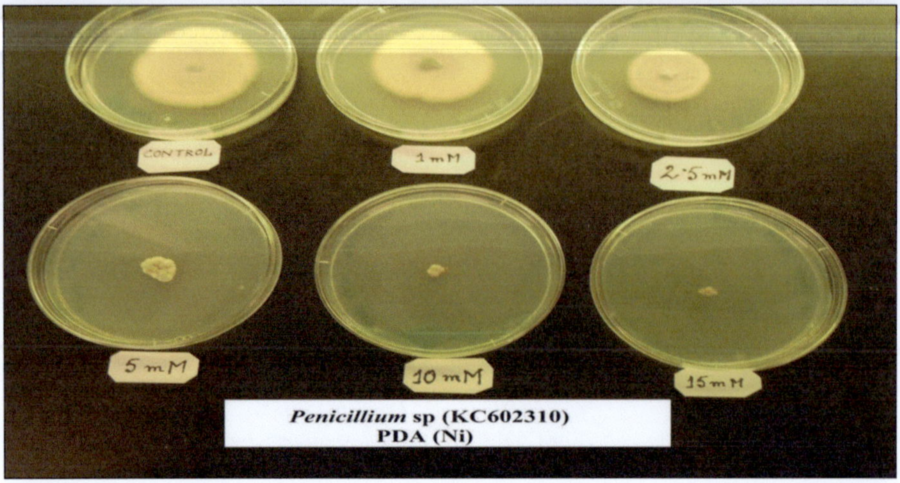

Plate 8.1 The growth of *Penicillium* sp. (KC602310*)* in nickel

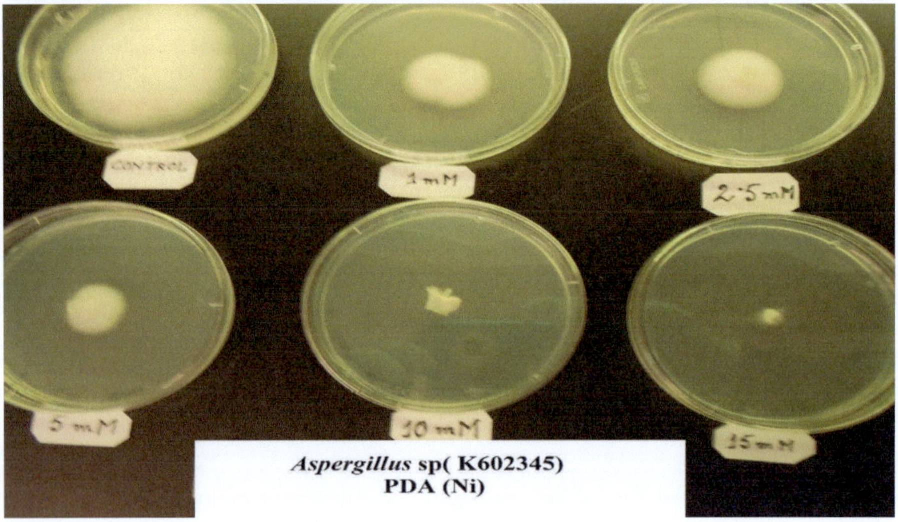

Plate 8.2 The growth of *Aspergillus* sp. (KC602345) in nickel

mycelium became blue green due to accumulation of Cu ions inside the cell wall of the tested fungi (Plate 8.3). The growth rate of fungi tested was reduced, and their conidiogenesis was also slowed down. Addition of copper sulphate to the PDA resulted in the growth of the isolated fungal strains and changed the colour and morphology of the mycelium. The mycelium of *Cladosporium* sp. (KC602374) secreted a deep brown substance (Plate 8.4), and the *Fusarium* isolate (KC602349) (Plate 8.5) secreted violet pigment due to the response to the metal stress.

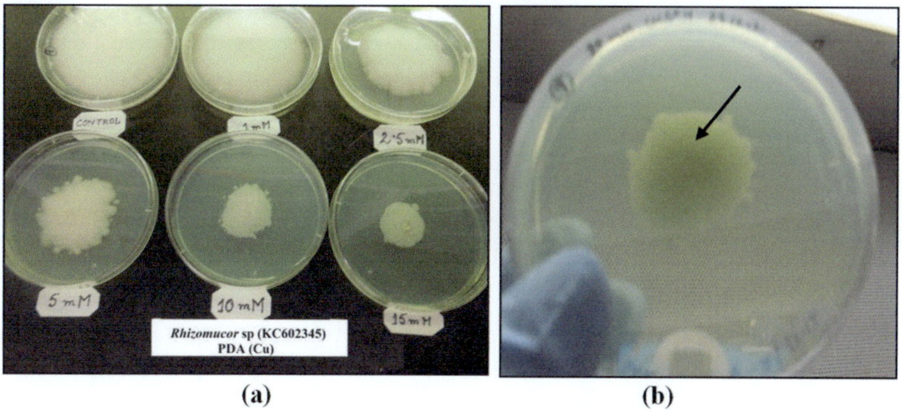

Plate 8.3 The growth of *Rhizomucor* sp. (KC602345) (**a** and **b**) in copper

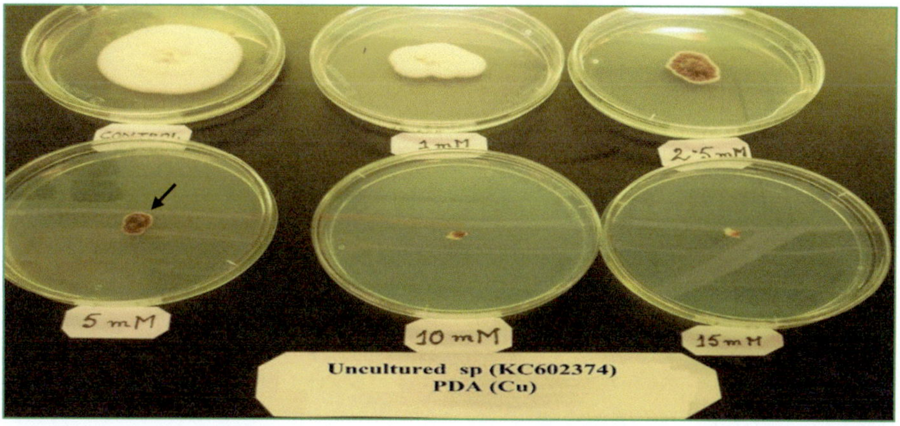

Plate 8.4 The growth of *Cladosporium* sp. (KC602374*)* in copper

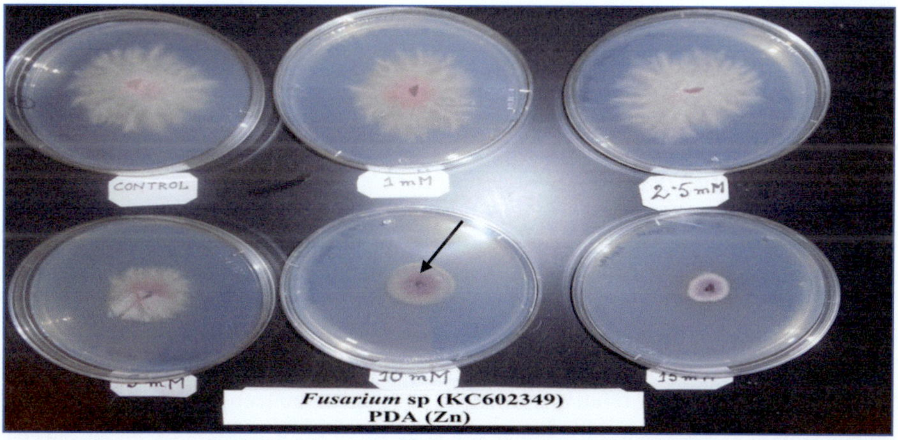

Plate 8.5 The growth of *Fusarium* sp. (KC602349) in copper

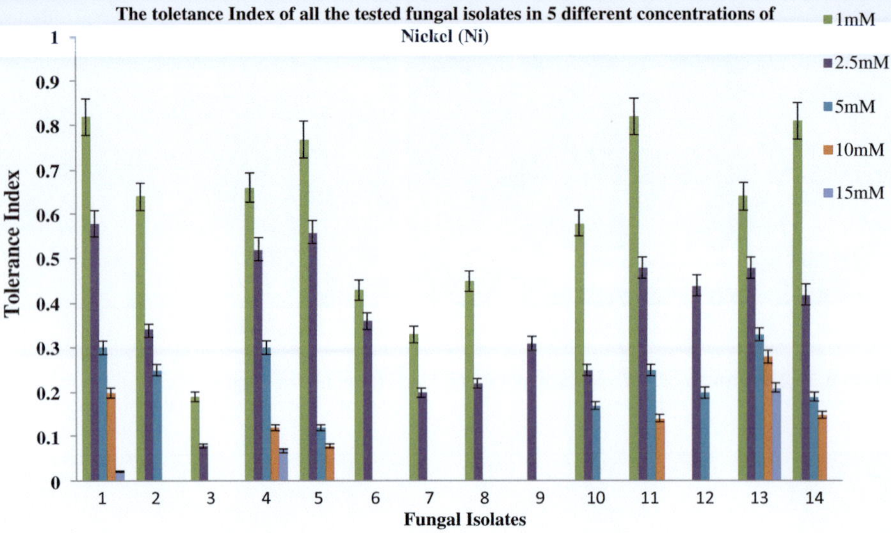

Fig. 8.1 The tolerance index of fungal strains in nickel (Ni)

In the presence of various concentrations of cadmium, the isolates *Trichoderma*, *Aspergillus* sp. and *Penicillium* sp. showed a high MIC with 15–20 mM. When the concentration of cadmium increased in the media, the absorbance of the fungal culture was found to be decreased. The most tolerant fungi which were found to grow in high concentration of the trace metals were *Penicillium* sp. (KC602310), *Trichoderma* sp. (KC602314), *Aspergillus* sp. (KC602350), *Fusarium* sp. (KC602349), *Hypocrea* sp. (KC602373), *Penicillium janthinellum* (KC602344) and *Cladosporium* (KC602374). The value of tolerance index of *Penicillium* sp. (KC602310 and KC602359), *Aspergillus* sp. (KC602350 and KC602371), *Rhizomucor* sp. (KC602345), *Fusarium* sp. (KC602349) and *Trichoderma* sp. (KC602314 and KC602331) showed a maximum value of 0.9 tested against all the metals, i.e. Ni, Cu, Zn and Cd (Figs. 8.1, 8.2, 8.3 and 8.4).

8.1.6 Identification and Characterization of Metal-Resistant Bacteria Isolates

Twenty-two bacterial isolates showed resistance to different trace metals, and the molecular characterization for these isolates was carried out (Table 8.3). The trace metals like Ni, Cu, Zn and Cd were selected in a concentration ranged from 0.1 to 4.0 mM for identification. Among the various genera, *Bacillus*, *Agromyces*, *Microbacterium*, *Arthrobacter*, *Chitinophaga*, *Rhizobium* and *Kribbella* were showing a range of 30–40% relative growth at the higher concentrations of all heavy metals tested. These bacterial isolates are capable to grow at higher concentrations of trace metals, and thus they were resistant to Ni, Cu, Zn and Cd. The species of

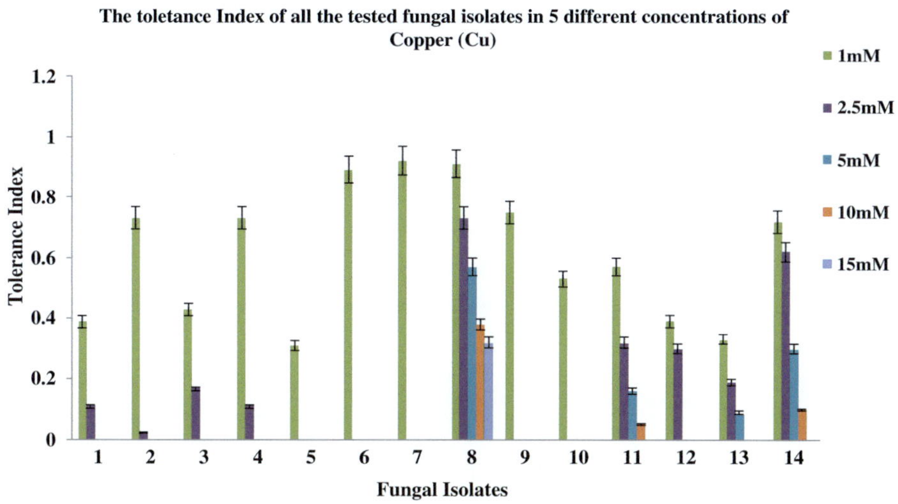

Fig. 8.2 The tolerance index of fungal strains in copper (Cu)

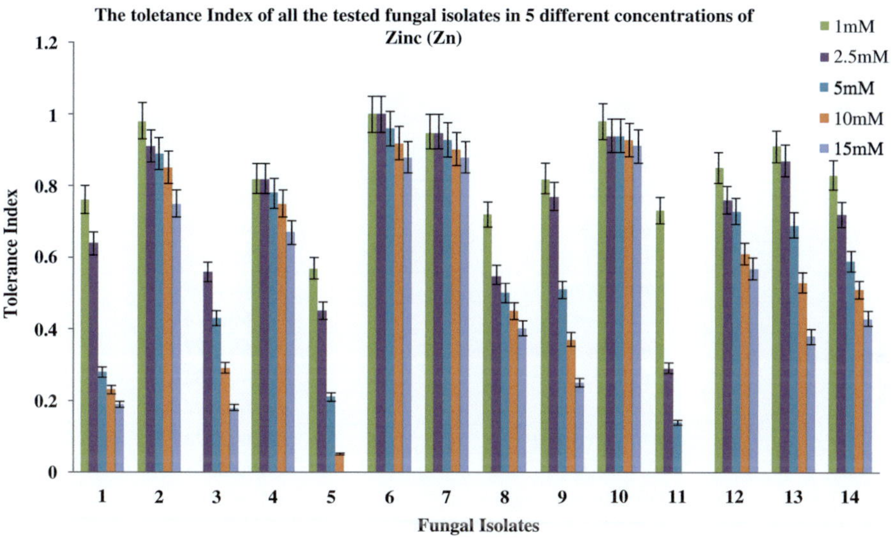

Fig. 8.3 The tolerance index of fungal strains in zinc (Zn)

Agromyces, *Bacillus*, *Chittinophaga* and *Kribbella* (isolates 1, 3, 4, 6, 10) showed significant relative growth values ranging from 40 to 70% at 2 mM and 4 mM concentrations of zinc. The species of *Rhizobium*, *Bacillus* and *Arthrobacter* showed a range of 20–60% of relative growth at 2 mM and 4 mM concentrations of nickel, copper and cadmium. The species of *Bacillus* and *Microbacterium* (isolates 18 and 19) showed a range of 20–30% of relative growth at 2 mM and 4 mM concentrations of cadmium.

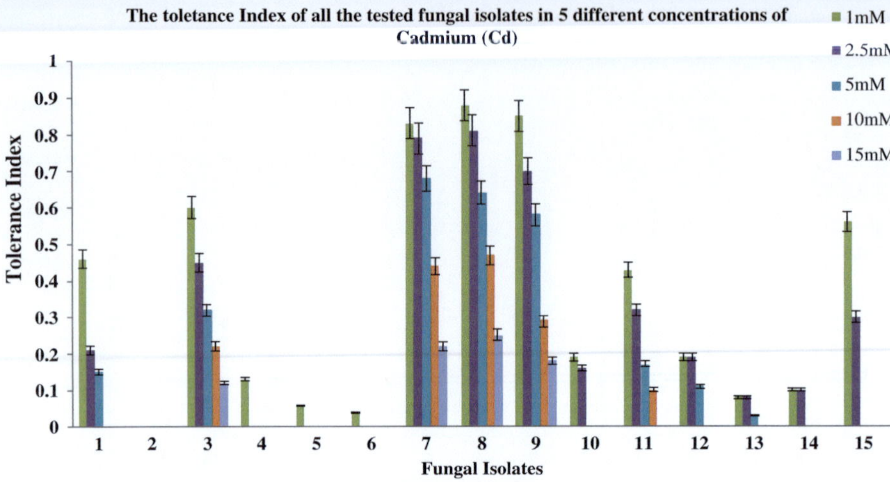

Fig. 8.4 The tolerance index of fungal strains in cadmium (Cd)

Table 8.3 List of bacterial strain tested for trace metal resistance and their accession numbers (NCBI)

Isolate no:	Name of the bacterial genus of max indent of 99%	Accession number
1.	*Agromyces* sp.	**KC602240**
2.	*Arthrobacter* sp.	**KC602245**
3.	*Bacillus cereus*	**KC602258**
4.	*Bacillus* sp.	**KC602265**
5.	*Chitinophaga* sp.	**KC602266**
6.	*Chitinophaga* sp.	**KC602269**
7.	*Rhizobium* sp.	**KC602276**
8.	*Microbacterium* sp.	**KC602277**
9.	*Bacillus* sp.	**KC602282**
10.	*Kribbella* sp.	**KC602294**
11.	*Arthrobacter* sp.	**KC602298**
12.	*Bacillus* sp.	**KC602301**
13.	*Arthrobacter oryzae*	**KC602305**
14.	*Arthrobacter nicotinovorans*	**KC602306**
15.	*Arthrobacter globiformis*	**KC602307**
16.	*Arthrobacter humicola*	**KC602308**
17.	*Arthrobacter* sp.	**KC602309**
18.	*Bacillus aryabhattai*	**KC602264**
19.	*Microbacterium* sp.	**KC602239**
20.	*Agromyces* sp.	**KC602270**
21.	*Bacillus drentensis*	**KC602283**
22.	*Bacillus* sp.	**KC602286**

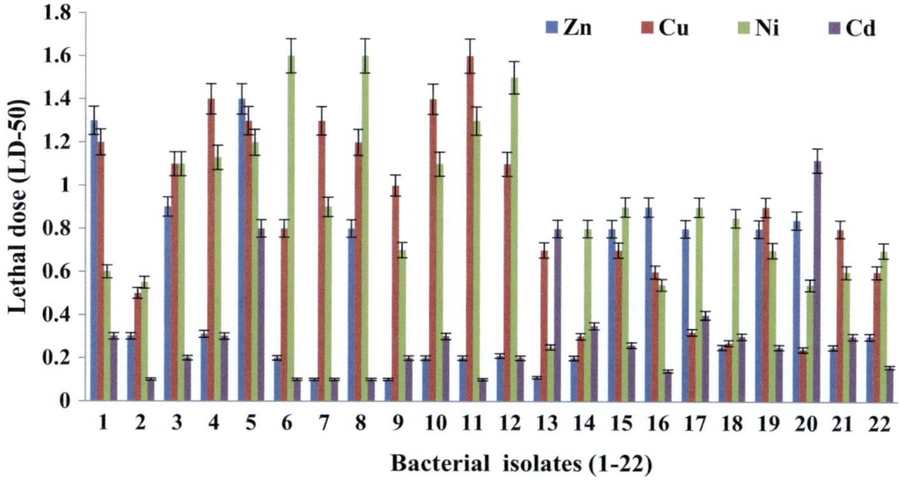

Fig. 8.5 Lethal dose (LD_{-50}) of bacteria isolates (1–22)

The isolated trace metal-resistant bacterial strains that were identified with their accession numbers are *Bacillus cereus* (KC602258), *Bacillus* sp. (KC602265), *Chitinophaga* sp. (KC602266), *Chitinophaga bacter* (KC602269), *Rhizobium* sp. (KC602276), *Microbacterium* sp. (KC602277), *Bacillus* sp. (KC602282), *Kribbella* sp. (KC602294), *Arthrobacter* sp. (KC602298), *Arthrobacter oryzae* (KC602305) and *Arthrobacter nicotinovorans* (KC602306) and were found to show positive test as tested against the 15 sugars, i.e. glucose, sucrose, xylose, maltose, rhamnose, rafffinose, cellobiose, dextrose, galactose, arabinose, lactose, sorbitol, melibiose, saccharose and trehalose. All the bacterial strains were tested for antibiotic sensitivity (Bauer 1996). Most of the isolates of *Bacillus*, *Agromyces*, *Microbacterium*, *Arthrobacter*, *Chitinophaga*, *Rhizobium*, *Brachybacterium* and *Kribbella* appeared to be inhibited by eight antibiotics and resistant to ampicillin, while *Chitinophaga* sp. (KC602269) was resistant to chloramphenicol (Adesoji et al. 2015). Among all the strains tested, the isolates (KC602240, KC602277, KC602301, KC602283 and KC602286) showed resistance to ampicillin, whereas the rest showed no inhibition. The antibiotics like streptomycin, polymyxin B, vancomycin, tetracycline, gentamicin, amikacin, ciprofloxacin and levofloxacin were found to be susceptible to all the 22 tested strains.

Lethal dose (LD-50) was estimated for all the tested bacterial isolates. The species of *Arthrobacter*, *Chitinophaga*, *Kribbella*, *Microbacterium*, *Bacillus*, *Agromyces* and *Rhizobium* showed a significant range of LD-50 values (0.2–1.8) tested against zinc, (0.3–1.6) for copper, (0.6–1.5) for nickel and (0.1–0.8) for cadmium. The highest LD-50 value of 1.8 was showed by the *Chitinophaga* sp. (KC602266), while the highest LD-50 value of 1.6 was showed by *Chitinophaga bacter* sp. (KC602269) and *Microbacterium* sp. (KC602277) against Ni. The highest LD-50 value of 0.8 was showed by the *Chitinophaga* sp. (KC602266), *Arthrobacter* (KC602303) and *Agromyces* (KC602270) against the metal Cd (Fig. 8.5).

The fungal and bacterial sequences were analysed by the Basic Local Alignment Tool (BLAST) for finding the closest homologous sequences. These sequences

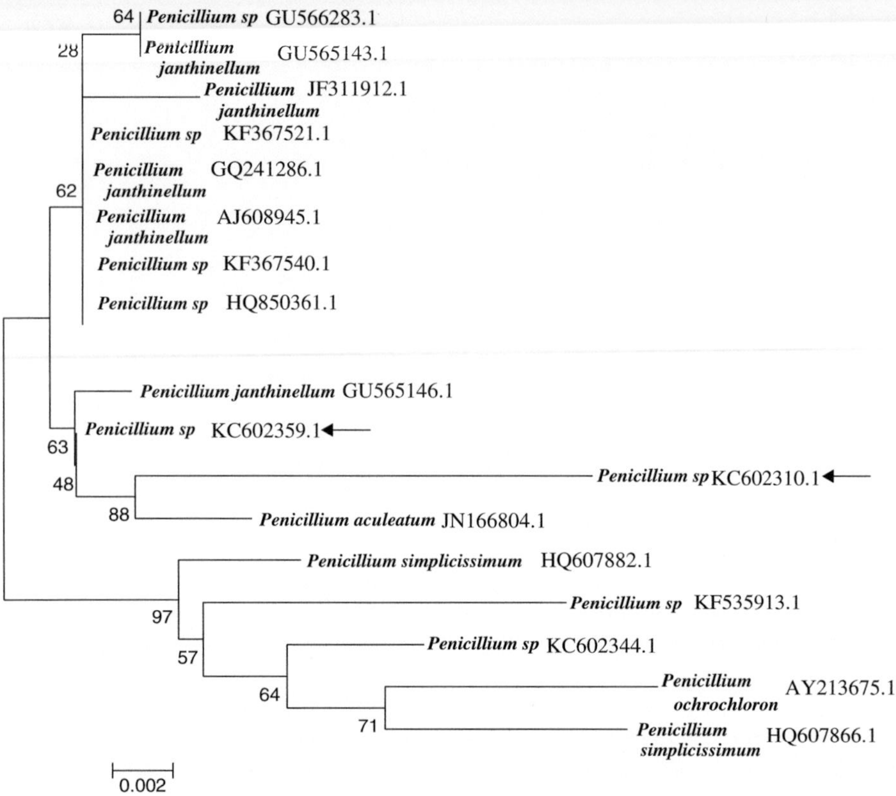

Fig. 8.6 Phylogenetic tree of *Penicillium* sp. (KC602310 and KC602359)

were saved in a fasta format (*.fasta or *.txt) and aligned with CLUSTAL X2. The output of CLUSTAL (i.e. *.aln file) was saved for the output of MEGA version 5. Phylogenetic and molecular evolutionary analysis was carried out by MEGA version 5 (Tamura et al. 2011). A distance matrix was made based on nucleotide sequence homology, and neighbour joining (NJ) consensus trees were obtained using Kimura-2 parameter substitution model (MEGA 5) (Saitou and Nie, 1987). The bootstrap values above 50% and the genetic distance scale are shown for the relationship of the isolated fungal (Figs. 8.6, 8.7, 8.8, 8.9 and 8.10) and bacterial strains (Figs. 8.11, 8.12, 8.13, 8.14, 8.15, 8.16 and 8.17) with their closely related neighbouring species.

The present experimental findings revealed the effects of trace metals on microbial diversity, i.e. fungi, bacteria and actinomycetes, in the polluted site of Hindustan Paper Corporation (HPC) paper mill. The diversity and abundance of soil microorganisms were found to be affected by naturally occurring environmental variables, including soil types, soil pH, moisture content and natural availability. Carson et al. (2010) and Stefanowicz et al. (2010) also reported that the soil microorganisms are affected positively by environmental factors. All the isolated strains of fungi,

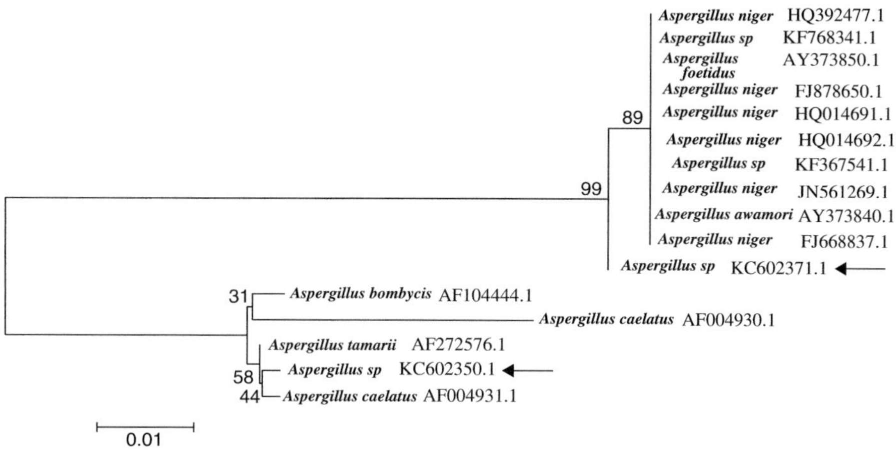

Fig. 8.7 Phylogenetic tree of *Aspergillus* sp. (KC602350 and KC602371)

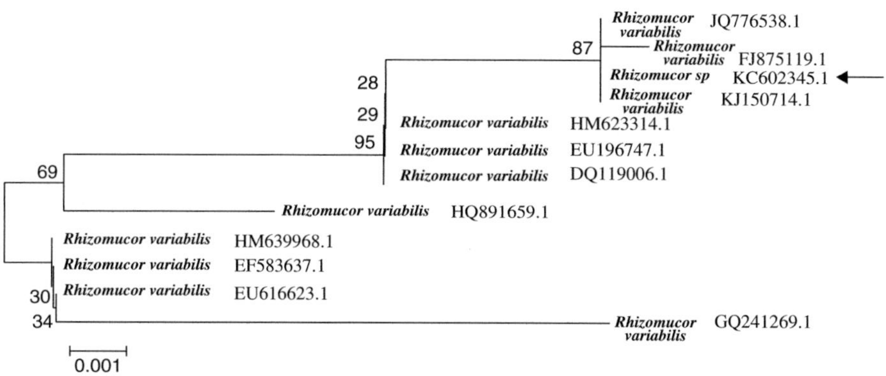

Fig. 8.8 Phylogenetic tree of of *Rhizomucor* sp. (KC602345)

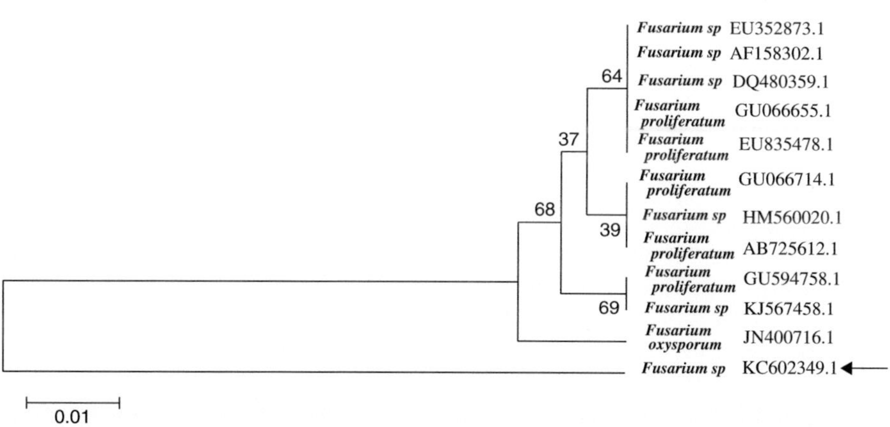

Fig. 8.9 Phylogenetic tree of *Fusarium* sp. (KC602349)

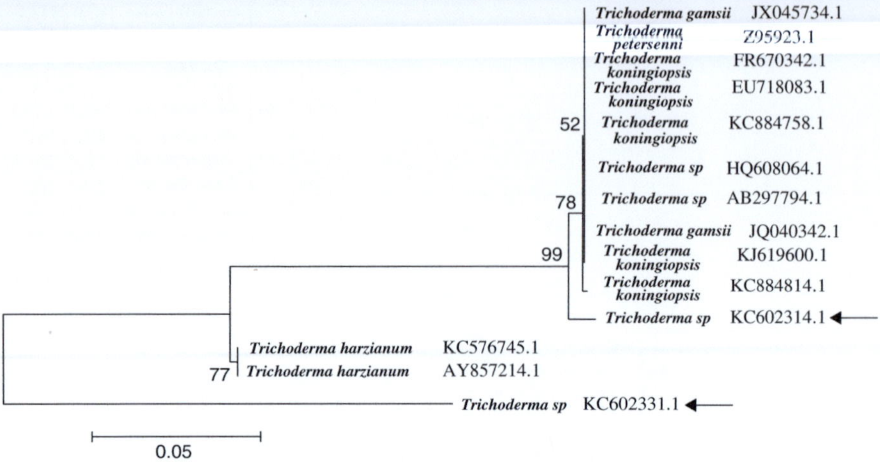

Fig. 8.10 Phylogenetic tree of *Trichoderma* sp. (KC602314 and KC602331)

bacteria and actinomycetes were found to be resistant to various trace metals at higher concentrations. Similar observations were observed by Freitas et al. (2009) and Appenroth (2010). Soil microbial populations were found to multiply even under metal-contaminated soil which in turn maintains the diversity of fungi and bacteria (Chen et al. 2014). The resistance of the selected strains to Cr^{6+}, Pb^{2+}, Zn^{2+} and Cu^{2+} was determined by the dilution method to calculate the tolerance index for all the tested fungi. *Penicillium* sp. (KC602310), *Trichoderma* sp. (KC602314), *Aspergillus* sp. (KC602350), *Fusarium* sp. (KC602349), *Hypocrea* sp. (KC602373), *Penicillium janthinellum* (KC602344) and *Cladosporium* (KC602374) were reported for their great importance in removal of trace metals from contaminated site. Some deuteromycetes have been reported by Ghorbani et al. (2007) and Zafar et al. (2007). Metals such as copper and zinc are essential to bioactivities; however, they tend to show toxicity after a certain level.

The fungal strains which were able to grow in 15–20 mM were *Trichoderma* sp., *Penicillium* sp., *Rhizomucor* sp., *Cladosporium* sp. *and Hypocrea* sp. The other tested strains like *Penicillium*, *Aspergillus* and *Cunninghamella* were also to grow in MIC of 10–15 mM (Table 8.2). Rao et al. (2005) and Sun and Shah (2007) also observed that with the increasing metal concentration of trace metals, the fungi *Aspergillus niger* and *Cunninghamella echinulata* can increase the rate of metal removal by saturation adsorbent concentrations by increasing mobilization of metal ions (Burford et al. 2003, Thippeswamy et al. 2012, 2014). *Penicillium* and *Aspergillus* showed a higher metal tolerance against nickel. Similar effects were also observed by Shivkumar et al. (2011) who discussed the high tolerance and bioaccumulation ability in *Penicillium* sp. and *Rhizopus* sp. against the various trace metal like Cu^{2+}, Zn^{2+}, Cd^{2+}, Ni^{2+} and Pb^{2+}. The growth of all fungi tested was decreased after addition of copper in high concentration in comparison with zinc, nickel and cadmium. All strains studied could not grow in higher concentrations except *Rhizomucor* sp. (KC602345) which showed the highest MIC of 15–25 mM. Van and Christov (2002) and Tripathi et al. (2007)

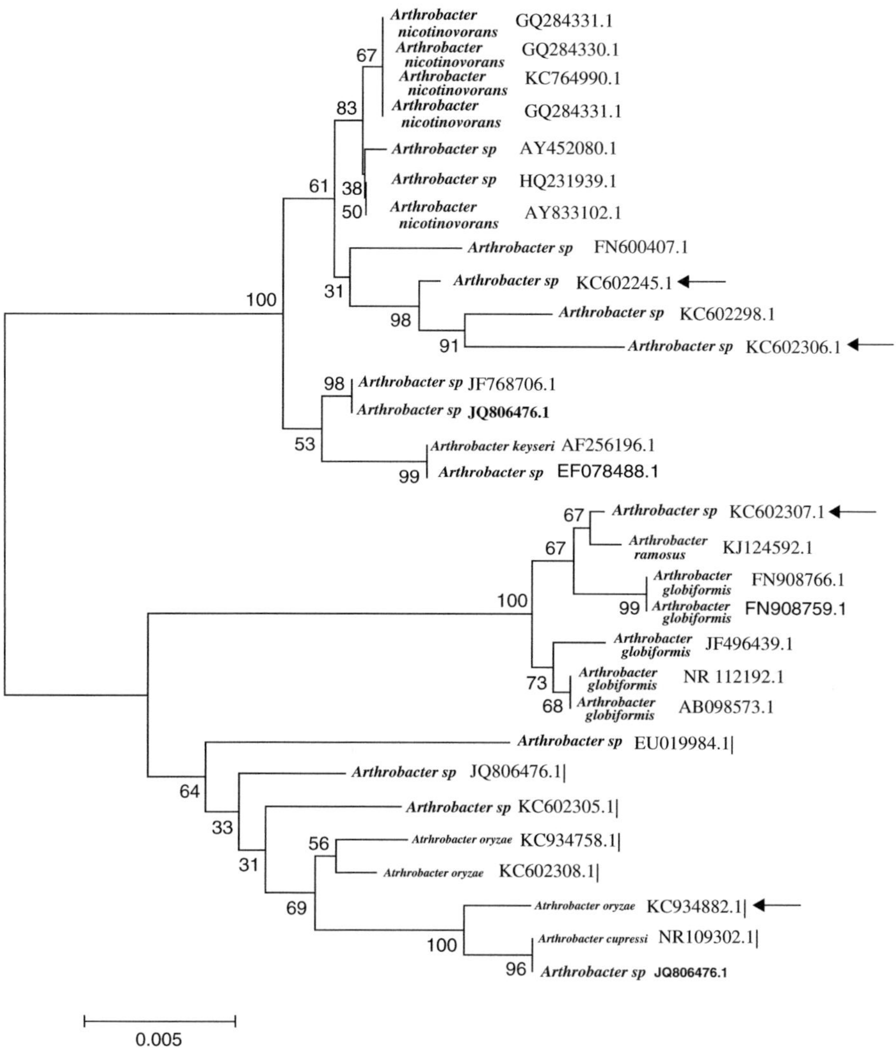

Fig. 8.11 Phylogenetic tree of *Arthrobacter* sp.

also observed that *Rhizomucor pusillus* adsorption capacity was isolated from effluent plant. Rouhollahi et al. (2014) studied the nickel biosorption capacity of *Rhizomucor pusillus* by enzymatic and alkali treatments. The white colour of the mycelium became blue green due to accumulation of Cu ions inside the cell wall of the tested fungi. Copper tolerance in fungi ascribed to diverse mechanisms also described by Cervantes and Gutierrez (1994). The most of the tested strains showed a very low MIC except *Cunninghamella* and *Cladosporium* where MIC range was 5–10 mM. The morphology of strains was highly affected by the presence of Cu. Their mycelia became diffused compared with the control. The growth rate of fungi tested was reduced, and their conidiogenesis was also slowed down. In *Cladosporium* sp. (KC602374), the

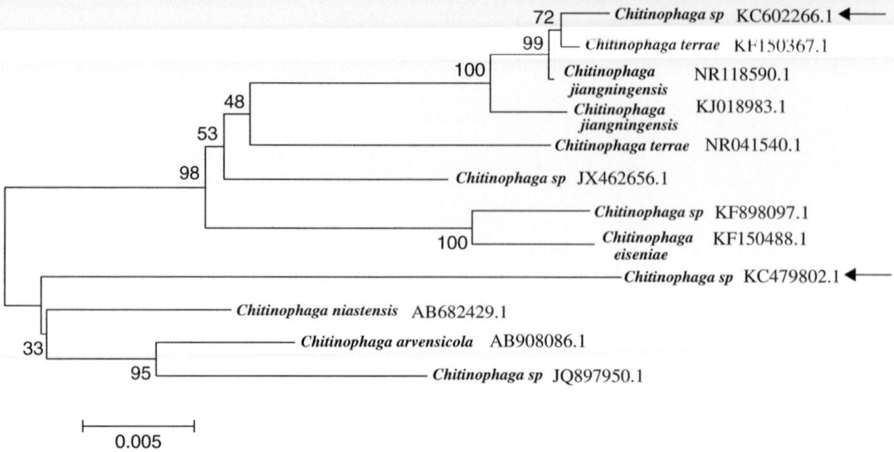

Fig. 8.12 Phylogenetic tree of *Chitinophaga* sp. (KC602266)

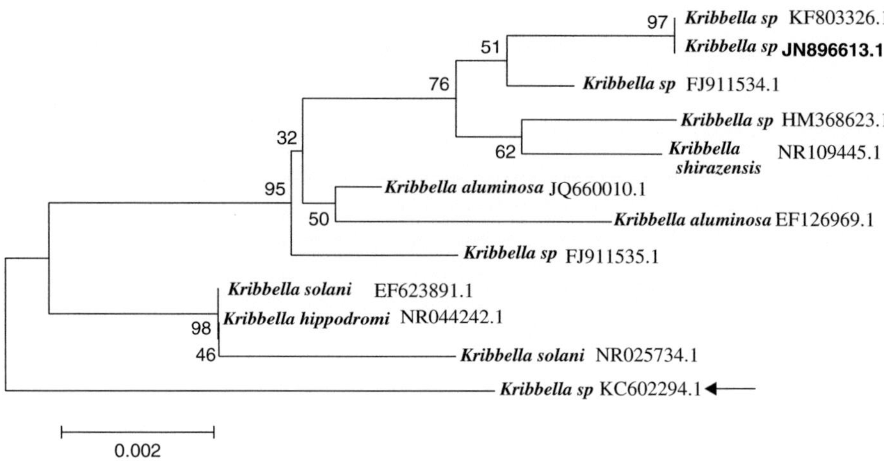

Fig. 8.13 Phylogenetic tree of *Kribbella* sp. (KC602294)

mycelia changed into deep brown colour in the high concentration of Cu. The tolerance of the tested fungi to high copper concentrations could be related to metallothioneins and other thiol compounds which may be promising detoxifying agents for copper as reported by Malik (2004) and Dusrun (2008). Similar biosorption mechanisms were also reported by Juliana et al. (2013) who discussed the biomass of *Cladosporium* as an efficient biosorbent of copper.

The fungal colour and morphology were both affected by high Zn concentrations in *Fusarium* sp. as the mycelium changed to violet pigment which is probably due to the stress imposed by the Zn. The zinc MIC was in the range 20–25 mM,

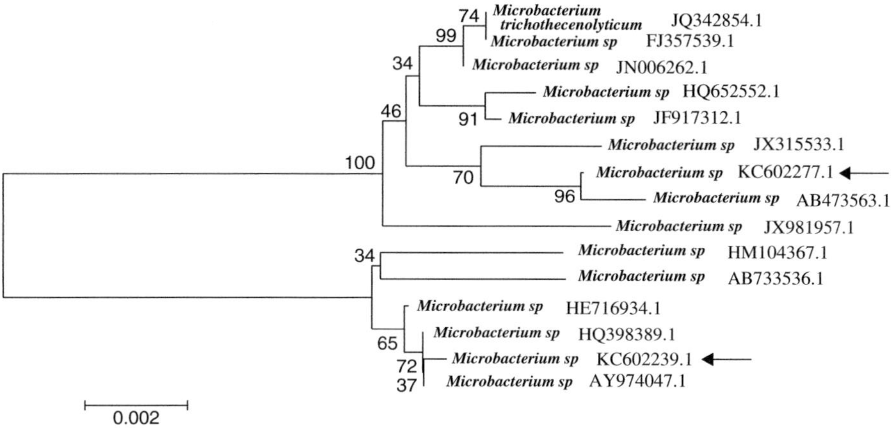

Fig. 8.14 Phylogenetic tree of *Microbacterium* sp. (KC602277 and KC602239)

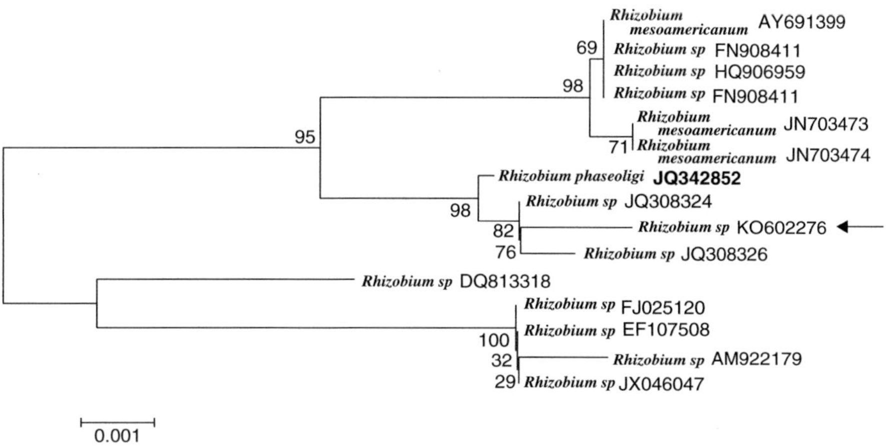

Fig. 8.15 Phylogenetic tree of *Rhizobium* sp. (KC602276)

15–20 mM, 10–15 mM and 5–10 mM for the *Fusarium* sp. Biosorption of various trace metals by *Fusarium* sp. was also reported earlier by Sen (2011), Zhang et al. (2012) and Verma et al. (2016).

The isolates *Trichoderma*, *Aspergillus niger*, *Cunninghamella* sp. and *Penicillium* sp. showed a high MIC with 15–20 mM in Cd-amended media. DeLima et al. (2011, 2013) also reported a higher potential of cadmium tolerance in the fungi *Trichoderma harzianum* and *Cunninghamella elegans*. The dominant genus of fungi identified and characterized were *Aspergillus*, *Penicillium*, *Fusarium*, *Cunninghamella*, *Trichoderma*, *Rhizomucor*, *Cladosporium* and *Hypocrea* by PCR with (forward) ITS1 and (reverse) ITS4 from the polluted soil. This may be due to the processes of valence transformation, active uptake, complexation, crystallization and biosorption of trace metals to the fungal cell walls (Jaeckel et al. 2005; Willie et al. 2007;

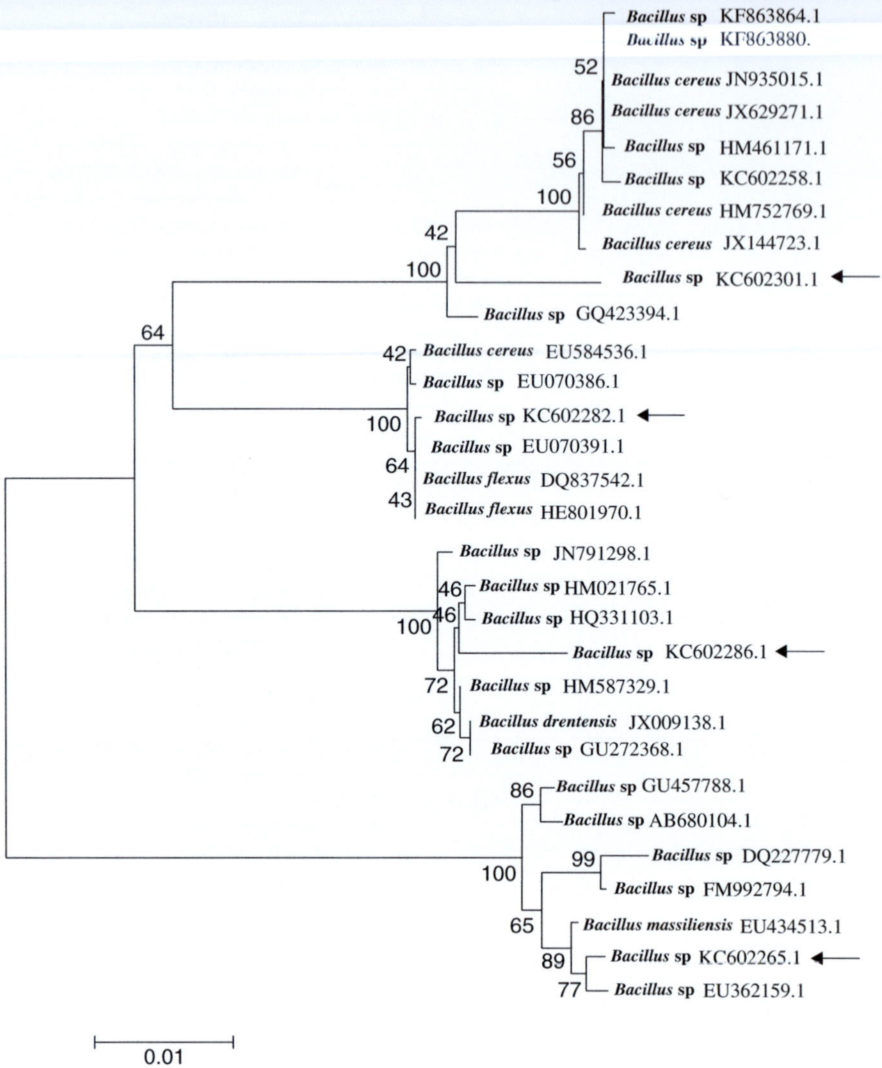

Fig. 8.16 Phylogenetic tree of *Bacillus* sp.

Palanivel et al. 2010; Anahid et al. 2011; Iram et al. 2012; Do Carmo et al. 2013; Rhodes 2013; Akhtar et al. 2013). Yazdani et al. (2009) and Malgorzata et al. (2014) found the application of *Trichoderma* sp. on various plant and found that this fungus has positive effects on increasing the biomass, soil parameters (C, N and P) and solubility of trace metals in soil, thereby enhancing phytoextraction in the plants. Copper tolerance of various *Trichoderma* sp. is also studied by Petrovic et al. (2014). Teng et al. (2015) also studied the phytoremediation in Cd-contaminated soil by *Trichoderma reesei* FS10-C strain.

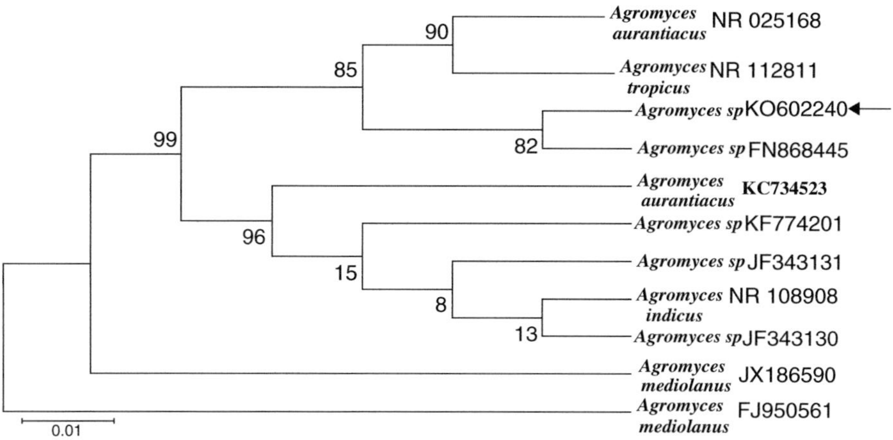

Fig. 8.17 Phylogenetic tree of *Agromyces* sp. (KC602240)

A total of 22 bacterial isolates exhibited resistance to different trace metals. These bacterial isolates were capable to grow at higher concentrations of trace metals and showing different degree of resistance to Ni, Cu, Zn and Cd. The major bacterial genera were identified as *Arthrobacter*, *Brachybacterium*, *Chitinophaga*, *Kribbella*, *Microbacterium*, *Bacillus*, *Agromyces* and *Rhizobium.* The resistance of these bacterial strains towards trace metal could be a result of the interaction between the metals and amphoteric groups such as the carboxyl and phosphoryl groups. In the present study, Gram-positive bacteria showed a major group for absorption capacity than the Gram-negative isolates as tested against different trace metals as Gram-positive bacteria have high chemisorption sites (Tunali et al. 2006, Long et al. 2012). The glycoproteins present on the outer site of Gram-positive bacterial cells have more potential binding sites than the Gram-negative bacteria having an outer layer of lipopolysaccharide (LPS), phospholipids and proteins (Gupta et al. 2012, 2016; Issazadeh et al. 2013).

The isolates of *Bacillus*, *Agromyces*, *Microbacterium*, *Arthrobacter*, *Chitinophaga*, *Rhizobium*, *Actinobacterium* and *Kribbella* showed positive activity towards urease, nitrate, H_2S production, citrate utilization, methyl red, malonate utilization, oxidase production, starch amylase and catalase activity. The Gram-positive isolates found to be positive against catalase and negative against oxidase activity were identified as *Brachybacterium*, *Agromyces*, *Arthrobacter*, *Kribbella* and *Microbacterium*. Similar observations were reported by different workers for these Gram-positive strains of same bacterial strain: *Agromyces* sp. (Chen et al. 2011; Thawai et al. 2011), *Arthrobacter* sp. (Elanvogvan et al. 2010; Rosales et al. 2012; Santa et al. 2013; Sahoo et al. 2014; Swer et al. 2016), *Chitinophaga* sp. (Gao et al., 2015), *Kribbella* sp. (Clara et al. 2008) and *Microbacterium* sp. (Mondani et al. 2012; Brown et al. 2012; Tappe et al.2013).

The strains of *Microbacterium* sp. showed a positive catalase activity and negative oxidase and H_2S production. Piccirillo et al. (2013) also observed similar biochemical activities in *Microbacterium oxydans* to be tolerant against Zn (II) and Cd (II). The isolated strains of *Brachybacterium* sp. were observed negative for catalase and oxidase activities and positive against starch hydrolysis and found to have relative growth range of 40–80% on higher concentrations of Zn, Cu, Ni and Cd at 2 mM and 4 mM (Wang et al. 2009; Park et al. 2011). Various strains of *Arthrobacter* sp. were isolated and found to be resistant against different trace metals (Paris and Blondeau 1999; Bafana et al. 2010; Inga 2013). The genus *Kribbella* sp. was isolated and showed 40–90% of relative growth on different concentrations of trace metals. Biochemical tests showed positive against catalase activity, nitrate reduction and H_2S production and negative against oxidase production. Similar chemotaxonomic characteristics were reported earlier by Carlsohn et al. (2007) who also reported the greater accumulation capacity of *Kribbella aluminosa* against the metal Pb, Fe, Zn and Cu when grown in medium with 200 ppm of Pb, Fe and Zn and 100 ppm of Cu.

Bacillus sp. was found to be resistant with a relative growth of 30–40% on higher concentrations of all the trace metals. The isolates were found to be positive against nitrate reduction, citrate utilization, oxidase production, starch amylase, methyl red test and catalase activity. Similar biochemical activities and multi-tolerance and bioremediation of trace metals in *Bacillus* strains were observed earlier by various workers (Rathnayake et al. 2009; Elsilk et al. 2014). *Chitinophaga* sp. was negative against oxidase, catalase and starch amylase tests. The similar biochemical characteristics were observed in *Chitinophaga* sp. by Lee et al. (2009) and Wang et al. (2014). *Rhizobium* sp. is a Gram-negative, aerobic, non-endospore-forming rods, showed positive results against nitrate and catalase test and negative against oxidase, indole, VP test and urease test (Kuykendall et al. 2005; Grison et al. 2015). *Rhizobium* sp. was found to grow with 20–40 % of relative growth of copper and nickel at higher concentrations. The resistance of *Rhizobium* towards trace metals can produce huge amount of extracellular polysaccharide (EPS) and lipopolysaccharide (LPS), which can attach most of the metals extracellularly, acting a first-defence barrier against trace metal stress (Mohamed et al. 2012; Mandal and Bhattacharyya 2012). Our results were supported by Reeve et al. (2002), Hemida et al. (2012) and Hao et al. (2014) who also observed that *Rhizobium* played a very important role of legume-rhizobia symbiosis in aiding phytoremediation of polluted site contaminated with trace metals (Mergeay et al. 2003; Piotrowska-Seget et al. 2005; Zhang et al. 2011; Aafi et al. 2012; Rajkumar et al. 2012; Yang et al. 2012; Adel et al. 2014).

8.2 Conclusion

The present study focused on the effect of trace metal on the diversity of microorganisms (fungi, bacteria, actinomycetes) in the Hindustan Paper Corporation (HPC), Cachar. The most tolerant fungi grown in high concentration of the trace metals were identified as *Penicillium* sp. (KC602310), *Trichoderma* sp. (KC602314),

Aspergillus sp. (KC602350), *Fusarium* sp. (KC602349), *Hypocrea* sp. (KC602373), *Penicillium janthinellum* (KC602344) and *Cladosporium* sp. (KC602374).

The most tolerant bacteria grown in high concentration of the trace metals were identified as *Bacillus cereus* (KC602258), *Bacillus* sp. (KC602265), *Chitinophaga* sp. (KC602266), *Chitinophaga bacter* (KC602269), *Rhizobium* sp. (KC602276), *Microbacterium* sp. (KC602277), *Bacillus* sp. (KC602282), *Kribbella* sp. (KC602294), *Arthrobacter* sp. (KC602298), *Arthrobacter oryzae* (KC602305) and *Arthrobacter nicotinovorans* (KC602306).

From the results of the present investigation, it can be concluded that biotic and abiotic stress in trace metal-polluted soil of the paper mill greatly influenced the enzyme activity, composition and function of the indigenous microorganisms (fungi, bacteria, actinomycetes). The current study clearly showed that the native dominant resistant indigenous fungal, bacterial isolates can be used as a biosensor to assess the trace metal toxicity in the polluted environment. Thus, future research may be proposed for further advances in microbial genetics by studying the mechanism of metal-microbe-plant interactions and their potential use as metal-resistant microbial inoculants in microbial-assisted phytoremediation.

8.2.1 Future Prospective

With the increased demand of paper, the treatment of effluents emerges as most pressing problem in environmental protection. The current study clearly showed that the native dominant resistant indigenous fungal, bacterial isolates can be used as a biosensor to assess the heavy metal toxicity in the polluted environment contaminated with paper mill effluents. A further understanding of metal-microbe-plant interactions will increase our knowledge to design microbial-assisted phytoremediation in the trace metal-contaminated sites.

Acknowledgements The authors are grateful to the Commonwealth Scholarship grant, Canada, for carrying out the present work in the Department of Biological Science, University De Montreal, Montreal, and Genome Quebec Innovation Centre, Canada, for getting the accession numbers of all isolated trace metal-tolerant microbial strains.

References

Aafi NE, Brhada F, Dary M, Maltouf AF, Pajuelo E (2012) Rhizostabilization of metals in soils using *Lupinus luteus* inoculated with the metal resistant rhizobacterium *Serratia* sp. MSMC 541. Int J Phytoremediation 14:261–274. doi:10.1080/15226514.2011.604693

Abdelatey LM, Khalil WKB, Ali TH, Mahrous KF (2011) Heavy metal resistance and gene expression analysis of metal resistance genes in Gram-positive and Gram-negative bacteria present in Egyptian soils. J Appl Sci Env San 6:201–211

Abou-Shanab RAI, Van Berkum P, Angle JS (2007) Heavy metal resistance and genotypic analysis of metal resistance genes in gram-positive and gram-negative bacteria present in Ni-rich serpentine soil and in the rhizosphere of *Alyssum murale*. Chemosphere 68:360–367. doi:10.1016/j.chemosphere.2006.12.051

Adel Al-Gheethi AS, Noril I, Lalung J, Azan AM, Nurfarehah ZA, Kadir AB (2014) Biosorption of heavy metals and cephalexin from secondary effluents by tolerant bacteria. Clean Technol Envir 16(1):137–148

Adesoji AT, Ogunjobi AA, Olatoye IO, Douglas DR (2015) Prevalence of tetracycline resistance genes among multi-drug resistant bacteria from selected water distribution systems in southwestern Nigeria. Ann Clin Microbiol Antimicrob 14:35. doi:10.1186/s12941-015-0093-1

Ahmed I, Hayat S, Ahmed A, Samiullah IA (2001) Metal and antibiotic resistance traits in Bradyrhizobium sp (cajanus) isolated from soil receiving oil refinery wastewater. World J Microbiol Biotechnol 17:379–384

Akhtar S, Mahmood-ul-Hassan M, Ahmad R, Suthor V, Yasin M (2013) Metal tolerance potential of filamentous fungi isolated from soils irrigated with untreated municipal effluent. Soil Environ 32:55–62

Akponah E (2013) Bacterial and fungal biosorbents of heavy metals associated with hydrocarbon contaminated rainforest soils of the Niger delta region of Nigeria. Eur J Exp Biol 3(3):95–100

Ali H, Khan E, Sajad MA (2013) Phytoremediation of metals-concepts and applications. Chemosphere 91:869–881. doi:10.1016/j.chemosphere.2013.01.075

Altschul SF, Gish W, Miller W, Meyer EW, Lipman DJ (1990) Basic local alignment search tool. J Mol Biol 215:403–410. doi:10.1016/S0022-2836(05)80360-2

Anahid S, Yaghmaei S, Ghobadinejad Z (2011) Heavy metal tolerance of fungi. Scientia Iranica 18(3):502–508

Anderson DI, Hughes D (2014) Microbiological effects of sublethal levels of antibiotics. Nature Rev Microbiol 12:465–478. doi:10.1038/nrmicro3270

Andrew B, Brown M, Steven DS, Peter HT (2013) Microbial community responses to anthropogenically induced environmental change: towards a systems approach. Ecol Lett 16(Supplement S1):128–139. doi:10.1111/ele.12109

Anyanwu CU, Nwankwo SC, Moneke AN (2011) Soil bacterial response to introduced metal stress. Int J Basic Appl Sci 11(1):109–115

Appenroth, K-J (2010) Defination of heavy metals and their role in biological systems. In: Sheramati I, Varma A (eds) Soil heavy metals. Soil biology, vol 19. Springer, Berlin, pp 19–29

Bafana A, Krishnamurthi K, Patil M, Chakrabarti T (2010) Heavy metal resistance in *Arthrobacter ramosus* strain G2 isolated from mercuric salt-contaminated soil. J Hazard Mater 177(1–3):481–486. doi:10.1016/j.saa.2009.12.058

Bargagli R (2000) Trace metals in Antarctica related to climate change and increasing human impact. Rev Environ Contam Toxicol 166:129–173

Bauer AW, Kirby WMM, Sherris JC (1996) Antibiotic susceptibility testing by a standardized single disc method. Am J Clin Pathol 45(4):493–496

Bello OA, Abdullahi OA (2016) Tolerance to heavy metals by some fungal isolates from petroleum refinery effluent in Kaduna, Nigeria. Br Microbiol Res J 12(6):1–8. doi:10.9734/BMRJ/2016/22728

Borjesson G, Menichetti L, Kirchmann H, Kätterer T (2012) Soil microbial community structure affected by 53 years of nitrogen fertilization and different organic amendments. Biol Fert Soils 48(3):245–257

Braud A, Jézéquel K, Bazot S, Lebeau T (2010) Enhanced phytoextraction of an agricultural Cr, hg^- and Pb^-contaminated soil by bioaugmentation with siderophore producing bacteria. Chemosphere 74:280–286. doi:10.1016/j.chemosphere.2008.09.013

Brown SD, Palumbo AV, Panikov N, Ariyawansa T, Klingeman DM, Johnson KM, Land ML, Utturkar SM, Epstein SL (2012) Draft genome sequence for *Microbacterium laevaniformans* strain OR221, a bacterium tolerant to metals, nitrate, and low pH. J Bacteriol 194(12):3279–3280. doi:10.1128/JB.00474-12

Burford EP, Fomina M, Gadd GM (2003) Fungal involvement in bioweathering and biotransformation of rocks and minerals. Mineral Mag 67:1127–1155. doi:10.1180/0026461036760154

Carlsohn MR, Groth I, Spror C, Schutze B, Saluz HP, Munder T, Stackebrandt E (2007) *Kribbella aluminosa* sp. nov., isolated from a medieval alum slate mine. Int J Syst Evol Microbiol 57:1943–1947

Carson JK, Gonzalez-Quinones V, Murphy DV, Hinz C, Shaw JA, Gleeson DB (2010) Low pore-connectivity increases bacterial diversity in soil. Appl Environ Microbiol 76:3936–3942. https://doi.org/10.1128/AEM.03085-09

Cervantes C, Gutierrez-Corona F (1994) Copper resistance mechanisms in bacteria and fungi. FEMS Microbiol Rev 14:121–137. doi:10.1111/j.1574- 6976.1994.tb00083.x

Cetin SC, Karaca A, Kizilkaya R, Turgay OC (2012) Role of plant growth promoting bacteria and fungi in heavy metal detoxification. In: Sheramati VA (ed) Detoxification of heavy metals. *Soil Biology*, vol 30. Springer-Verlag, Berlin, pp 369–388

Chen J, Chen HM, Zhang YQ, Wei YZ, Li QP, Liu HY, Jing S, Zhang YQ, Yu LY (2011) *Agromyces flavus* sp. nov., an actinomycete isolated from soil. Int J Syst Evol Microbiol 61:1705–1709. doi:10.1099/ijs.0.023242-0

Chen J, He F, Zhang X, Sun X, Zheng J, Zheng J (2014) Heavy metal pollution decreases microbial abundance, diversity and activity within particle-size fractions of a paddy soil. FEMS Microbiol Ecol 87:164–181. https://doi.org/10.1111/1574-6941.12212

Clara U, Filomena DL, Schumann P (2008) *Kribbella catacumbae* sp. nov. and *Kribbella sancticallisti* sp. nov., isolated from whitish-grey patinas in the catacombs of St Callistus in Rome, Italy. Int J Syst Evol Microbiol 58:2090–2097. doi:10.1099/ijs.0.65613-0

de Lima AF, de Moura GF, de Lima MA, de Souza PM, da sliva CA, De Campos GM, Do Nascimento AE (2011) Role of the morphology and polyphosphate in *Trichoderma harzianum* related to cadmium removal. Molecules 16(3):2486–2500. doi:10.3390/molecules16032486

de Lima MAB, de Franco, Luciana O, de Souza, Patrícia M, do Nascimento, Aline E, da Silva, Carlos AA, de Maia, Rita CC, Rolim, Hercília ML, Galba, MC (2013). Cadmium tolerance and removal from *Cunninghamella elegans* related to the polyphosphate metabolism. Int J Mol Sci 14(4):7180–7192. doi:https://doi.org/10.3390/ijms14047180

Do Carmo JR, Pimenta CJ, da Silva JF, de Souza SMC (2013) Recovery of copper (II) absorbed in biomass of *Cladosporium cladosporioides*. SciAgric 70:3. org:10.1590/S0103-90162013000300002

Dursun AY (2008) A comparative study on determination of the equilibrium, kinetic and thermodynamic parameters of biosorption of copper (II) and lead (II) ions onto pretreated *Aspergillus niger. Biochem.* Eng J 28:187–195. doi:org/10.1016/j.bej.2005.11.003

Elanvogvan R, Philip L, Chandraraj K (2010) Hexavalent chromium reduction by free and immobilized cell-free extract of *Arthrobacter rhombi*-RE. Appl Biochem Biotechnol 160(1):81–97. doi:org/10.1007/s12010-008-8515-6

El-Morsy EM (2004) *Cunninghamella echinulata* a new biosorbent of metal ions from polluted water in Egypt. Mycologia 96:1183–1189

Elsilk SE, Abd El-Raheem R, El-Shanshoury APS (2014) Accumulation of some heavy metal resistant avirulent *Bacillus anthracis* PS2010 isolated from Egypt. Afr J Micro Res 8(12):1266–1276. doi:10.5897/AJMR2013.6551

Essam F, Jumaily BH, Saleh BH, Shalal M, Hussain SM (2013) Determination of lethal dose (LD50)of exotoxin a from Pseudomonasaeruginosa26A in mice histopathology. IOSR J Pharm Biol Sci 7(6):49–54

Ezzourhi L, Castro E, Moya M (2009) Heavy metal tolerance of filamentous fungi isolated from polluted sites in tangier, Morocco. Afr J Microbiol Res 3:35–48. doi.org/10.7745/KJSSF.2012.45.4.565

Freitas AC, Ferreira F, Costa AM, Pereira R, Antunes SC, Gonçalves F, Rocha-Santos TAP, Diniz MS, Castro L, Peres I, Duarte AC (2009) Biological treatment of the effluent from a bleached kraft pulp mill using basidiomycete and zygomycete fungi. Sci Total Environ 407:3282–3289. doi:org/10.1016/j.scitotenv.2009.01.054

Gao Y, Miao C, Xia J, Mao L, Wang Y, Zhou P (2012) Plant diversity reduces the effect of multiple heavy metal pollution on soil enzyme activities and microbial community structure. Front Environ Sci En 6(2):213–223. doi:10.1007/s11783-011-0345-z

Gao S, Zhang WB, Sheng XF, He LY, Huang Z (2015) *Chitinophaga longshanensis* sp. nov., a mineral-weathering bacterium isolated from weathered rock. Int J Syst Evol Microbiol 65:418–423. doi:10.1099/ijs.0.067249-0

Ghorbani Y, Oliazadeh M, Shahvedi A (2007) Use of some isolated fungi in biological leaching of aluminum from low grade bauxite. Afr J Biotech 11:1284–1288. doi:10.5897/AJB2007.000-2178

Gilman JC (1957) A manual of soil fungi. Iowa State Univ Press, Ames, p 450

Grison CM, Jackson S, Merlot S, Dobson A, Grison C (2015) *Rhizobium metallidurans* sp. nov., a symbiotic heavy metal resistant bacterium isolated from the *Anthyllis vulneraria* Zn-hyperaccumulator. Int J Syst Evol Microbiol 65:1525–1530. doi:10.1099/ijs.0.000130

Gupta K, Chatterjee C, Gupta B (2012) Isolation and characterization of heavy metal tolerant Gram-positive bacteria with bioremedial properties from municipal waste rich soil of Kestopur canal (Kolkata)est Bengal, India. Biologia 67(5):827–836. doi:10.2478/s11756-012-0099-5

Gupta R, Kumar T, Mittal A (2016) Isolation, identification and characterization of heavy metal resistant bacteria from soil of an iron industry, Haryana (India). IJPSR 7(3):1308–1313. doi:10.13040/IJPSR.0975-8232.7(3).1308-13

Hao X, Taghavi S, Xie P, Orbach J, Alwathnani HA, Rensing C, Wei G (2014) Phytoremediation of heavy and transition metals aided by legume-rhizobia symbiosis. Int J Phytoremed 16(2):179–202. https://doi.org/10.1080/15226514.2013.773273

Harriso JJ, Ceri H, Raymond JT (2007) Multimetal resistance and tolerance in microbial biofilms. Nat Rev Microbiol 5:928–938. doi:10.1038/nrmicro1774

Hayat S, Ahmed I, Ahmed A, Samiullah IA (2002) Effect of long-term application of oil refinery wastewater on soil health with special reference to microbiological characteristics. Bioresour Technol 84:159–163. doi:10.1016/S0960-8524(02)00027-5

Hemida M, Morsy FM, El-Enany AE, Ohyama T (2012) Isolation and characterization of a heavy-metal-resistant isolate of *Rhizobium leguminosarum viciae* potentially applicable for biosorption of Cd^{2+} and Co^{2+}. IntBiodeterBiodegr 67:48–55. doi:10.1016/j.ibiod.2011.10.008

Hijri M, Bell TH, Hassan S, Lauron-Moreau A, Al-Otaibi F (2014) Linkage between bacterial and fungal rhizosphere communities in hydrocarbon-contaminated soils is related to plant phylogeny. ISME J 8:331–334. doi:10.1038/ismej.2013.149

Hirak RD, Das S (2014). Bioremediation potential of mercury by *Bacillus* species isolated from marine environment and wastes of steel industry. Biorem J.18(3):204–212. doi.org/10.1080/10889868.2014.899555

Inga Z (2013) Study of the interaction of metals (Hg, Cr, Zn, Ag, Au) with *Arthrobacter* genera and *Spirulina platensis*. PhD thesis

Iram S, Parveen K, Usman J, Nasir K, Akhtar N, Arouj S, Ahmed I (2012) Heavy metal tolerance of filamentous fungal strains isolated from soil irrigated with industrial wastewater. Biologija 58(3):107–116. doi:org/10.6001/biologija.v58i3.2527

Issazadeh K, Jahanpour N, Pourghorbanali F, Raeisi G, Faekhondeh J (2013) Heavy metals resistance by bacterial strains. Ann Biol Res 4(2):60–63

Jaeckel P, Krauss GJ, Krauss G (2005) Cadmium and zinc response of the fungi *Heliscus lugdunensis* and *Verticillium cf. alboatrum* isolated from highly polluted water. *Sci.* Total Environ 346:274–279. doi:10.1016/j.scitotenv.2004.12.082

Juliana RDC, Pimenta CJ, Silva JF, Souza SMC (2013) Recovery of copper (II) absorbed in biomass of Cladosporium cladosporioides. Sci Agric 70(3):147–151. doi:org/10.1590/S0103-90162013000300002

Kanmami P, Aravind J, Preston D (2012) Remediation of chromium contaminants using bacteria. Int J Environ Sci Technol 9(1):183–193. doi:10.1007/s13762-011-0013-7

Karelová E, Harichová J, Stojnev T, Pangallo D, Ferianc P (2011) The isolation of heavy-metal resistant culturable bacteria and resistance determinants from a heavy-metal contaminated site. Biologia 66:18–26. doi:10.2478/s11756-010-0145-0

Karn SK, Reddy S (2012) Degradation of 2, 4,6-trichlorophenol by bacteria isolated from secondary sludge of a pulp and paper mill. J Gen Appl Microbiol 58:413–420. doi:org/10.2323/jgam.58.413

Khan AG (2000) Role of soil microbes in the rhizospheres of plants growing on trace metal contaminated soils in phytoremediation. JTraceElemMedBiol 18:355–364. doi:10.1007/s11356-015-4471-1

Kumar R, Singh P, Dhir B, Sharma AK, Mehta D (2014) Potential of some fungal and bacterial species in bioremediation of heavy metals. Journal of Nuclear Physics, Material Sciences, Radiation and Application 1(2):212–223

Kuykendall LD, Young J, Martinez-Romero E, Kerr A, Sawada H (2005) Genus I. *Rhizobium*. In: Garrity GM, Brenner DJ, Krieg NR, Staley J (eds) Bergey's manual of systematic bacteriology, vol 2. Springer: Verlag, New York, pp 325–340

Lane DJ (1991) 16S/23S rRNA sequencing. In: Stackebrandt E, Goodfellow M (eds) Nucleic acid techniques in bacterial systematics (modern microbiological methods). Wiley, New York, pp 115–175

Lee DW, Lee JE, Lee SD (2009) *Chitinophaga rupis* sp. nov., isolated from soil. Int J Syst Evol Microbiol 59:2830–2833. doi:10.1099/ijs.0.011163-0

Lee S, Cho K, Lim J, Kim W, Hwang S (2011a) Acclimation and activity of ammonia-oxidizing bacteria with respect to variations in zinc concentration, temperature, and microbial population. Bioresour Technol 102:4196–4203. doi:10.1016/j.biortech.2010.12.035

Lee M, Ten LN, Woo SG, Park J (2011b) *Agromyces soli sp. nov*., isolated from farm soil. *Int.* J Syst Evol Microbiol 61:1286–1292. doi:10.1155/2013/831308

Long F, Su CC, Lei HT, Bolla JR, Do SV, Yu EW (2012) Structure and mechanism of the tripartite CusCBA heavy-metal efflux complex. Philos Trans R Soc Lond Ser B Biol Sci 367:1047–1058. doi:10.1098/rstb.2011.0203

Ma Y, Oliveira RS, Freitas H, Zhang C (2016) Biochemical and molecular mechanisms of plant-microbe-metal interactions: relevance for phytoremediation. FrontPlant Sci 7:918. doi:org/10.3389/fpls.2016.00918

Madhaiyan M, Poonguzhali S, Sa T (2007) Metal tolerating methylotrophic bacteria reduces nickel and cadmium toxicity and promotes plant growth of tomato (*Lycopersicon esculentum* L.) Chemosphere 69:220–228. doi:10.1016/j.chemosphere.2007.04.017

Malgorzata JK, Rosikon K., Fijalkowski K, Grobelak A (2014).The effect of *Trichoderma* on heavy metal mobilityand uptake by *Miscanthus giganteus*, *Salix* sp., *Phalaris arundinacea*and *Panicumvirgatum.*Appl Environ Soil Sci.1–10. doi.org/10.1155/2014/506142

Malik A (2004) Metal bioremediation through growing cells. Environ Int 30:261–278. doi:10.1016/j.envint.2003.08.001

Mandal SM, Bhattacharyya R (2012) *Rhizobium*-legume Symbiosis: a model system for the recovery of metal-contaminated agricultural land. In: Zaidi A, Wani PA, Khan MS (eds) Toxicity of heavy metals to legumes and bioremediationol.12.Springer, New York, pp 115–127

Mergeay M, Monchy S, Vallaeys T, Auquier V, Benotmane A, Bertin P, Taghavi S, Dunn J, van der Lelie D, Wattiez R (2003) *Ralstonia metallidurans*, a bacterium specifically adapted to toxic metals: towards a catalogue of metal-responsive genes. FEMS Microbiol 9:1181–1191. doi:10.1016/S0168-6445(03)00045-7

Mindlin S, Petrenko A, Kurakov A, Beletsky A, Mardanov A, Petrova M (2016) Resistance of permafrost and modern *Acinetobacter lwoffii* strains to heavy metals and arsenic revealed by genome analysis. Biomed Res Int. doi:org/10.1155/2016/3970831

Mohamed HA, Morsy FM, Enamy AWE, Ohyama (2012) Isolation and characterization of a heavy-metal-resistant isolate of *Rhizobium leguminosarum* bv. *viciae* potentially applicable for biosorption of Cd^{2+} and Co^{2+}. Int Biodeter Biodegr 67:48–55. doi:10.1016/j.ibiod.2011.10.008

Mondani L, Piette L, Christen R, Bachar D, Berthomieu C, Chapon V (2012) *Microbacterium lemovicicum* sp. nov., a bacterium isolated from a natural uranium-rich soil. IntJSystEvolMicrobiol 63:2600–2606. doi:10.1007/s00203-015-1183-3

Nies DH (1995) The cobalt, zinc, and cadmium efflux system CzcABC from *Alcaligenes eutrophus* functions as a cation-proton antiporter in *Escherichia coli*. J Bacteriol 177:2707–2712

Nies DH (1999) Microbial heavy metal resistance. ApplMicrobial Biotechnol 51:730–750. doi:10.1007/s002530051457

Olukoya DK, Smith SI, Hori MO (1997) Isolation and characterization of heavy metal resistant bacteria from Lagos lagoon folia. Microbiol 42(5):441–444. doi:10.1007/BF02826550

Osborn AM, Bruce KD, Strike P, Ritchie DA (1997) Distribution, diversity and evolution of the bacterial mercury resistance (mer) operon. FEMS Microbiol Rev 19:239–262. doi:http://dx.doi.org/10.1111/j.1574-6976

Pal A, Choudhuri P, Dutta S, Mukherjee PK, Paul AK (2004) Isolation and characterization of nickel-resistant microflora from serpentine soils of Andaman. World J Microbiol Biotech 20:881–886. doi:10.1007/s11274-004-2776-1

Palanivel VP, Shim J, You Y, Choi S, Seralathan KK, Lee KJ, Kim HJ, Oh BT(2010). Removal of zinc by live, dead, and dried biomass of *Fusarium* spp. isolated from the abandoned-metal mine in South Korea and its perspective of producing nanocrystals. J Haz Mat. 182(1–3):317–324.doi:org/10.1016/j.jhazmat.2010.06.032

Paris D, Blondeau R (1999) Isolation and characterization of *Arthrobacter* sp. from activated sludge of a pulp and paper mill. Water Res 33(4):947–950. doi:org/10.1016/S0043-1354(98)00298-X

Park SK, Kim MS, Jung MJ, Nam YD, Park EJ, Roh SW, Bae JW (2011) Brachybacterium squillarum sp. nov., isolated from salt-fermented seafood. Int J Syst Evol Microbiol 61:1118–1122. https://doi.org/10.1099/ijs.0.022517-0

Pattus F, Abdallah M (2000) Siderophores and iron-transport in microorganisms: review. J Chin Chem Soc 47:1–20. doi:10.1002/jccs.200000001

Petrovic JJ, Danilovic G, Curcic N, Milinkovic M, Stosic N, Pankovic D, Raicevic V (2014) Coppertolerance of Trichoderma species. Arch Biol Sci Belgrade 66(1):137–142. https://doi.org/10.2298/ABS1401137J

Piccirillo C, Pereira SIA, Marques APGC, Pullar RC, Tobaldi DM, Pintado ME, Castro PML (2013) Bacteria immobilization on hydroxyapatite surface for heavy metals removal. J Environ Manag 121:87–95. doi:org/10.1016/j.jenvman.2013.02.036

Piotrowska-Seget Z, Cycon, M, Kozdroj J (2005) Metal tolerant bacteria occurring in heavily polluted soil and mine spoil. Appl Soil Ecol. 28:237–246. doi:org/10.1016/j.apsoil.2004.08.001

Rajbanshi A (2008) Study on heavy metal resistant bacteria in Guheswori sewage treatment plant. Our Nature 6:52–57. doi:10.3126/on.v6i1.1655

Rajkumar M, Ae N, Prasad MNV, Freitas H (2010) Potentialof siderophore-producing bacteria for improving heavy metal phytoextraction. Trends Biotechnol. 28:142–149. doi:org/10.1016/j.tibtech.2009.12.002

Rajkumar M, Sandhya S, Prasad MNV, Freitas H (2012) Perspectives of plant-associated microbes in heavy metal phytoremediation. Biotechnol Adv. 30(6):1562–1574.doi:org/10.1016/j.biotechadv.2012.04.011

Rao KR, Rashmi K, Latha JNL (2005) Bioremediation of toxic metal ions using biomass of *Aspergillus fumigates* from fermentative waste. Indian J Biotechnol 4:139–143. http://hdl.handle.net/123456789/5627

Raper KB, Fennell DJ (1965) The genus *Aspergillus*. Williams and Wilkins, Baltimore, p 686

Rathnayake VN, Megharaj M, Bolan N, Naidu R (2009) Tolerance of heavy metals by gram positive soil bacteria. WASET 53:1185–1189

Reeve WG, Tiwari RP, Kale NB, Dilworth MJ, Glenn AR (2002) ActP controls copper homeostasis in *Rhizobium leguminosarum* bv. *viciae* and *Sinorhizobium meliloti* preventing low pH-induced copper toxicity. Mol Biol 43(4):981–991. doi:10.1046/j.1365-2958.2002.02791.x

Rhodes CJ (2013) Applications of bioremediation and phytoremediation. Sci Prog. 96(4): 417–427.doi:org/10.3184/003685013X13818570960538

Rosales E, Pazos M, Sanroman MA, Tavares T (2012) Application of zeolite-*Arthrobacter viscosus* system for the removal of heavy metal and dye: chromium and azure B. Desalination. 284:150–156. doi:org/10.1016/j.desal.2011.08.049

Rouhollahi F, Zamani A, Karimi K, Etesami N (2014) Enhancement of nickel biosorption on fungal biomass by enzymatic and alkali pretreatments. Int J Environ Sci Technol. doi:10.1007/s13762-014-0624-x

Sahoo KN, Pakshirajan K, Ghosh PK (2014) Evaluation of 4-bromophenol biodegradation in mixed pollutants system by *Arthrobacter chlorophenolicus* A6 in an upflow packed bed reactor. Biodegradation 25(5):705–718. doi:10.1007/s10532-014-9693-2

Saitou N, Nei M (1987) The neighbour-joining method: a new method for reconstruction phylogenetic trees. Mol Biol Evol 4:406–425

Santa CL, Gonzalez-Lopez J, Manzanera M (2013) *Arthrobacter siccitolerans* sp. nov., a highly desiccation-tolerant, xeroprotectant-producing strain isolated from dry soil. Int J Syst Evol Microbiol 63:4174–4180. doi:10.1099/ijs.0.052902-0

Schalk IJ, Hannauer M, Braud A (2011) New roles for bacterial siderophores in metal transport and tolerance. Environ Microbiol 13:2844–2854. doi:10.1111/j.1462-2920.2011.02556.x

Sen M (2011) Biosorption of Cr (VI) by resting cells of *Fusarium solani* Iran. J Environ Health SciEng 8(2):117–120. doi:10.4236/ojapps.2012.23021
Shivkumar CK, Thippeswamy B, Krishnappa M, Ananthamurthy KS (2011) Bioaccumulation potency of ecotoxic heavy metals by indigenous Rhizopus and Penicillium sp: a comparative study. J Pure. Appl Microbiol 6(1):231–239
Sessitsch A, Kuffner M, Kidd P, Vangronsveld J, Wenzel WW, Fallmann K, Puschenreiter M (2013) The role of plant-associated bacteria in the mobilization and phytoextraction of trace elements in contaminated soils. Soil Biol Biochem 60(100):182–194. doi:10.3389/fpls.2016.00918
Smith KS, Balistrieri LS, Todd AS (2015) Using biotic ligand models to predict metal toxicity in mineralized systems. Appl Geochem 57:55–72. doi:10/1016/j.apgeochem.2014.07.005
Song N, Ma Y, Zhao Y, Tang S (2015) Elevated ambient carbon dioxide and *Trichoderma* inoculum could enhance cadmium uptake of *Lolium perenne* explained by changes of soil pH, cadmium availability and microbial biomass. Appl Soil Ecol 85:56–64. doi:10.1016/j.apsoil.2014.09.007
Stan V, Gament E, Cornea CP, Voaideş C, Duşa M, Plopeanu G (2011) Effects of heavy metal from polluted soils on the *Rhizobium* diversity. Not Bot Hort Agrobot Cluj 39(1):88–95. doi:http://dx.doi.org/10.15835/nbha3916081
Stefanowicz AM, Niklińska M, Kapusta P, SzarekŁukaszewska G, Grodzińska K (2010).Metal pollution, soil properties and plant diversity as determinants of bacterial and fungal performance in soils. In: 15th ICHMET, pp 708–712
Sun F, Shao Z (2007) Biosorption and bioaccumulation of lead by *Penicillium sp.* Psf-2 isolated from the deep sea sediment of the Pacific Ocean. Extremophiles 11:853–858. doi:10.1007/s00792-007-0097-7
Swer PB, Joshi SR, Acharya C (2016) Cesium and strontium tolerant *Arthrobacter* sp. strain KMSZP6 isolated from a pristine uranium ore deposit. AMB Expr 6:69. doi:10.1186/s13568-016-0247-3
Tamura K, Peterson D, Peterson N, Stecher G, Nei M, Kumar S (2011) MEGA5: molecular evolutionary genetics analysis using maximum likelihood, evolutionary distance, and maximum parsimony methods. Mol Biol Evol 28:2731–2739. doi:10.1093/molbev/msr121.pmid: 21546353
Tappe W, Herbst M, Hofmann D, Koeppchen S, Kummer S, Thiele B (2013) Degradation of sulfadiazine by *Microbacterium lacus* strain SDZm4, isolated from lysimeters previously manured with slurry from sulfadiazine-medicated pigs. Appl Environ Microb 79(8):2572–2577. doi:10.1128/AEM.03636-12 pmid:23396336
Teng Y, Luo Y, Ma W, Zhu L, Ren W, Luo Y. Li, Z (2015). *Trichoderma reesei* FS10-C enhances phytoremediation of cd-contaminated soil by *Sedum plumbizincicola* and associated soil microbial activities. Front Plant Sci. 9: 220. doi.org/10.3389/fpls.2015.00438
Thakre NA, Shanware AS (2015) Promising biological indicator of heavy metal pollution: bioluminescent bacterial strains isolated and characterized from marine niches of Goa, India. Indian J Microbiol 55(3):327–332. doi:10.1007/s12088-015-0531-y
Thawai C, Tanasupawat S, Suwanborirux K, Kudo T (2011) *Agromyces tropicus* sp. nov., isolated from soil. Int J Syst Evol Microbiol 61:605–609. doi:10.1099/ijs.0.021774-0
Thippeswamy B, Shivakumar CK, Krishnappa M (2012) Accumulation potency of heavy metals by *Saccharomyces* sp. indigenous to paper mill effluent. J Env Res Develop 6(3):439–445
Thippeswamy B, Shivakumar CK, Krishnappa M (2014) Optimization of heavy metals bioaccumulation in *Aspergillus niger* and *Aspergillus flavus*. Int J Environ Biol 4(2):188–195
Tripathi AK, Harsh NSK, Gupta N (2007) Fungal treatment of industrial effluents: a mini-review. Life Science Journal 4(2):78–81
Tunali S, Cabuk A, Akar T (2006) Removal of lead and copper ions from aqueous solutions by bacterial strain isolated from soil. ChemEngJ 115(3):203–211. doi:10.1016/j.cej.2005.09.023
Turner S, Pryer KM, Miao VPW (1999) Investigating deep phylogenetic relationships among cyanobacteria and plastids by small subunit rRNA sequence analysis. J Eukaryot Microbiol 46:327–338. doi:10.1111/j.1550-7408.1999.tb04612.x
Ullah A, Heng S, Munis MFH, Fahad S, Yang X (2015) Phytoremediation of heavy metals assisted by plant growth promoting (PGP) bacteria: a review. Environ Exp Bot 117:28–40. doi:10.1016/j.envexpbot.2015.05.001

Valix M, Tang JY, Malik R (2001) Heavy metal tolerance of fungi. Miner Eng 14:499–505. doi:10.1016/S0892-6875(01)00037 1

Van DB, Christov L (2002) Adsorption of colour from bleach plant effluent using biomass and cell wall fractions from *Rhizomucor pusillus*. J Chem Technol Biotechnol 77:155–158. doi:10.1002/jctb.535

Verma P, Singh S, Verma RK (2016) Heavy metal biosorption by *Fusarium* strains isolated from iron ore mines overburden soil. IntJEnvironSci Toxicol Res 4(4):61–69

Volesky B, Holan ZR (1995) Biosorption of heavy metals. Biotechnol Prog 11:235–250. doi:10.1021/bp00033a001

Wang YP, Shi JY, Wang H, Lin Q, Chen XC, Chen YX (2007) The influence of soil heavy metals pollution on soil microbial biomass, enzyme activity and community composition near a coppersmelter. Ecotoxicol Environ Saf 67:75–81. doi:10.1016/j.ecoenv.2006.03.007

Wang W, Shao Z, Liu Y, Wang G (2009) Removal of multi-heavy metals using biogenic manganese oxides generated by a deep-sea sedimentary bacterium - *Brachybacterium* sp. strain Mn32. Microbiology 155:1989–1996. doi:10.1099/mic.0.024141-0

Wang Q, Cheng C, He LY, Huang Z (2014) *Chitinophaga jiangningensis* sp. nov., a mineral-weathering bacterium. Int J Syst Evol Microbiol 64:260–265. doi:10.1099/ijs.0.067249-0

White TJ, Bruns T, LeeS TJ (1990) Amplification and direct sequencing of fungal ribosomal RNA genes for Phylogenetics. In: Innis MA, Gelfand DH, Sninsky JJ, White TJ (eds) PCR protocols: a guide to methods and applications. Academic, New York, pp 315–322

Willie JG, Peijnenburg M, Zablotskaja M, Martina GV (2007) Monitoring metals in terrestrial environments within a bioavailability framework and a focus on soil extraction. Ecotox Environmen Safe 67(2):163–179. doi:10.1016/j.ecoenv.2007.02.008

Wu SC, Cheung KC, Luo YM, Wong MH (2006) Effects of inoculation of plant growth-promoting rhizobacteria on metal uptake by *Brassica juncea*. Environ Pollut 140:124–135. doi:10.1016/j.envpol.2005.06.023

Wuana RA, Okieimen FE (2011) Heavy metals in contaminated soils: a review of sources, chemistry, risks and best available strategies for remediation. ISRN Ecology. doi: org/10.5402/2011/402647

Yang HB, Tan N, Wu FJ, Liu HJ, Sun M, She ZG, Lin YC (2012) Biosorption of uranium (VI) by mangrove endophytic fungus *Fusarium* sp. ZZF51 from the South China Sea. J Radioanal Nucl Chem 292(3):1011–1016. doi:10.1007/s10967-011-1552-6

Yasir M, Chung EJ, Song GC, Bibi F, Jeon CK, Chung YR (2011) *Chitinophaga eiseniae* sp. nov., isolated from vermicompost. Int J Syst Evol Microbiol 61:2373–2378. doi:10.1099/ijs.0.040014-0

Yazdani MF, Abdullah CKY, Tan SG (2009) Trichoderma atroviride as a bioremediatorof cu pollution: an in vitro study. Toxicol Environ Chem 91(7):1305–1314. doi:10.1080/02772240802616510

Yu X, Li Y, Zhang C, Liu H, Liu J, Zheng W, Kang X, Qiang C (2014) Culturable heavy metal-resistant and plant growth promoting bacteria in V-Ti magnetite mine tailing soil from Panzhihua, China. PLoS One 9(9):e106618. doi:10.1371/journal.pone.0106618

Zafar S, Aqil F, Ahmad I (2007) Metal tolerance and biosorption potential of filamentous fungi isolated from metal contaminated agriculture soil. Bioresour Technol 98:2557–2561. doi:10.1016/j.biortech.2006.09.051

Zhang YF, He LY, Chen ZJ, Zhang WH, Wang QY, Qian M (2011) Characterization of lead-resistant and ACC deaminase-producing endophytic bacteria and their potential in promoting lead accumulation of rape. J Hazard Mater 186:720–725. doi:10.1016/j.jhazmat.2010.12.069

Zhang X, Lin L, Chen M, Zhu Z, Yang W, Chen B, Yang X, An Q (2012) A nonpathogenic Fusarium oxysporum strain enhances phytoextraction of heavy metals by the hyperaccumulator sedum alfredii Hance. J Hazard Mater 229-230:361–370. doi:10.1016/j.jhazmat.2012.06.013

Zhou Q, Chen Y, Yang M, Li W, Deng L (2013) Enhanced bioremediation of heavy metal from effluent by sulfate-reducing bacteria with copper-iron bimetallic particles support. Bioresour Technol 136:413–417. doi:10.1016/j.biortech.2013.03.047

Dhritiman Chanda completed his M.Sc. and M. Phil. in (Botany, Specialization: Microbiology) from Gauhati University and Ph.D. (Microbiology) from Assam university, Silchar, Assam, India. He was awarded Common Wealth Fellow and worked under Canada Genome Project funded by Genorem Quebec, McGill University and University de Montreal at Institut de recherche en biologie végétale (IRBV), from January, 2012 to May, 2013 under the supervision of Prof. Mohamed Hijri and Prof. Marc St-Arnaud, IRBV, University De Montreal, Montreal, Quebec, Canada (www.hijrilab.org). His current research interests are: Environmental Microbiology/toxicology, Microbial ecology, Metagenomics and Phytoremediation.

G.D. Sharma Professor, Vice-Chancellor, Bialspur University, His current research interests are: Microbial diversity, microbial resources, mycorrhiza, microbial interactions, plant pathogens, microbial degradation.

D.K. Jha Professor, Department of Botany, Gauhati University. His current research interests are: plant-microbe interaction, AM fungi, Microbial diversity, Bioremediation.

Mohamed Hijri Professor, Department of Biological Sciences, University De Montreal, Montreal, Canada. His research interestare: Molecular genetics, genomics and evolution of arbuscular mycorrhizal fungi, Environmental genomics and soil microbial biodiversity.

Transformation, Purification, and Quantification of Soy Isoflavone from *Lactobacillus* sp. and *Bifidobacterium* sp.

9

V. Usha Rani and B.V. Pradeep

Abstract

Lactobacillus and *Bifidobacterium* are capable of hydrolyzing glycosidic bond of glycosidic isoflavone in soy extract (milk) during microbial fermentation. Each strain has proved enzymatic ability in converting the glucoside isoflavones to increased bioactive aglycone concentration in fermented soy milk. Various chromatographic techniques (TLC, HPLC, and LC) were used to study the analyte to know its range of applications.

Keywords

Lactobacillus • *Bifidobacterium* • Transformation • Isoflavones • Genistein • Daidzein • Quantification

9.1 Introduction

Concern about soy isoflavones has enlarged and escalating worldly, as the consumption of soy isoflavones (Rostagno et al. 2004) is associated with lowering the incidence of certain types of cancers (Messina and Barnes 1991; Ingram et al. 1997; Watanabe and Koessel 1993; Choi et al. 2000; Strauss et al. 1998) and lessening the menace of innumerable sicknesses including cardiovascular problems (Clarkson et al. 1995) and osteoporosis (Alekel et al. 2000; Knight and Eden 1996). To current soy isoflavones have been increasingly used in traditional hormone replacement therapy (HRT) for women particularly with estrogen deficiency (Ettinger et al. 2003; Lawton et al. 2003). Isoflavones are a group of phytoestrogens structurally

V.U. Rani • B.V. Pradeep (✉)
Department of Microbiology, Karpagam University (Karpagam Academy of Higher Education), Coimbatore 641 021, Tamil Nadu, India
e-mail: usmicro07@gmail.com; bvpradeepku@gmail.com

P. Shukla (ed.), *Recent Advances in Applied Microbiology*,
DOI 10.1007/978-981-10-5275-0_9

similar to 17β-estradiol (E2) and have weak estrogenic activities that are able to duplicate the actions of endogenous estrogens with high-affinity binding to estrogen receptors (ERs) (Amnon and Ayelet 1999).

This synergetic outcome of isoflavone has relieved estrogen-deficient diseases and involved in decreasing the HRT-induced risk factors relating to hormone-induced cancers (Cornwell et al. 2004; Adlercreutz and Mazur 1997). Through binding estrogen receptor sites with genistein, daidzein, and glycitein, the biologically active estrogen-like isoflavone mimics the functions of human estradiol (Setchell and Cassidy 1999). Epidemiological review states that isoflavone consumption reduces risk factors of hormone-associated health issues that are more predominant in present Western culture (Kurzer 2000; Anthony 2000).

King and Bignell in the year 2000 have said that in nonfermented soy foods, about 80%–95% of the isoflavone concentrations are found as inactive biological glucoside conjugates. The metabolic and bioavailability of isoflavones plays an important function in intestinal microflora projection. The β-galactosidase activity of intestinal flora projects key factor in metabolising and hydrolyzing bioactive isoflavone by biotransformation. In the fermentation of soy milk with LAB *Bifidobacteria*, isoflavone glycosides are biotransformed into bioactive aglycone. The enzyme possesses an ability to split the β-glycosidic linkage. The fermentation should be used as a starter in the hydrolysis of soybean isoflavone glycosides by some lactic and bacteria which was described in previous study (Lee et al. 1993). During fermentation volatile compounds such as aldehydes and flatulence caused high amount of oligosaccharides reducing the undesirable beany flavor). Predominant members of the intestinal microflora are *Bifidobacteria* which is gram positive, bacilli, anaerobic in nature. An optimum growth parameter such as temperature at 37 °C and pH 6.8 in anaerobic enevironment are 37 °C and pH 6.8, respectively (Ballongue 1999). When live microbial feed is supplemented, *Bifidobacteria*, a probiotic, supply useful effects to the host by developing intestinal microbial balance. Bifidobacteria are combined with probiotic bacterium *Lactobacillus acidophilus* and are implemented in dairy preparation because of their health benefits. The positive benefits paved the path for microbial-based biotransformation of soybean substrates with *Lactobacillus* and *Bifidobacterium*. Isoflavones from the fermented soy represent an adjunctive therapeutic property as good anticancer agent targeting the cell proliferation, induction of apoptosis, and exertion of inhibitory effects on carcinogenesis (Usha Rani and Pradeep 2014).

9.2 Isoflavones

Isoflavones are intensely seized attention in regular dietary for its immense nutritious load of bioactive molecules. Isoflavonoids or isoflavones are leguminous phytochemical of many plants. In soybeans isoflavones exist in various chemical forms like aglycones, malonyl/acetyl, and glucoside conjugates. Setchell (2000) and Otieno et al. (2006) studied the hydrolysis of isoflavones, by converting the glucoside conjugates of isoflavones into aglycones, by β-glucosides of intestinal bacteria.

Fig. 9.1 Comparative structures of genistein, daidzein, and 17β-estradiol

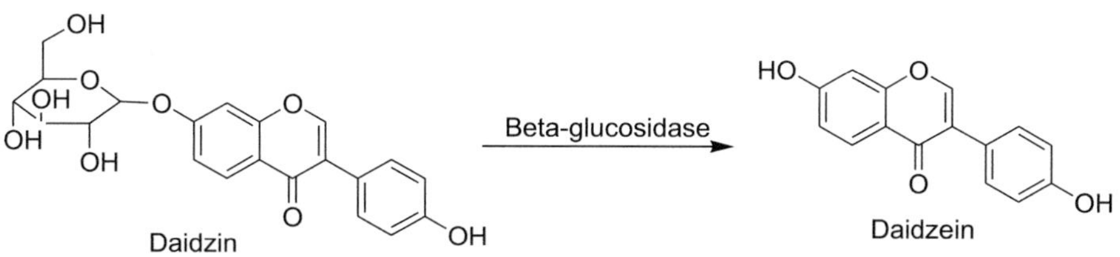

Fig. 9.2 Biotransformation by glucosidase – isoflavone (daidzein) to bioactive aglycone (daidzein)

Enzyme β-glucosidase is mainly responsible for the conversion of isoflavones present in fermented soybean foods. To enrich isoflavone aglycones in fermented soy milk, β-glucosidase induces the conversion of acetyl glycoside and malonyl glucoside to their aglycones (Tsangalis 2002). The level of aglycones is high in fermented products. Based on the processing techniques, isoflavones possess different forms and concentration in soybean products. To produce yogurt-like fermented soybean milk, which contains high concentrations of isoflavone aglycones, *Bifidobacterium breve*, *Bifidobacterium bifidum*, and *L. casei* subsp. *rhamnosus* strains are used. Genistein and daidzein have been documented to exhibit additional benefits having anticancer, antioxidant, anti-osteoporosis, anti-inflammation, anti-cardiovascular, and enzyme inhibitory effects (Wang and Murphy 1994) (Fig. 9.1).

9.3 β-Galactosidase

β-Glucosidases (EC 3.2.1.21) are heterogeneous in nature. They have the capability to split β-glucoside linkages of di- and oligosaccharide or other glucose conjugates. The biological activity of soy milk is improved by effectively converting isoflavone to aglycones (Pandjaitan et al. 2000; Otieno et al. 2006). β-Glucosidase enzyme plays a vital role in transforming isoflavone glucoside into bioactive isoflavone aglycones. By using genetic engineering approaches, we can also enhance the production of β-glucosidase enzyme in probiotics (Yadav et al. 2017b). According to the outcome of earlier reports, the strains of *Lactobacillus* and *Bifidobacterium* spp. contains β-glucosidase (Yadav et al. 2016a, b) (See Fig. 9.2).

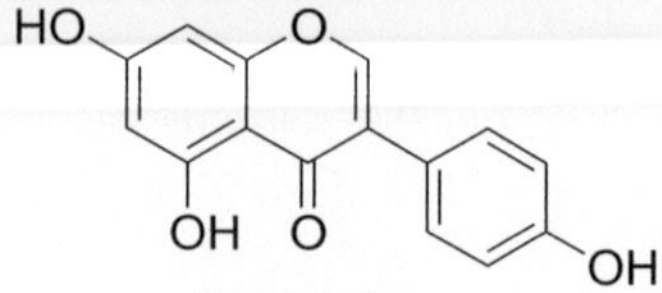

Fig. 9.3 Structure – genistein

9.4 Genistein

Important isoflavone found in a variety of human foods made of soybeans is genistein (4′,5,7-trihydroxyisoflavone). Genistein is a colorless crystal with needle shape having a melting point of 296–298 °C, $C_{15}H_{10}O_5$ as molecular formula, and molecular weight of 270. Its solubility is high in ether and hot ethanol, whereas its solubility is weak in glacial acetic acid, and cold ethanol. When it is mixed with ethanolic iron chloride (III) solution, the solution turns to be dark red and yellow, while genistein is dissolving in alkali. Genistein has been confirmed to promote benefits to humans by reducing the occurrence of specific chronic diseases, namely, cancer and atherosclerosis (Onozawa et al. 1998). Genistein reduces the growth of various cancer cells. Modification in genes that are intimately related in regulating cell cycle leads to programmed cell death. When administered in dietary form, genistein helps boost the antioxidant enzymes present in organs like the liver, small intestine, and kidney. Growth of leukemia, lymphoma, ovarian, cervical, leiomyoma, melanoma, neuroblastoma, gastric, pancreatic, breast, and prostate cancer cells has been suppressed by genistein (See Fig. 9.3).

The gene expression is regulated by genistein when it binds to estrogen receptors present in the cell, which exhibit agonist and antagonist properties of estrogen. Genistein competes with ATP and binds to tyrosine kinase. This interferes with tyrosine kinase cascade activated by mitogens showing its protein tyrosine kinase inhibitor activity. It has also shown to inhibit DNA topoisomerase activity thereby repressing angiogenesis and inducing apoptosis (Dixona and Ferreira 2002).

9.5 Daidzein

Daidzein (4′,7-dihydroxyisoflavone) is the subclass of flavonoids found in soybeans, fruits, nuts, and soy-based products. Daidzein is a colorless crystal with columnar shape having 315–320 °C as its melting point, having $C_{15}H_{10}O_4$ as its molecular formula, and having a molecular weight of 254. Daidzein is insoluble in water and soluble in methanol, ethanol, and acetone. Daidzein can be deducted under the fluorescence by UV radiation; it turns yellow after dissolving in alkali. It decomposes with alkali to form formic acid and p-hydroxybenzoate. The nontoxic compounds of daidzein are capable of causing cell death in various cancer cells. The pro-apoptotic effects of daidzein contribute to cancer chemotherapy and inhibit tumor growth (See Fig. 9.4).

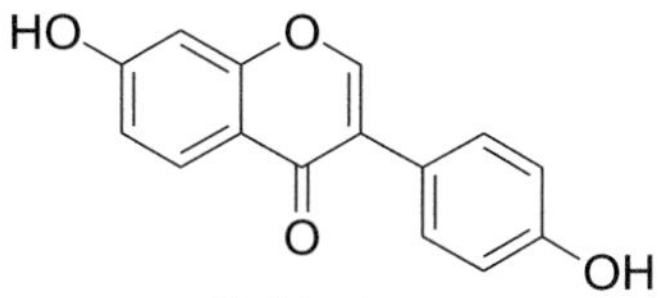

Fig. 9.4 Structure -daidzein

9.6 Strains and Culture Conditions

Since 1900, in the field of industrial microbiology, interest in microorganism from food source has increased because of the potential of new bacterial species and strains (Singh and Sharma 2009). Among various food sources, milk plays a major role as nutritious for humans and animals. In addition to nutrition, the existence of specific components or beneficial bacteria called probiotic is gaining scientific credibility at a rapid pace (Bhardwaj et al. 2012). LAB are members of the microflora present in the intestine. These microflora will inhibit the bacteria that are harmful. They assist in digestion and boost immune functions thereby resistance to various infections is promoted (Yadav et al. 2016a, b). Colonic bacteria like *Lactobacillus* spp., *Bacteroides* spp., and *Bifidobacterium* spp. have the ability to produce β-glucosidase enzyme. Lactobacillus a group of microaerophilic organism showing gram positive with rod shape, non-spore former, and non-pigmented class which is considered as GRAS (Yadav et al. 2017a, b). Reports of *Lactobacilli* show its diverse behavior as potential source of new application for enzyme production. Bile salt hydrolase enzyme produced by *Lactobacillus plantarum* RYPR1 also shows probiotic property (Yadav et al. 2016a, b). Members of *Bifidobacterium* are the most common microorganisms comprising up to 3% in the human gut especially in infant gut. The total microflora in breast-fed babies is 91%, but in formula-fed infants, it shows 75% a drop in microflora. *Bifidobacterium* spp. have various beneficial aspects like improving the immunity by maintaining the intestinal flora count, thereby it assists in diarrheal treatment. They also help in relieving symptoms of lactose intolerance and lowering the cholesterol level in the serum, reducing the risk of colon cancer. The interest of using bifidobacterial strains in food industries as additives is rapidly growing. *Bifidobacteria* produce acetic and lactic acids that lower the pH of the colon and inhibit the rise in of pathogens. There has been increasing evidence supporting that probiotic cultures may have the ability to modulate the composition of the intestinal microbiota and deliver a series of health benefits (Yadav and Shukla 2015). *Bifidobacterium* sp. and *Lactobacillus* sp. are capable of producing a positive effect on human health. These two strains have been widely used as excellent source of probiotics, as they retain enzymes such as β-glucosidase and β-galactosidase (Sieladie et al. 2011; Tochikura et al. 1986).

9.7 Fermentation

Isoflavone glucosides can be hydrolyzed through β-glucosidases enzyme activity in soy milk which also enhances its biological activity. During fermentation, probiotic organisms such as *Bifidobacterium*, *Lactobacillus acidophilus*, and *L. casei* increase the convergence of isoflavone aglycones in soy milk which is bioactive in nature. During biotransformation of isoflavone glycosides, β-glucosidases elevate the concentration of bioactive isoflavone aglycones. The commonly used probiotic in industry is *Lactobacilli* strains. β-Galactosidase, an intracellular enzyme extracted from lactic acid bacteria, is not discharged under conventional fermentation conditions to the outside of the cells (Yadav et al. 2016a, b). These organisms play a role in shuffling the soy constituents during fermentation. Thus, the profile or constituent alteration is due to the potential hydrolyzation by probiotic group during fermentation that causes the major bioactive isoflavone aglycones to inflate. The undesirable beany taste caused by n-hexanal and penta is removed as a result and improves the nutritional characteristics of soy milk (Scalabrinia et al. 1998). Fermentation with *Bifidobacteria* also makes the proteins present more digestible and brings down the quantity of soy oligosaccharides, stachyose, and raffinose that are prone to cause digestive issues. Culture conditions of fermentation have a large effect on the growth activity of *Lactobacilli* (Gilliland 1985). The content and composition of bioactive compounds are modified by physical parameters like pH and temperature (Rickert et al. 2004). The idea of associating soy milk and probiotic bacteria could be a unique multifunctional food.

9.8 Preparation and Fermentation of Soy Milk

The amazing nutritive significance and favorable characteristics of soy milk consumed become a very interesting food. The nutritional advantages of soy milk over dairy milk include condensed cholesterol level and saturated fat as well as the absence of lactose. It consists of proteins, unsaturated fatty acids, and soluble and insoluble dietary fibers which form a significant part of everyday diet. Whole soya beans are soaked overnight in distilled water after washing. The water is then emptied, and soya beans are blended for 3 min with distilled water (1:6 w/v). A double-layered cheesecloth is used to filter the remnants in order to yield soy milk. 50 mL of soy milk is discharged into screw cap containers and sterilized at 121 °C for 15 min.

9.9 Extraction of Soy Isoflavones

Different modes have been investigated for extracting isoflavones from soy. As described by Luthria et al. (2007), four different extraction methodologies using acetonitrile: water (58:42, v/v); ethanol: water (70:30, v/v); methanol: water (90:10, v/v); and superheated pressurized water can be deployed for extraction of

Table 9.1 Abstract: Procedures of extraction for isoflavones

S.No	References	Extraction	Solvent
1	Lin and Giusti (2005)	Six different extraction were executed (RT for 2 h) using solvents like ACN (83%), acidified ACN (83%), MeOH (80%), acidified MeOH (80%), ACN (58%), and acidified ACN (58%)	ACN (58%)
2	Lee et al. (2004)	Shaking the sample for 2 h at RT using acidified ACN (58%)	One solvent
3	Penalvo et al. (2004)	Time intervals (30–240 min); extraction and hydrolysis were compared with three different solvents EtOH (80%), acidified EtOH (80%), and MeOH (80%) at different temperatures (60–100 °C)	EtOH (80%) with HCl (1 M)
4	Murphy et al. (2002)	Four various acidified solvents ACN (53%), MeOH (53%), EtOH (53%), and acetone (53%) blended at RT for 2 h	ACN (53%)
5	Achouri et al. (2005)	ACN (80%), MeOH (80%), and EtOH (80%), by stirring and sonication	Yields of all the three solvents were similar
6	Rostagno et al. (2004)	Pressurized liquid extractor in different proportions of aqueous EtOH (30–80%) and MeOH solvent mixture	One solvent
		Varied temperatures, pressures, and solid to solvent ratios	
7	Li-Hsun et al. (2004)	High-pressure hot water process was considered as 110 and 641 psig over 2.3 h of extraction	MeOH (90%)
8	Klejdus et al. (2004, 2005, b)	PLE; varied extraction time, temperature, pressure, cycles, solvents, and sonication and Soxhlet	MeOH (90%)

isoflavone. Yoo et al. (2004) worked on extracting the freeze-dried fermented soybean in 80% ethanol (10 g/100 mL) and stored at 25 °C for 24 h. The preparation is centrifuged at 4 °C and 15,000× *g* for a duration of 30 min. The supernatant is subsequently evaporated in a vacuum and the resulting residue was freeze-dried to be made into a powder form (See Table 9.1).

9.10 Purification of Fermented Soy Isoflavones

TLC is a simple and inexpensive procedure used to give a satisfactory separation of components in a mixture.

9.10.1 Various Methodologies of TLC Technique

The sample and the standard genistein were spotted near the bottom of the TLC plate and placed inside the TLC chamber containing mobile phase. The mobile phase is left to elevate the TLC plate by capillary action. TLC plates were dried and visualized as fluorescent spots under UV light (255 nm). The spots were marked and the Rf values were calculated. Rf value of compound is compared with the Rf value of standard genistein. The extracted isoflavone is further purified by TLC and HPLC. Silica-Gel 60 is precoated on TLC plate and solvent of chloroform/methanol (20/1, v/v) as the mobile phase for the development of spots. Sulfuric acid (50%) spray with heat (120 °C) is used to identify organic compound. Jyoti et al. (2015) performed TLC for separating and identifying genistein from alcoholic extraction. Chloroform/methanol (10:1, v/v) is used as the developing solution, and Rf value for standard genistein 0.50 has been obtained. Hu et al. (2009) studied the purification of isoflavones by acidifying the extract to pH 3 with hydrochloric acid (2 N) and extracted with EtOAc (2 × 10 ml). One ml of MeOH is used to dry and dissolve the extracts. The MeOH extracts are utilized for analysis using thin layer chromatography. Under 254 nm UV light and by spraying them, the developed plates are monitored by fluorescence quenching. The Pauley's reagent is an amalgamation of solution A, $NaNO_2$ 0.5%; solution B, sulfanilic acid 0.5% in 2% HCl; and solution C, 5% NaOH in 50% EtOH. Freshly prepared mixtures of solution A and B are sprayed at equal volumes on to the developed TLC plates after which solution C is sprayed. Burnt orange colors is seen when it is warmed with heat gun. The substrate controls are then treated with 2 N HCl to hydrolyze the isoflavone glycosides into their aglycones. Orange color spots are observed in the cultures extracts that are different than the controls Rf values. Yuan et al. (2006) studied purification of soy semen sojae praeparatum ultrasonically extracted with 10 mL methanol for 20 min, and at 3500 RPM for 5 min, it was centrifuged. The residues extracted by evaporating the combined supernatants are dissolving in 2 ml methanol for the TLC analysis. For the TLC analysis, HSGF254 plate's development was carried out in a solvent system of toluene-ethyl acetate-acetone-formic acid (20:4:2:1). It can be detected directly at UV-254 nm or at UV-365 nm after exposure to fumes of concentrated ammonia solution. The volume applied was 10 μL, and methanolic solutions of daidzein (0.5 mg/mL^{-1}) and genistein (0.5 mg/mL^{-1}) were used as reference substances.

9.10.2 Analytical Methods for Soy Isoflavones

The isoflavone and its nutrious, metabolism, bioavailabilty of flavonoids can be accessed by in vitro and in vivo values methods are required affordably and accurately to quantify isoflavone in fermented soy milk. These methods include high-pressure liquid chromatography (HPLC) and liquid chromatography and mass spectroscopy (LC-MS).

9.11 HPLC Analysis

HPLC can be directly resolved and quantified spectrometry, when isoflavone are isolated from the matrix (Aaron et al. 2011; Hubert et al. 2005). This allows quantification of complex mixtures of phytoestrogens present in the fermented soy milk. HPLC is a process that requires simple pre-analysis sample preparation, for measurement of all isoflavone forms. The high-efficient, reproducible technique will directly resolve and quantify isoflavone. Both the conjugated and the unconjugated forms of the flavonoids are measured by this approach. Isoflavones, as weak acidic in nature, could dissociate the analyte in a solvent system with enhanced resolution of chromatographic peak. Jyoti et al. (2015) used C18 column with methanol (B) and water (A) (80:20), mobile phase at a flow rate of 1.0 mL/min to analyze isoflavones genistein. The sample volume of 10 μL sample could be injected under the optimized conditions such as column temperature of 40 °C and wavelength of 260 nm; the absorbance of standard and sample peak areas can be quantified. By comparing the retention time, the genistein was identified. Preliminary peak identity was based on a comparison of retention times of standard genistein and unknown peaks in the sample extracts. Jackson et al. (2003) studied various solvent ratios such as mobile phases solvent (A), acetic acid 0.1% (v/v) filtered in Milli-Q water, and (B) acetic acid 0.1% (v/v) in acetonitrile to elude the isoflavone and solvent B with increasing percentage of 15 to 25% for 35 min, followed by 26.5% within the next 12 min, and finally to 50% within 50 mins and held at that percentage for the next 14.50 min. The flow rate was 1 mL/min maintained up to 48 min and increased to 1.3 mL/min from 48.50 min till 63 min. A water 996 series photodiode array detector (Millipore Corp, Marlborough, MA) was monitored from 200 to 350 nm. 100 ng/mL of genistein and 185 ng/mL of daidzein are the minimum concentrations that were detectable (see Fig. 9.5 and Table 9.2).

Hubert et al. (2008) studied lactic acid bacteria fermentation of soy germ under analytical column at 30 °C with the flow rate of 1.5 mL per min. The 0.05% trifluoroacetic acid solvent system in water (solvent A) and acetonitrile (solvent B) as mobile phase developed the chromatogram with increased aglycone after 48 h incubation at 37 °C with the strains A, B, and C; the chromatogram revealed the quantification of aglycones accounted for 62.7%, 75.2%, and 75.3% of total isoflavones, respectively. Hu et al. (2009) analyzed fermented soy extract using a AichromBond-AQ C18 column with mobile phase as methanol-water-acetic acid (66:33:1, v/v) at a flow rate of 0.5 ml/min with 248 and 260 nm for daidzein and genistein, respectively. Chen et al. (2007) studied the comparative efficiency of acetonitrile, ethanol, and methanol, for the highest recovery process by sonication at 25 °C for 40 min and reported acetonitrile elution of isoflavone range with 92.4–111.0%, followed by ethanol and methanol. As of the demand for isoflavones with enormous potential leads to new advances chromatography technique, with less than 1 min. The mobile and stationary phase compositions are noted, and optimization of mobile phase parameters is always considered first, since this is much easier and convenient than stationary phase optimization.

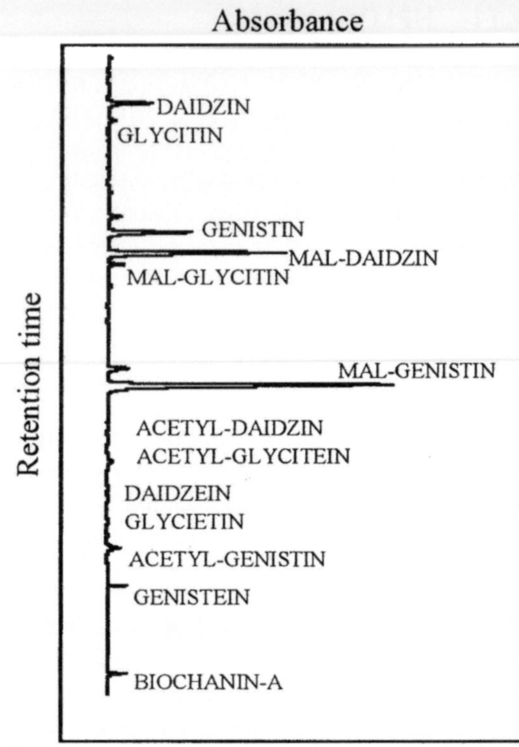

Fig. 9.5 Soybean reverse-phase – the 12 isoflavones' HPLC chromatogram

9.12 Liquid Chromatography Mass Spectrophotometry

The reversed phase (RP) mode is generally used in LC-MS analysis of flavonoids on C18- or C8-bonded silica columns which ranges in length from 250 mm and diameter of 3.9–4.6 mm internally. The mobile phases have been acetonitrile and/or methanol in collaboration with water consisting acetate or formate buffer. LC-MS is commonly carried out at room temperature, by 40 °C, and suggested to shorten the time of evaluation because of thermostated columns. For clinical analysis of isoflavones, liquid chromatography and mass spectrophotometry (LC-MS) has been mainly used. LC-MS's efficient separation capability and structural characterization part with high sensitivity of MS are combined by LC-MS. The identity of flavonoide in a sample is facilitated by MS in connection with UV detectors with the help of standard and reference data. It can also be used with tandem mass spectrometry (MS/MS or MSn) technique in the detection of unknown samples but needs more absorption. With the characterization on structure of flavonoids, LC-MS transmits report about:

(i) Aglycone structure
(ii) Type of carbohydrates or other substituent present
(iii) The part of glycan sequence

Table 9.2 Summary of various HPLC conditions for purifying isoflavones

References	Extraction	Separation condition	HPLC column
Klejdus et al. (2008)	Ethanol (80%) and elution with methanol (80%)	Acetic acid (0.3%) (A) in water-methanol (B); flow rate – 0.8 mL/min 270 nm UV	BEH C18 column
Zheng and Row (2007)	Isoflavone standards	Acetic acid (0.1%) in water (A) acetic acid (0.1%) in acetonitrile (B); flow rate: 1 mL/min; at UV 254 nm	RS-tech C18
Qu et al. (2007)	Supercritical fluid extraction, refluxing with ethanol, drying, redissolving, chromatographed	Methanol-water-acetic acid (30: 70: 2, v/v/v) (glycoside fraction); 45: 55: 2 (v/v/v) (aglycone fraction); flow rate: 1 mL/min; UV at 260 nm	Lichrospher C18 column
Rostagno et al. (2007)	Sonicated with ethanol (50%) in water – centrifuged	Acetic acid (0.1%) in water (A), acetic acid (0.1%) in methanol (B); flow rate: 5 mL/min; UV at 254 nm	Chromolith monolithic RP-18e column
Chen et al. (2007)	Methyl *tert*-butyl ether, centrifuged, dried, and redissolved in methanol-0.05 mol/L ammonium formate (50:50, v/v)	0.05 mol/L ammonium formate (pH 4) (A) acetonitrile (B); SIM/MS	Luna C18 column
Chan et al. (2006)	Hydrolyzed, extracting in a SPE cartridge with formic acid (0.1%) in methyl *tert*-butyl ether and methanol (80/20, v/v) as elution solvent, drying and redissolving in methanol-water (50/50, v/v)	Acetic acid (0.3%) (A) in water acetic acid (0.3%) (B) in acetonitrile; flow rate: 0.2 mL/min; SIM/MS	Agilent Zorbax bonus-RP C18 column
Klejdus et al. (2005a)	Two-step controlled conditions: (1) Preheated and extracted twice with hexane, followed by subjecting to pressurized nitrogen (2) Preheated and extracted twice with 90% methanol, followed by subjecting to pressurized nitrogen. Centrifuged, dried, and redissolved in methanol and filtered	Acetic acid (0.1%) (A) in water (pH 3.75), methanol (B); flow rate: 0.35 mL/min; DAD at 254 nm	Atlantis C18 column

(continued)

Table 9.2 (continued)

References	Extraction	Separation condition	HPLC column
Kao and Chen (2002) Hsieh et al. (2004)	Acetone in hydrochloric acid (0.1 M) (5: 1, v/v), shaken (2 h), centrifuged, supernatant dried, redissolved in methanol, and filtered	Water (A) and acetonitrile (B); flow rate: 2.0 mL/min; UV at 262 nm	Vydac 201TP54 C18 column
Twaddle et al. (2002)	Diluted serum/plasma poured into the SPE cartridge for elution with acetonitrile, dried, and redissolved in methanol	Formic acid (0.1%) in water (A); acetonitrile (B); flow rate of 0.2 mL/min; ESI/MS/MS	Ultracarb ODS column
Krenn et al. (2002)	Methanol (80%) (pH 3 with TFA), filtered, dried, redissolved in dimethyl sulfoxide	A: water (adjusted to pH 2.7 with sulfuric acid). B, acetonitrile; flow rate, 1 mL/min; UV at 254 nm.	Hypersil BDS-C18 column
Satterfield et al. (2001)	Urine sample poured into a SPE C18 cartridge for elution with water-methanol (85: 15, v/v) and subjected to solid-phase microextraction (SPME)	Methanol (A) and water (B); flow rate: 0.1 mL/min; ESI/MS	Waters symmetry C18
Thomas et al. (2001)	Plasma or urine hydrolyzed and extracted with methyl *tert*-butyl ether	Ammonium formate (0.05 M) (A); methanol-acetonitrile (50: 50, v/v) (B)	Luna phenyl-hexyl column
Griffith and Collison (2000)	Acetonitrile and deionized water, diluting with water to make the final acetonitrile concentration to 50% (v/v)	Acetic acid (0.1%) in water (A) acetic acid (0.1%) in acetonitrile (B); flow rate of 0.65 mL/min	Standard method: YMC ODS-AM rapid method, Alltima C18 rocket column
Hutabarat et al. (2000)	HCl (2 M) and ethanol (96%)	Acetonitrile (A) and water (B); flow rate of 1 mL/min; DAD (200–400 nm)	Phenyl Nova-Pak column
Franke et al. (1999)	Aqueous methanol (80%) by sonication and stirring, followed by centrifuging and diluting (1: 1) with 0.2 M acetate buffer	Acetic acid water (10: 90, v/v) (A) methanol-acetonitrile dichloromethane (10: 5: 1, v/v/v) (B); flow rate of 0.8 mL/min; DAD at 260 and 280 nm	Nova-Pak C18 column
Cimino et al. (1999)	Diethyl ether (twice) – dried, redissolved in methanol (25%) containing ammonium acetate (10 mM) and triethylamine (71 mM)	Methanol (25%) containing 10 mM ammonium acetate and 71 mM triethylamine (pH 4.5) (A) methanol (95%) containing 10 mM ammonium acetate and 71 mM triethylamine (pH 4.5) (B); flow rate at mL/min; 1.0 MS	Discovery RPamide-C16 column

(iv) Interglycosidic linkages
(v) Substituents attachment points to the aglycone

The two interfaces used are electrospray ionization (ESI) and atmospheric pressure ionization interfaces (APCI) to show higher ionization stability and more sensitivity than any other interfaces. In the application of positive ionization (PI) as well as negative ionization (NI) for ESI and APCI, the best sensitivity was achieved at NI mode.

Aaron et al. (2011) analyzed the LC-MS evaluation by using P4000 quaternary gradient pump, AS3000 autosampler with sn4000 system controller having uv 6000 diode array detector and an SCM1000 solvent degassing module. Setchell et al. (2002) reported the spectra of daidzein and genistein, characterized their molecular ions and stated that the spectra can be used for high-sensitivity selected ion monitoring detection of daidzein because of the presence of abundant ions which arises from fragmentation and cleavage of the glycoside moiety. Chedea et al. (2009) by LC-ESI-MS analysis studied LE extract of soybean and examine the presence of large spectra of the phenolic compounds at m/z 100 to m/z 1000. The validations were hardly ever discussed as far in the previous review articles on isoflavones and related compounds.

9.13 Conclusion

With the added benefits, the present chapter specifies the transformation of isoflavones into aglycones using β-glucosidase produced by *Bifidobacterium* and *Lactobacillus* and its biological activity. The field of pharmaceutical analysis has gained a great deal of attention due to the development in the analytical methods. This helps in identification and quantification of isoflavones from the fermented soy foods which are also considered as functional foods. This review describes HPLC method development and validation in general way. A general and very simple approach for the HPLC and LC-MS gives an opportunity to determine the phytoestrogens even if they are in the conjugated form.

References

Aaron P, Griffith MW, Collison (2011) Improved methods for the extraction and analysis of isoflavones from soy-containing foods and nutritional supplements by reversed-phase high-performance liquid chromatography and liquid chromatography–mass spectrometry. J Chromatogr A 913:397–313

Achouri A, Boye JI, Belanger D (2005) Soybean isoflavones: efficacy of extraction conditions and effect of food type on extractability. Food Res Int 38(10):1199–1204

Adlercreutz H, Mazur W (1997) Phyto-oestrogens and western diseases. Ann Med 29(2):95–120

Alekel DL, Germain AS, Peterson CT, Hanson KB, Stewart JW, Toda T (2000) Isoflavone rich soy protein isolate attenuates bone loss in the lumbar spine of perimenopausal women. Am J Clin Nutr 72:844–852

Amnon B, Ayelet D (1999) Phytoestrogens the natural selective estrogen receptor modulators. Eur J Obstet Gynecol Reprod Biol 85:47 41

Anderson EL, Van Voris LP, Bartram J, Hoffman HE (1987) Pharmacokinetics of a single dose of rimantadine in young adults and children. Antimicrob Agents Chemother 3:1140–1142

Anderson JW, Johnstone BM, Cook Newell ME (1995) Meta analysis of the effects of soy protein intake on serum lipids. N Engl J Med 333:276–282

Anthony MS (2000) Soy and cardiovascular disease cholesterol and beyond. J. Nutri 130:662–663

Ballongue J (1999) Bifidobacteria and probiotic action. In: Salminen S, Von Wright A (eds) Lactic Acid Bacteria. Marcel Dekker, New York, pp 519–587

Bhardwaj A, Puniya M, Sangu KPS, Kumar S, Dhewa T (2012) Isolation and biochemical characterization of *Lactobacillus* species isolated from Dahi Research & Reviews. J Dairy Sci Technol 1(2):18–31

Chan SA, Lin SW, Yu KJ, Liu TY, Fuh MR (2006) Quantitative analysis of isoflavone aglycones in human serum by solid phase extraction and liquid chromatography-tandem mass spectrometry. Talanta 69:952–956

Chedea VS, Echim C, Vicas SI (2009) Composition in isoflavones of a soybean enzymatic extract. Bull UASVM Agri 66:259–264

Chen L, Zhao X, Fang L, Games DE (2007) Quantitative determination of acetyl glucoside isoflavones and their metabolites in human urine using combined liquid chromatography-mass spectrometry. J Chromatogr A 1154:103–110

Choi YH, Lee WH, Park KY, Zhang L (2000) p53-independent induction of p21 (WAF1/CIP1), reduction of cyclin B1 and G2/M arrest by the isoflavone genistein in human prostate carcinoma cells. Jap J Can Res 91:164–173

Cimino CO, Shelnutt SR, Ronis MJJ, Badger TM (1999) An LC-MS method to determine concentrations of isoflavones and their sulfate and glucuronide conjugates in urine. Clin Chim Acta 287:69–82

Clarkson TB, Anthony MS, Hughes CL (1995) Estrogenic soybean isoflavones and chronic disease: risks and benefits. Trends Endocrinol Metab 6:11–16

Cornwell T, Cohick W, Raskin I (2004) Dietary phytoestrogens and health. Phytochemistry 65:995–1016

Dixona RA, Ferreira D (2002) Molecules of interest genistein. Phytochemistry 60:205–211

Ettinger B, Grady D, Tostesonm ANA, Pressman A, Macer JL (2003) Effect of the Women's health initiative on women's decisions to discontinue postmenopausal hormone therapy. Obstet Gynecol 102:1225–1232

Franke AA, Hankin JH, Yu MC, Maskarinec G, Low SH, Custer LJ (1999) Isoflavones levels in soy food consumed by multiethnic populations in Singapore and Hawaii. J Agric Food Chem 47:977–986

Gilliland SE (1985) Concentrated starter cultures. In: Gilliland SE (ed) Bacterial starter culture for food. CRC Press, Inc., Boca Raton, pp 145–157

Griffith AP, Collison MW (2000) Improved methods for the extraction and analysis of isoflavones from soy containing foods and nutritional supplements by reversed-phase high-performance liquid chromatography and liquid chromatography-mass spectrometry. J Chromatogr A 913:397–313

Hsieh HC, Kao TH, Chen BH (2004) A fast HPLC method for analysis of isoflavones in soybean. J Liq Chromatogr Relat Technol 27:315–324

Hu, SC, Hong, K, Song, YC, Liu, JY. & Tan, RX 2009, 'Biotransformation of soybean isoflavones by a marine'. *World J Microbiol Biotechnol*. vol, *25*. pp. 115–121 Streptomyces sp. 060524 and cytotoxicity of the products

Hubert J, Berger M, Dayde J (2005) Validation of a HPLC–UV method to quantify soy sapogenols A and B in soy germs from different cultivars and in soy isoflavone-enriched supplements. J Food Sci 70(7):471–477

Hubert J, Berger M, Nepveu F, Paul F, Daydé J (2008) Effects of fermentation on the phytochemical composition and antioxidant properties of soy germ. Food Chem 109(4):709–721

Hutabarat LS, Greenfield H, Mulholland M (2000) Quantitative determination of isoflavones and coumestrol in soybean by column liquid chromatography. J Chromatogr A 886:55–63

Ingram D, Sanders K, Kolybaba M, Lopez D (1997) Case-control study of phyto-oestrogens and breast cancer. Lancet 350:990–994

Jyoti ASS, Saxena S, Sharma A (2015) Phytoestrogen "genistein": Its extraction and isolation from soybean seeds. Int J Pharmaco Phytochem Res 7(6):1121–1126

Kao TH, Chen BH (2002) An improved method for determination of isoflavones in soybean powder by liquid chromatography. Chromatography 56:423–430

King RA, Bignell CM (2000) Concentrations of isoflavone phytoestrogens and their glucosides in Australian soya beans and soya foods. Aust J Nutri Diet 57(2):70–78

Klejdus B, Mikelova R, Adam V, Zehnalek J, Vacek J, Kizek R, Kuban V (2004) Liquid chromatographic–mass spectrophotometric determination of genistein and daidzein in soybean food samples after accelerated solvent extraction with modified content of extraction cell. Anal Chim Acta 517:1–11

Klejdus B, Mikelova R, Petrolova J, Potesil D, Adam V, Stiborova M, Hodek P, Vacek J, Kizek R, Kuban V (2005a) Evaluation of isoflavone aglycon and glycoside distribution in soy plants and soybeans by fast column high-performance liquid chromatography coupled with a diode-array detector. J Agric Food Chemi 53:5848–5852

Klejdus B, Mikelova R, Petrlova J, Potesil D, Adam V, Stiborova M, Hodek P, Vacek J, Kizek R, Kuban V (2005b) Determination of isoflavones in soy bits by fast column high-performance liquid chromatography coupled with UV-visible diode-array detection. J Chromatogr A 1084:71–79

Klejdus B, Vacek J, Lojkova L, Benesova L, Kuban V (2008) Ultrahigh-pressure liquid chromatography of isoflavones and phenolic acids on different stationary phases. J Chromatogr A 1195:52–59

Knight DC, Eden JA (1996) A review of the clinical effects of phytoestrogens. Obstet Gynecol 87:897–804

Krenn L, Unterrieder I, Ruprechter R (2002) Quantification of isoflavones in red clover by high-performance liquid chromatography. J Chromatogr B 777(1-2):123–128

Kurzer MS (2000) Hormonal effects of soy isoflavones studies in premenopausal and postmenopausal women. J. Nutri 130:660–661

Lawton B, Rose S, McLeod D, Dowell A (2003) Changes in use of hormone replacement therapy after the report from the women's health initiative cross sectional survey of users. Br Med J 327:845–846

Lee J, Ametani A, Enomoto A, Sato Y, Motoshima H, Ike F, Kaminogawa S (1993) Screening for the immune potent activity of food microorganisms and enhancement of the immune response by *Bifidobacterium adolescentis* M101-4. Biosci Biotechnol Biochem 57:2127–2132

Lee JH, Renita M, Fioritto RJ, Martin SK, Schwartz SJ, Vodovotz Y (2004) Isoflavone characterization and antioxidant activity of Ohio soybeans. J Agric Food Chem 52:2647–2651

Li-Hsun C, Ya-Chuan C, Chieh-Ming C (2004) Extracting and purifying isoflavones from defatted soybean flakes using superheated water at elevated pressures. Food Chem 84:279–285

Lin F, Giusti MM (2005) Effects of solvent polarity and acidity on the extraction efficiency of isoflavones from soybeans (Glycine max). J Agric Food Chem 53:3795–3800

Luthria D, Biswas R, Natarajan S (2007) Comparison of extraction solvents and techniques used for the assay of isoflavones from soybean. Food Chem 105(1):325–333

Messina M, Barnes S (1991) The role of soy products in reducing risk of cancer. J Nat Cancer Inst 83(8):541–546

Murphy PA, Barua K, Hauck CC (2002) Solvent extraction selection in the determination of isoflavones in soy foods. J Chromatogr B 777:129–138

Onozawa M, Fukuda K, Ohtani M, Akaza H, Sugimura T, Wakabayashi K (1998) Effects of soybean isoflavones on cell growth and apoptosis of the human prostatic cancer cell line LNCaP. *Jpn J Clin Onco*l 28:360–63

Otieno D, Ashton JF, Shah NP (2006) Evaluation of enzymic potential for biotransformation of isoflavone phytoestrogen in soymilk by *Bifidobacterium animalis*, *Lactobacillus acidophilus* and *Lactobacillus casei*. Food Res Int 39:394–307

Pandjaitan N, Hettiarachchy N, Ju ZY (2000) Enrichment of Genistein in soy protein concentrate with β-glucosidase. Food Chem Toxicol 65(3):403–407

Penalvo JL, Nurmi T, Adlercreutz H (2004) A simplified HPLC method for total isoflavones in soy products. Food Chem 87:297–205

Qu LP, Fan GR, Peng JY, Mi HM (2007) Isolation of six isoflavones from semen sojae praeparatum by preparative HPLC. Fitoterapia 78:200–204

Rickert DA, Meyer MA, Murphy PA (2004) Effect of extraction pH and temperature on isoflavone and saponin partitioning and profile during soy protein isolate production. J Food Sci 69(8):623–631

Rostagno MA, Palma M, Barroso CG (2004) Pressurized liquid extraction of isoflavones from soybeans. Anal Chim Acta 522:169–177

Rostagno MA, Palma M, Barroso CG (2007) Fast analysis of soy isoflavones by high-performance liquid chromatography with monolithic columns. Analas Chim Acta 582:243–249

Satterfield M, Black DM, Brodbelt JS (2001) Detection of the isoflavone aglycones genistein and daidzein in urine using solid-phase micro extraction high performance liquid chromatography electrospray ionization mass spectrometry. J Chromatogr B 759:33–41

Scalabrinia P, Rossib M, Spettolia P, Matteuzzib D (1998) Characterization of Bifidobacterium strains for use in soymilk fermentation. Int J Food Microbiol 39:213–219

Setchell KDR (2000) Absorption and metabolism of soy isoflavones from food to dietary supplements and adults to infants. J Nutr 130:654S–655S

Setchell KD, Cassidy A (1999) Dietary isoflavones biological effects and relevance to human health. J Nutr 129:758–767

Setchell KDR, Brown NM, Nechemias LZ, Brashear WT, Wolfe BE, Kirschner AS, Heubi JE (2002) Evidence for lack of absorption of soy isoflavone glycosides in humans, supporting the crucial role of intestinal metabolism for bioavailability1–3. Am J Clin Nutr 76:447–453

Sieladie DV, Zambou NF, Kaktcham PM, Cresci A, Fonteh F (2011) Probiotic properties of lactobacilli strains isolated from raw cow milk in the western highlands of Cameroon. Innov Rom Food Biotechnol 9:12–28

Singh GP, Sharma RR (2009) Dominating species of lactobacilli and Leuconostocs present among the lactic acid bacteria of milk of different cattles. Asian J Exp Sci 23:173–179

Strauss L, Santti R, Saarinen N, Streng T, Joshi S, Makela S (1998) Dietary phytoestrogens and their role in hormonally dependent disease. Toxicol Lett 102-103:349–354

Thomas BF, Zeisel SH, Busby MG, Hill JM, Mitchell RA, Scheffler NM, Brown SS, Bloeden LT, Dix KJ, Jeffcoat AR (2001) Quantitative analysis of the principle soy isoflavones genistein, daidzein, and glycitein, and their primary conjugated metabolites in human plasma and urine using reversed-phase high-performance liquid chromatography with ultraviolet detection. J Chromatogr B Biomed Sci Appl 760:191–105

Tochikura, Kerlji S, Takashi F, Tachiki T, Kumagai H (1986) ρ-nitrophenyl glycoside hydrolyzing activities in *Bifidobacteria* and characterization of β-d-galactosidase of *Bifidobacterium longum* 401. Agri Biol Chem 50(9):2279–2286

Tsangalis D, Ashton JF, Mcgill AEJ, Shah NP (2002) Enzymic transformation of isoflavone phytoestrogens in soymilk by β-Glucosidase producing *Bifidobacteria*. J Food Sci 67(8):3104–3113

Twaddle NC, Churchwell MI, Doerge DR (2002) High-throughput quantification of soy isoflavones in human and rodent blood using liquid chromatography with electrospray mass spectrometry and tandem mass spectrometry detection. J of Chromatogr B Anal Technol Biomed Life Sci 777:139–145

Usha Rani V, Pradeep BV (2014) Application of anticancer agents from fermentation of soy using Bifidobacterium and Lactobacillus sp., a review. Acad J Cancer Res 7(3):185–192
Wang H, Murphy PA (1994) Isoflavone content in commercial soybean foods. J Agric Food Chem 42:1666–1673
Watanabe S, Koessel S (1993) Colon Cancer: an approach from molecular epidemiology. J Epidemiol 3:47–61
Yadav R, Shukla P (2015) An overview of advanced technologies for selection of probiotics and their expediency: a review. Crit Rev Food Sci Nutr 57:3233–3242
Yadav R, Puniya AK, Shukla P (2016a) Probiotic properties of *Lactobacillus plantarum* RYPR1 from an indigenous fermented beverage Raabadi. Front Microbiol 7(1683):1–9
Yadav R, Singh PK, Puniya AK, Shukla P (2016b) Catalytic interactions and molecular docking of bile salt hydrolase (BSH) from *L. plantarum* RYPR1 and its prebiotic utilization. Front Microbiol 7(2116):1–9
Yadav R, Singh PK, Shukla P (2017a) Metabolic engineering for probiotics and their genome-wide expression profiling. Curr Protein Pept Sci 18
Yadav R, Kumar V, Baweja M, Shukla P (2017b), Gene editing and genetic engineering approaches for advanced probiotics: a review. *Crit Rev Food Sci Nutr*
Yuana D, Chena Y, Baia X, Pana Y, Kanob Y (2006) TLC and HPLC analysis of soy isoflavones in semen sojae praeparatum. Asian J Tradi Med 1:1–7
Zheng JZ, Row KH (2007) Optimum of mobile phase condition for resolving isoflavones in RP-HPLC. Chin J Chem Eng 15:291–295

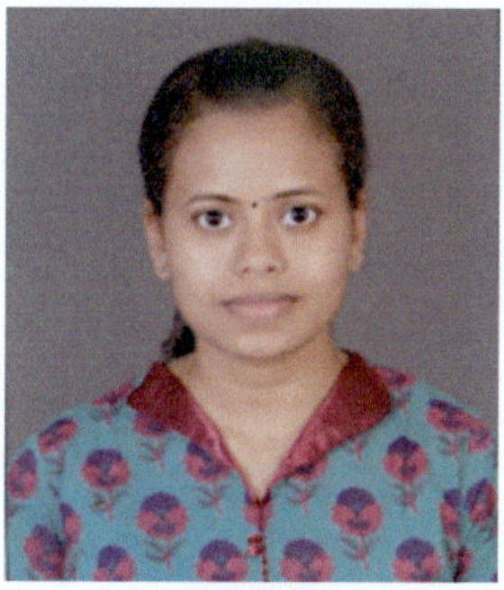

V. Usha Rani completed her M.Sc. (Microbiology) and M.Phil (Microbiology) from Bharathiar University, and Ph.D. at the Karpagam University (Karpagam Academy of Higher Education). Her area of research is on isoflavone production from soy fermentation using *Lactobacillus* sp., and *Bifidobacterium* sp., and its application in cancer studies.

B.V. Pradeep is M.Sc. (Applied Microbiology) from The University of Madras, Chennai, Tamil Nadu, India and Ph.D. (Microbiology) from Bharathiar University, Coimbatore, Tamil Nadu, India. His current research interests are microbial pigments and probiotics.

Burkholderia to *Paraburkholderia*: The Journey of a Plant-Beneficial-Environmental Bacterium

10

Chandandeep Kaur, G. Selvakumar, and A.N. Ganeshamurthy

Abstract

The genus *Burkholderia* is a versatile member of class *Proteobacteria* with over a hundred validly described species. Though endowed with a vast ecological diversity and metabolic versatility, the agro-biotechnological use of members of this genus has remained highly restricted over the past few decades owing to the pathogenic nature of nearly twenty species classified as the *Burkholderia cepacia* complex (Bcc), *B. pseudomallei* the causative agent of melioidosis, *B. mallei* the causative agent of glanders disease in equines and a few plant pathogenic species. Despite the presence of several environmental isolates with beneficial traits, they were overshadowed by their pathogenic relatives. Though initial attempts were made to segregate the clinical and environmental isolates based on the 16S rRNA gene sequences and multilocus sequence typing (MLST), they have failed to remove the stigma associated with the genus. In order to enable the utilization of this genus in agro-biotechnological applications, attempts were made to bifurcate the genus based on phylogenetic evidence. While an earlier attempt to this effect was unsuccessful, the attempt in describing the novel genus *Paraburkholderia* based on the presence of conserved sequence indels and its subsequent taxonomical validation have opened up the possibilities of utilizing this genus that largely remains untainted by any pathogenic potential. But the widespread use of members of this novel genus has to follow a cautious path in order to eliminate any possibility of mammalian pathogenicity and the possible transfer of virulence genes from the members of genus *Burkholderia*. If such concerted steps are taken up, we shall be adding one more potential genus for agro-biotechnological applications.

C. Kaur • G. Selvakumar (✉) • A.N. Ganeshamurthy
ICAR- Indian Institute of Horticultural Research, Hessaraghatta Lake Post, Bengaluru 560089, India
e-mail: gselva74@rediffmail.com

P. Shukla (ed.), *Recent Advances in Applied Microbiology*,
DOI 10.1007/978-981-10-5275-0_10

Keywords
Burkholderia • *Paraburkholderia* • Plant-beneficial bacterium

10.1 Introduction

The genus *Burkholderia*, named after the American phytopathologist Walter Burkholder, is a large genus with more than a hundred validly described species till date. *Burkholderia* belongs to class Betaproteobacteria and family Burkholderiaceae (Palleroni 2005; Parte 2013). The genus was first described by Burkholder who described *Phytomonas caryophylli* and *Phytomonas alliicola* as pathogens of carnation and onion, respectively (Burkholder 1942). In 1992, Yabuuchi proposed the genus *Burkholderia* to encompass most of the former rRNA group II of *Pseudomonas* sp. with the exception of *Pseudomonas pickettii* and *Pseudomonas solanacearum* (Yabuuchi et al. 1995). Ever since a number of novel *Burkholderia* species have been identified and duly validated in the International Journal of Systematic and Evolutionary Microbiology (IJSEM). Though a huge diversity exists within genus *Burkholderia*, the *B. cepacia* complex (Bcc) comprising of twenty species that are widely distributed in soil and clinical environments has received widespread attention owing to their ability to cause opportunistic infections in cystic fibrosis patients. The other cluster that has been well studied within this genus is the *B. pseudomallei* group which includes *B. pseudomallei* the causative agent of melioidiosis (Cheng and Currie 2005) and *B. mallei* that causes glanders disease in equines (Nierman et al. 2004; Whitlock et al. 2007). The transfer of plant pathogenic *Pseudomonas* species, viz. *P. andropogonis*, *P. gladioli*, *P. cepacia*, *P. glumae* and *P. plantarii* to the genus *Burkholderia* (Yabuchi et al. 1992; Urakami et al. 1994; Coenye et al. 2001), has led to the genus attaining a plant pathological hue, which overshadowed its widespread diversity and utility in natural environments.

There was an attempt to bifurcate the harmful and environmental strains present within this genus, and hence the groups A and B were delineated based on 16S rRNA gene sequence data and multilocus sequence analysis (Estrada-de los Santos et al. 2001; Caballero-Mellado et al. 2004; Reis et al. 2004; Perin et al. 2006b). While Group A predominantly comprised of the plant-beneficial-environmental species (Suárez-Moreno et al. 2012), Group B comprised of the plant, human and animal pathogens (Estrada-de los Santos et al. 2013). But this grouping failed to remove the stigma around this species, owing to the cross group presence of member of the Bcc complex in Group A and vice versa. Later developments in this regard were the proposal of the genus *Caballeronia* (Gyaneshwar et al. 2011) to accommodate non-clinical isolates and the bifurcation of the genus *Burkholderia* into *Paraburkholderia* by Sawana et al. (2014), both of which were primarily aimed at accommodating plant-beneficial- environmental species within the genus *Burkholderia*. The latest development in this direction is the valid description of genus*Caballeronia* by (Dobrista and Samadpour 2016). In this chapter, we will discuss the genesis of the genus *Paraburkholderia* and its possible implications for utilization in agro-biotechnology.

10.2 Classification of *Burkholderia*

10.2.1 Earlier Scheme of Classification of *Burkholderia*

The earlier scheme of classification derived from sequence analysis and multilocus sequence typing (MLST) of *Burkholderia* species mainly classified *Burkholderia's* into two groups that were reflective of their functional roles.

Group A The Group A comprised of non-pathogenic *Burkholderia* species associated with plants and/or the environment. This group was derived from consistent multilocus sequence typing data and was placed at a distance from the *Burkholderia cepacia* complex (Bcc) and *B. pseudomallei* group (Spiker et al. 2009; Ussery et al. 2009; Vanlaere et al. 2009). Several species from this group have been reported to diazotrophic abilities and the potential to degrade recalcitrant compounds and thereby survive in environments with limited nutrient availability (Estrada -de los Santos et al. 2001; Martinez-Aguilar et al. 2008). Since some species possessed plant growth-promoting traits while others were useful in phytoremediation and biocontrol, therefore this group was referred to as "plant-beneficial-environmental (PBE) *Burkholderia* group". Members of this group have been found in various ecological niches and unrelated locations. *Burkholderia caribensis* strains were first isolated from a vertisol in Martinique and later from nodules of *Mimosa* spp. in Taiwan and China (Achouak et al. 1999; Liu et al. 2011). Similarly *Burkholderia graminis* strains have been reported from rhizospheric soils of France and Australia (Viallard et al. 1998). The transcontinental occurrence of strains of *Burkholderia tropica* from the rhizospheres of sugarcane, maize and teosinte from locations as diverse as Mexico, Brazil and South Africa has been reported (Perin et al. 2006b). The rhizosphere appears to be the most important habitat for PBE strains, and most strains including *B. unamae*, *B. xenovorans*, *B. tropica*, *B. silvatlantica*, *B. heleia*, *B. bannensis*, *B. acidipaludis*, *B. oxyphila* and *B. ginsengisoli* have been recovered from the rhizosphere (Caballero-Mellado et al. 2004, 2007; Estrada-de los Santos et al. 2001; Perin et al. 2006a; Aizawa et al. 2010a, b; Lim et al. 2008). It has been postulated that the rhizospheric association confers the ability to degrade root exudates and other rootderived compounds (Chain et al. 2006). Some species were also isolated from the bulk soil, e.g. *B. xenovorans* was isolated from polychlorobiphenyl (PCB)-polluted soil as well as coffee and tomato rhizospheres (Bopp 1986; Caballero-Melato et al. 2007; Goris et al. 2004). *B. sartisoli* has been isolated from polycyclic aromatic hydrocarbon (PAH)-polluted soils, rhizosphere of maize and compost heaps (Vanlaere et al. 2008). *B. caledonica*, which exists in sandy soils, has been reported from the rhizosphere of *Vitis vinifera* in Scotland (Coenye et al. 2001). The legume-nodulating *B. phymatum* and the soil inhabiting *B. terrae* have been reported to colonize the rhizosphere of tomato (Wong-Villarreal et al. 2010). To address potential safety issues, a study of the pathogenic potential of several plant-associated symbiotic *Burkholderia* species was conducted, which resulted in the conclusion that it is highly unlikely that they could infect mammals as they lacked the type 3 secretion system, and no mortality or cell lysis was observed with such strains on *Caenorhabtidis elegans* and HeLa cells models (Angus et al. 2014).

Group B This group comprised the species pathogenic to plant, humans and animals (Estrada-de los Santos et al. 2013). During the early 1980s, strains of *B. cepacia* were consistently recovered from respiratory tract cultures of cystic fibrosis patients (Isles et al. 1984). Subsequent studies showed that some patients remained infected with *B. cepacia* without specific symptoms, but other patients produced necrotizing pneumonia and sepsis, a condition referred to as the cepacia syndrome (CS). Polyphasic taxonomy studies revealed that *B. cepacia* was comprised of five closely related but distinct genomic species referred to as the *Burkholderia cepacia* complex (Bcc). Each species was initially designated as a genomovar but was later accorded binomial nomenclature (Vandamme et al. 1997). At present twenty closely related species comprise the Bcc. *Burkholderia pseudomallei*, a species related to the Bcc, causes melioidosis, a lethal septic infection contributing to nearly 20% of all community acquired septicaemias (White 2003; Limmathurotsakul and Peacock 2011). *Burkholderia mallei* is a species related to the Bcc and causes glanders disease in equines, while species like *Burkholderia glumae* and *Burkholderia gladioli* infect a wide variety of crops (Whitlock et al. 2007; Nandakumar et al. 2009). Due to their clinical importance, the species comprising the Bcc, *B. pseudomallei*, *B. mallei* and the plant pathogenic *Burkholderia species* have been extensively studied by several researchers (Coenye 2010; Gilad et al. 2007; Mahenthiralingam et al. 2008; Mahenthiralingam et al. 2005).

An interesting feature of this method of classification was the cross presence of some species with beneficial traits in Group B. For example, *B. vietnamensis* a member of Bcc is able to fix atmospheric nitrogen (Gillis et al. 1995). Similarly the genome of a *B. cenocepacia*, which is generally considered to be the most troublesome in cystic fibrosis patients, with extensive plant-beneficial traits has been reported (Ho and Huang 2015). This cross presence of members with contrasting traits leads to further confusion and had further strengthened the need for the separation of the genus.

10.2.2 Proposals for Bifurcation of Genus *Burkholderia*

Subsequent to the classification of *Burkholderia* into Groups A and B, there was an attempt to rename the group containing the plant-beneficial-environmental *Burkholderia* species as *Caballeronia*, based on the results of phylogenetic studies performed using 16S rRNA, *rec*A, *gyr*B, *rpo*B and *act*S gene sequences of different *Burkholderia* species and comparison of genome sequences (Gyaneshwar et al. 2011). But, this was not formally approved due to the absence of distinctive phenotypic features among different phylogenetic clusters and limited phylogenetic support that was required for the description of a new genus. Later based on the comparative analysis of protein sequences, Sawana et al. (2014) described the existence of forty two highly specific molecular markers in the form of conserved sequence indels (CSIs) found in specific *Burkholderia* groups. The conserved sequence indels represent inherited characters of gene insertion or deletion by various descendant species in essential genes/proteins from a common ancestor and serve as an important molecular marker besides providing a means to understand the evolutionary relationships between different closely related species (Gupta 1998; Rokas and Holland 2000; Gogarten et al. 2002; Gupta and Griffiths 2002; Gao and Gupta 2012). Using these markers they divided the genus *Burkholderia* to two clades; clade I is characterized by the presence of six highly specific CSI's comprised members of Bcc, *B. pseudomallei* and plant pathogenic *Burkholderia* species, while clade II is characterized by the presence of two highly specific CSI's comprised members of the environmental origin. Additionally within the first clade, they identified three CSIs specific to the Bcc group, four CSIs specific to the *B. pseudomallei* group and five CSIs specific to the plant pathogenic group. These were in addition to twenty-two CSIs that differentiate two groups within clade II. Based on this phylogenetic evidence, Sawana et al. (2014) proposed the division of the genus *Burkholderia* into two genera, viz. an emended genus *Burkholderia* containing clinically important and phytopathogenic members of the genus (clade I) and a new genus *Paraburkholderia* gen. nov. harbouring the environmental species (clade II). Due to the large number of unsequenced *Burkholderia* species present in the 16S rRNA database, the accurate identification of the groups within the clades IIa and IIb within clade II as proposed by Sawana et al. (2014) becomes difficult (Fig. 10.1). This suggestion to split the genus *Burkholderia* has met with criticism from scientists who were of the opinion that separation of the genus should be advantageous in future exploration, without dealing with the safety issues regarding human infections caused by members of this genus (Vandamme and Peeters 2014).

Though *Paraburkholderia* has been recognized as a validly published species, the separation of erstwhile *Burkholderia* species as good and bad purely based on taxonomy has been questioned (Eberl and Vandamme 2016).The argument put forth by these authors is that though most *Paraburkholderia* strains do not possess types III, IV and VI secretion systems which are mostly found in the *Burkholderia* sensu stricto clade, the lack of suitable experimental evidence in mammalian models is a major deterrent in establishing the non-pathogenic nature of *Paraburkholderia* species. Further the existence of a transition group between the Groups A and B as revealed by phylogenetic analysis of the 16S rRNA gene has questioned the need to bifurcate the genus *Burkholderia.* This transition group shares a similarity greater than 96% with Group B and predominantly consists of isolates from soil, water and rhizosphere and is associated with plants/ fungi (Estrada-de Los Santos et al. 2016). Despite agreements and counter agreements, the genus *Paraburkholderia* and forty-six of its species have been validated by including them in the International Journal of Systematic and Evolutionary Microbiology (IJSEM) validation list nos. 164 and 165 (Oren and Garrity 2015a, b).The morphological and metabolic characteristics of *Paraburkholderia* are similar to the genus *Burkholderia.* The G + C content of *Paraburkholderia* members ranges from 61.4 to 65.0 mol %, while the G + C content of *Burkholderia* spp. is in the range of 65.7 to 68.5 mol% (Sawana et al. 2014). Though members of genus *Paraburkholderia* are not associated with humans,

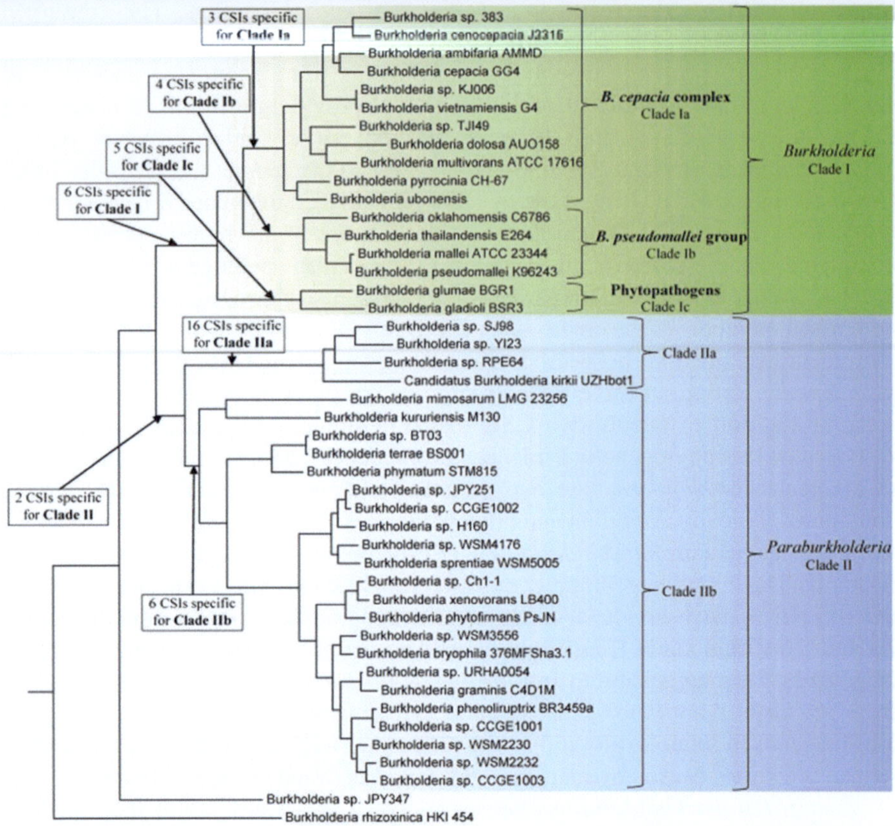

Fig. 10.1 Maximum likelihood phylogenetic tree based on concatenated sequences of 21 conserved proteins of genus *Burkholderia*. The major clades are indicated by brackets (Reproduced from Sawana et al. 2014)

isolation sources may be diverse due to their ecological versatility. Most members of the PBE group have a strict respiratory metabolism with oxygen as terminal electron acceptor with the exception of *B. sartisol* (reclassified as *P. sartisoli*), which reduces nitrate to nitrite (Vanlaere et al. 2008). With continuous developments in bacterial taxonomy and molecular markers used in identification, it can be concluded that *Burkholderia* taxonomy is an ever-evolving area, and many newer species are waiting to be discovered or realigned in tune with modern taxonomical trends.

10.2.3 Revival of the Genus *Caballeronia*

During the time period (2014–2015) between the effective and valid publications of genus *Paraburkholderia*, 16 novel species of *Burkholderia* were described, but only two of them could be accommodated in *Burkholderia* based on the emended description of the genus. This led to the accommodation of 11 novel species in the genus *Paraburkholderia* and the transfer of the remaining three species that form a distinctive clade in the phylogenetic tree to the novel genus *Caballeronia*, along with several other novel species (Dobrista and Samadpour 2016). With this development the process of seperation of genus *Burkholderia* that commenced in 2012 has reached its logical conclusion.

10.3 Agro-biotechnological Potential of *Burkholderia*

Though *Burkholderia* species were described from the rhizosphere and plant-associated environments in the past, two discoveries that had a strong ecological perspective were the ability of several species such as *B. brasiliensis* to fix atmospheric nitrogen (Los Santos et al. 2001) and the description of legume-nodulating *Burkholderia* (Moulin et al. 2001). Both these developments led to an explosion of information on the plant-associated *Burkholderia* and the establishment of the endophytic nature of the genus of the bacterium (da Silva et al. 2014; Carrell and Frank 2015; Banik et al. 2016). But much prior to these developments, *Burkholderia*-based products were used in pest control, plant growth promotion, production of industrial compounds and degradation of toxic molecules (Jaeger et al. 1999; Van et al. 2000; Hussain et al. 2007; Li et al. 2013). The United States Environmental Protection Agency (USEPA) registered several products such as Deny ®, Blue Circle® and Intercept ® that contained Bcc strains as their active ingredient. But these products were withdrawn after risk assessment studies showed that they posed a threat to humans on account of their opportunistic pathogenecity (http://www.gpo.gov/fdys/pkg/FR-2004/09/29/pdf/04-21695.pdf). The question that arises now is that whether *Paraburkholderia*-based products can substitute the earlier Bcc products. Literature search, genome mining and experimental evidence have revealed that only certain species, viz. *P .megapolitina* and *P. bryophila*, originally isolated from mosses, possess the ability to produce antifungal compounds (Vandamme et al. 2007), while this property is widespread among Bcc and plant pathogenic species. Hence the prospects of utilizing *Paraburkholderia* species with the exception of the above species for biological control are quite remote at present. But the utilization of *Paraburkholderia* species as biofertilizers especially as nitrogen fixers, growth promoters and bioremediation agents is quite bright. A list of *Paraburkholderia* species that were earlier classified as *Burkholderia* and possess agro-biotechnological traits is presented in Table 10.1. Recently Kaur et al. (2016) reported the 8.9 Mb draft genome sequence of phosphate-solubilizing bacterium *Paraburkholderia tropica* strain P-31, isolated from pomegranate (*Punica granatum*) rhizosphere. The draft genome sequence consists of 8,881,246 base pairs with a G + C content of 64.7%, 8039 protein-coding genes and 49 RNAs (DDBJ/ENA/GenBank under the accession LXGI00000000).

Table 10.1 *Paraburkholderia* species having agro-biotechnological potential

Species	Source	New combination	Characteristics	Original References
Burkholderia acidipaludis	Endophyte, rhizosphere	*Paraburkholderia acidipaludis* comb. nov.	Nitrogen fixing, aluminium-tolerant species	Aizawa et al. (2010b)
Burkholderia bannensis	Rhizosphere	*Paraburkholderia bannensis* comb. nov.	Nitrogen fixing species	Aizawa et al. (2011)
Burkholderia bryophila	Associated with mosses	*Paraburkholderia bryophila* comb. nov.	Growth promotion observed in lettuce, possesses antifungal activity	Vandamme et al. (2007)
Burkholderia caballeronis	Rhizoplane	*Paraburkholderia caballeronis* comb. nov.	Nitrogen fixing species	Martínez-Aguilar et al. (2013)
Burkholderia caledonica	Rhizosphere, sandy soil	*Paraburkholderia caledonica* comb. nov.	Possesses the *braI/R* quorum-sensing system and ACC deaminase activity	Coenye et al. (2001)
Burkholderia caribensis	Nodules, vertisols	*Paraburkholderia caribensis* comb. nov.	Nitrogen fixing and nodulating species with high EPS production	Achouak et al. (1999)
Burkholderia. ferrariae	Soil	*Paraburkholderia ferrariae* comb. nov.	Nitrogen fixing species with phosphate-solubilizing ability	Valverde et al. (2006)
Burkholderia fungorum	Fungal endosymbiont	*Paraburkholderia fungorum* comb. nov.	Possesses the *braI/R* quorum-sensing system, ACC deaminase activity and ability to degradearomatic compounds	Coenye et al. (2001)
Burkholderia graminis	Rhizosphere	*Paraburkholderia graminis* comb. nov.	Possesses the *braI/R* quorum-sensing systemand ACC deaminase activity may induce systematic tolerance to salt and drought	Barriuso et al. (2008);Viallard et al. (1998)
Burkholderia ginsengisoli	Rhizosphere	*Paraburkholderia ginsengisoli* comb. nov.	Possesses Beta-galactosidase activity	Kim et al. (2006)
Burkholderia heleia	Rhizosphere	*Paraburkholderia heleia* comb. nov.	Nitrogen-fixing species that grows in acidic environments	Aizawa et al. (2010a)
Burkholderia kururiensis	Endophyte, rhizosphere, TCE3 soil	*Paraburkholderia kururiensis* comb. nov.	Nitrogen-fixing species possessing *braI/R* quorum-sensing system and ACC deaminase activity	Anandham et al. (2009); Baldani et al. (1997); Zhang et al. (2000)

Burkholderia megapolitana	Associated with mosses	*Paraburkholderia megapolitana* comb. nov.	Possesses plant growth promotion and antifungal activity	Vandamme et al. (2007)
Burkholderia mimosarum	Nodules	*Paraburkholderia mimosarum* comb. nov.	Nitrogen-fixing and nodulating species occurs as the main endosymbiont of *Mimosa* spp.	Chen et al. (2006)
Burkholderia nodosa	Nodules	*Paraburkholderia nodosa* comb. nov.	Nitrogen-fixing species *nod*+, nodulates *Mimosa* roots	Chen et al. (2007)
Burkholderia oxyphila	Acidic soil	*Paraburkholderia oxyphila* comb. nov.	Catabolizes (+)-catechin into taxifolin	Otsuka et al. (2011)
Burkholderia phenazinium	Soil associated with mosses	*Paraburkholderia phenazinium* comb. nov.	Possesses the *braI/R* quorum-sensing systemand produces ionidin, acidophilic in nature	Izumi et al. 2010; Vanhaverbeke et al. 2003; Viallard et al. 1998
Burkholderia phenoliruptrix	Isolated from a chemostat with 2,4,5 Trichlorophenoxyacetic acid	*Paraburkholderia phenoliruptrix* comb. nov.	Possesses the *braI/R* quorum-sensing systemand ACC deaminase activity degrades halogen-phenol substituted compounds	Coenye et al. (2004); Kellogg et al. (1981); Kilbane et al. (1982)
Burkholderia phymatum	Nodules	*Paraburkholderia phymatum* comb. nov.	Nitrogen-fixing and nodulating species that possesses the *braI/R* quorum-sensing system and ACC deaminase activity. N-fixationin in vivo and *ex planta conditions*	Talbi et al. (2010); Vandamme et al. (1997)
Burkholderia phytofirmans	Endophyte, rhizosphere	*Paraburkholderia phytofirmans* comb. nov.	Possesses the *braI/R* quorum-sensing system, ACC deaminase activity, plant growth promotion and antifungal activity	Frommel et al. (1991); Vandamme et al. (2002); Sessitsch et al. (2005)
Burkholderia sabiae	Nodules	*Paraburkholderia sabiae* comb. nov.	Nitrogen fixing and nodulating species can producePHA	Chen et al. (2008)
Burkholderia saccharii	Soil	*Paraburkholderia saccharii* comb. nov.	Possesses the *braI/R* quorum-sensing system and can producePHA	Brämer et al. (2001)
Burkholderia sartisoli	Rhizosphere, PAH applied soil, compost	*Paraburkholderia sartisoli* comb. nov.	Ability to degrade aromatic compounds	Vanlaere et al. (2008)
Burkholderia sediminicola	Water sediments	*Paraburkholderia sedimonicola* comb. nov.	PHA production	Lim et al. (2008)

(continued)

Table 10.1 (continued)

Species	Source	New combination	Characteristics	Original References
Burkholderia silvatlantica	Rhizosphere, endosphere	*Paraburkholderia silvatlantica* comb. nov.	Nitrogen fixing species Possesses the *braI/R* quorum-sensing system, ACC deaminase activity and plant growth promotion ability	Perin et al. (2006a)
Burkholderia terrae	Soil, rhizosphere	*Paraburkholderia terrae* comb. nov.	Nitrogen fixing speciesPossesses the *braIR* quorum-sensing system	Kim et al. (2006); Wong-Villarreal et al. (2010); Yang et al. (2006)
Burkholderia terricola	Soil	*Paraburkholderia terricola* comb. nov.	Possesses the *braI/R* quorum-sensing system and ACC deaminase activity besides the acceptor of the plasmids pJP4.pEMT1	
Burkholderia tropica	Rhizosphere, endophyte	*Paraburkholderia tropica* comb. nov.	Nitrogen fixing speciesPossesses the *braI/R* quorum-sensing system and plant growth promotion abilities	Reis et al. (2004)
Burkholderia tuberum	Nodules	*Paraburkholderia tuberum* comb. nov.	Nitrogen fixing and nodulating species Possesses the *braI/R* quorum-sensing system and nodulates several plants	Moulin et al. (2001); Vandamme et al. (2002)
Burkholderia unmae	Endophyte, roots, rhizosphere	*Paraburkholderia unmae* comb. nov.	Nitrogen fixing species Possesses the *braI/R* quorum-sensing system, ability to degrade phenol and benzene and promote plant growth	Coenye et al. (2001); Caballero- Mellado et al. (2004)
Burkholderia xenovorans	Rhizosphere, PCB-polluted soil	*Paraburkholderia xenovorans* comb. nov.	Nitrogen fixing and nodulating species Possesses the *braI/R* quorum-sensing system Capable of PCB degradation, found in association with plants	Boop (1986)

10.4 Future of *Paraburkholderia* in Agro-biotechnology

Though it is amply clear that the newly created genus *Paraburkholderia* contains a vast number of species of environmental and plant origin, it would be wise to add an element of caution before being put to widespread agro-biotechnological use. Though the bifurcation of this genus from the parent *Burkholderia* genus has helped to some extent in removing opportunistic pathogenic aspersions that have pre-empted the widespread commercialization of erstwhile *Burkholderia* species of agro-biotechnological importance, the biosafety of potential *Paraburkholderia* species has to be established on mammalian models, in order to clear any element of suspicion about these isolates. This is absolutely necessary since the biosafety data generated at present is mainly on nematode and cell lines that are quite different from the mammalian system. It would be also wise to analyse the selected isolates for their ability to acquire genes from the Bcc complex, in order to eradicate the possibility of such isolates acquiring virulence-associated genes in the future. Both these measures would improve the confidence of the regulatory bodies and the public at large on the safety and utility of *Paraburkholderia* species. If such concerted efforts are made, it will be possible to utilize *Paraburkholderia* as a potentially useful microorganism in agriculture and environmental cleanup.

10.5 Conclusion

Nature has endowed us with a limitless possibility of microbial wealth, a large proportion of which await discovery and sustainable usage. But in the endeavour to scout for beneficial microbes, we frequently encounter bacterial genera and species that are well-known pathogens or have gained the opportunistic pathogen tag. This pre-empts the further exploration of such genera and species though they contain several other species with beneficial traits and very often entire genera are ignored. The genus *Burkholderia* is one such example wherein the presence of known pathogens has excluded the utilization of several potentially beneficial species within the genus. The bifurcation of the genus into genus *Paraburkholderia* and the latest description of genus *Caballeronia* are positive attempts in bringing several plant-beneficial-environmental strains out of the ambit of the genus *Burkholderia*. But the utilization of species presently classified as *Paraburkholderia* and *Caballeronia* requires stringent regulation in view of their past history. It can be said for sure that the recent developments in the taxonomical realignments of genus *Burkholderia* have provided mankind with a treasure trove of novel microbes for sustainable crop production and environmental cleanup.

Acknowledgements Chandandeep Kaur was supported by a grant from the Department of Science and Technology, Ministry of Science and Technology, Government of India, under the WOS-A scheme.

References

Achouak W, Christen R, Barakat M et al (1999) *Burkholderia caribensis* sp.nov. an exopolysaccharide-producing bacterium isolated from vertisol micro aggregates in Martinique. Intl J Sys Evol Microbiol 49:787–794

Aizawa T, Ve NB, Nakajima M et al (2010a) *Burkholderia heleia* sp.nov. a nitrogen-fixing bacterium isolated from an aquatic plant, Eleocharis dulcis, that grows in highly acidic swamps in actual acid sulphate soil areas of Vietnam. Intl J Sys Evol Microbiol 60:1152–1157

Aizawa T, Ve NB, Vijarnsorn P et al (2010b) *Burkholderia acidipaludis* sp.nov. aluminium-tolerant bacteria isolated from Chinese water chestnut (Eleocharis dulcis) growing in highly acidic swamps in South-East Asia. Intl J Sys Evol Microbiol 60:2036–2041

Aizawa T, Vijarnsorn P, Nakajima M, Sunairi M (2011) *Burkholderia bannensis* sp. nov., an acid-neutralizing bacterium isolated from torpedo grass (*Panicum repens*) growing in highly acidic swamps. Int J Syst Evol Microbiol 61(7):1645–1650

Anandham R, Gandhi PI, Kwon SW et al (2009) Mixotrophic metabolism in *Burkholderia kururiensis* subsp. *thiooxidans* subsp. nov., a facultative chemolithoautotrophic thiosulfate oxidizing bacterium isolated from rhizosphere soil and proposal for classification of the type strain of *Burkholderia kururiensis* as *Burkholderia kururiensis* subsp. *kururiensis* subsp. nov. arch. Microbiol 191(12):885–894

Angus AA, Agapakis CM, Fong S et al (2014) Plant associated symbiotic *Burkholderia* species lack hallmark strategies required in mammalian pathogenesis. PLoS One 9(1):e83779

Baldani J, Caruso L, Baldani VL et al (1997) Recent advances in BNF with non-legume plants. Soil Biol Biochem 29(5):911–922

Banik A et al (2016) Characterization of N -fixing plant growth promoting endophytic and epiphytic bacterial community of Indian cultivated and wild rice (*Oryza* spp.) genotypes. Planta 243(3):799–812

Barriuso J, Ramos Solano B, Fray RG et al (2008) Transgenic tomato plants alter quorum sensing in plant growth promoting rhizobacteria. Plant Biotech J 6(5):442–452

Bopp LH (1986) Degradation of highly chlorinated PCBs by *Pseudomonas* strain LB400. J Ind Microbiol Biotechnol 1:23–29

Brämer CO, Vandamme P, da Silva LF et al (2001) Polyhydroxyalkanoate-accumulating bacterium isolated from soil of a sugarcane plantation in Brazil. Intl J Sys Evol Microbiol 51(5):1709–1713

Burkholder WH (1942) Three bacterial plant pathogens:*Phytomonas caryophylli*, sp.nov. *Phytomonas alliicola* sp.nov. and *Phytomonas manihotis* (Artaud-Berthet et Bondar) Viegas. Phytopathology 32:141–149

Caballero-Mellado J, Martínez-Aguilar L, Paredes-Valdez G et al (2004) *Burkholderia unamae* sp. nov., an N – fixing rhizospheric and endophytic species. Int J Syst Evol Microbiol 54:1165–1172

Caballero-Mellado J, Onofre-Lemus J, Estrada-de los Santos P et al (2007) The tomato rhizosphere, an environment rich in nitrogen-fixing *Burkholderia* species with capabilities of interest for agriculture and bioremediation. Appl Environ Microbiol 73(16):5308–5319

Carrell AA, Frank AC (2015) Bacterial endophyte communities in the foliage of coast redwood and giant sequoia. Frontiers Microbiol 6:1008

Chain PS, Denef VJ, Konstantinidis KT et al (2006) *Burkholderia xenovorans* LB400 harbours multireplicon, 9.73-Mbp genome shaped for versatility. Proc Natl Acad Sci U.S.A 103:15280–15287

Chen WM, James EK, Coenye T et al (2006) *Burkholderia mimosarum* sp.nov. isolated from root nodules of *Mimosa* spp. from Taiwan and South America. Intl J Sys Evol Microbiol 56:1847–1851

Chen WM, De Faria SM, James EK et al (2007) *Burkholderia nodosa* sp.nov. isolated from root nodules of the woody Brazilian legumes *Mimosa bimucronata* and *Mimosas cabrella*. Intl J Sys Evol Microbiol 57:1055–1059

Chen WM, de Faria SM, Chou JH et al (2008) *Burkholderia sabiae* sp. nov., isolated from root nodules of *Mimosa caesalpiniifolia*. Intl J Sys Evol Microbiol 58:2174–2179

Cheng AC, Currie BJ (2005) Melioidosis: epidemiology, pathophysiology, and management. Clin Microbiol Rev 18(2):383–416

Coenye T (2010) Social interactions in the *Burkholderia cepacia* complex: biofilms and quorum sensing. Future Microbiol 5(7):1087–1099

Coenye T, Laevens S, Willems A et al (2001) *Burkholderia fungorum* sp. nov. and *Burkholderia caledonica* sp. nov., two new species isolated from the environment, animals and human clinical samples. Intl J Sys Evol Microbiol 51:1099–1107

Coenye T, Henry D, Speert DP et al (2004) *Burkholderia phenoliruptrix* sp. nov. to accommodate the 2,4,5-trichlorophenoxyacetic acid and halophenol-degrading strain AC1100. Syst Appl Microbiol 27:623–627

da Silva DA, Cotta SR, Vollú RE (2014) Endophytic microbial community in two transgenic maize genotypes and in their near-isogenic non-transgenic maize genotype. BMC Microbiol 14(1):1

Dobritsa AP, Samadpour M (2016) Transfer of eleven *Burkholderia* species to the genus *Paraburkholderia* and proposal of *Caballeronia* gen. nov., a new genus to accommodate twelve species of *Burkholderia* and *Paraburkholderia*. Intl J sys Evol Microbiol 66(8):2836–2846

Eberl L, Vandamme P (2016) Members of the genus *Burkholderia*: good and bad guys. F1000Research 5. doi:10.12688/f1000research.8221.1

Estrada-de los Santos P et al (2001) *Burkholderia*, a genus rich in plant-associated nitrogen fixers with wide environmental and geographic distribution. Appl Environ Microbiol 67(6):2790–2798

Estrada-de los Santos P, Vinuesa P, Martínez-Aguilar L et al (2013) Phylogenetic analysis of *Burkholderia* species by multilocus sequence analysis. Curr Microbiol 67:51–60

Estrada-De Los Santos P, Rojas-Rojas FU, Tapia-García EY (2016) To split or not to split: an opinion on dividing the genus *Burkholderia*. Ann Microbial 66:1303–1314

Frommel MI et al (1991) Growth enhancement and developmental modifications of in vitro grown potato (*Solanum tuberosum* spp. *tuberosum*) as affected by a non fluorescent *Pseudomonas* sp.Plant. Physiol 96:928–936

Gao B, Gupta RS (2012) Microbial systematics in the post-genomics era. Antonie Van Leeuwenhoek 101:45–54

Gilad J, Harary I, Dushnitsky T et al (2007) *Burkholderia mallei* and *Burkholderia pseudomallei* as bioterrorism agents: national aspects of emergency preparedness. Isr Med Assoc J 9(7):499

Gillis M, Van Van T, Bardin R et al (1995) Polyphasic taxonomy in the genus *Burkholderia* leading to an emended description of the genus and proposition of *Burkholderia vietnamiensis* sp. nov. for N -fixing isolates from rice in Vietnam. Intl J Sys Evol Microbiol 45(2):274–289

Gogarten JP et al (2002) Prokaryotic evolution in light of gene transfer. Mol Biol Evol 19:2226–2238

Goris J, Dejonghe W, Falsen E et al (2004) Classification of the biphenyl and polychlorinated biphenyl-degrading strain LB400T and relatives as *Burkholderia xenovorans* sp.nov. Intl J Sys Evol Microbiol 54:1677–1681

Gupta RS (1998) Protein phylogenies and signature sequences: are appraisal of evolutionary relationships among archae bacteria,eubacteria,and eukaryotes. Microbiol Mol Biol Rev 62:1435

Gupta RS, Griffiths E (2002) Critical issues in bacterial phylogeny. Theor Popul Biol 61:423–434

Gyaneshwar P, Hirsch AM, Moulin L, Chen WM, Elliott GN, Bontemps C, Estrada-de Los Santos P, Gross E, Dos Reis FB et al (2011) Legume-nodulating beta Proteobacteria: diversity, host range, and future prospects. Mol Plant Microbe Interact 24:1276–1288

Ho YN, Huang CC (2015) Draft genome sequence of *Burkholderia cenocepacia* strain 869T2, a plant-beneficial endophytic bacterium. Genome Announc 3(6):e01327–e01315

Hussain S, Arshad M, Saleem M et al (2007) Biodegradation of α-and β-endosulfan by soil bacteria. Biodegradation 8(6):731–740

Isles A, Maclusky I, Corey M et al (1984) *Pseudomonas cepacia* infection in cystic fibrosis: an emerging problem. J Pediatr 104(2):206–210

Izumi H, Cairney JW, Killham K et al (2010) Bacteria associated with ectomycorrhizas of slash pine (*Pinus elliotti*) in south-eastern Queensland, Australia. FEMS Microbiol Lett 282:196–204

Jaeger K, Dijkstra B, Reetz M (1999) Bacterial biocatalysts: molecular biology, three-dimensional structures, and biotechnological applications of lipases. Annu Rev Microbiol 53:315–351

Kaur C et al (2016) Draft genome sequence of phosphate-solubilizing bacterium *Paraburkholderia tropica* strain P-31 isolated from pomegranate (*Punica granatum*) rhizosphere. Genome Announc 4(4):e00844–e00816

Kellogg ST et al (1981) Plasmid assisted molecular breeding: new technique for enhanced biodegradation of persistent toxic chemicals. Science 214:1133–1135

Kilbane JJ, Chatterjee DK, Karns JS et al (1982) Biodegradation of 2,4,5-trichlorophenoxyacetic acid by a pure culture of *Pseudomonas cepacia*. Appl Environ Microbiol 44:72–78

Kim H et al (2006) *Burkholderia ginsengisoli* sp.nov. a β-glucosidase-producing bacterium isolated from soil of a ginseng field. Intl J Sys Evol Microbiol 56:2529–2533

Li X, Prescott M, Adler B et al (2013) Beclin 1 is required for starvation-enhanced, but not rapamycin-enhanced, LC3-associated phagocytosis of *Burkholderia pseudomallei* in RAW 264.7 cells. Infect Immun 81(1):271–277

Lim HB, Park MJ, Yang HC et al (2008) *Burkholderia sediminicola* sp.nov., isolated from fresh water sediment. Intl J Sys Evol Microbiol 58:565–569

Limmathurotsakul D, Peacock SJ (2011) Melioidosis: a clinical overview. Br Med Bull 99(1):125–139

Liu Y, Chen D, Yan Y et al (2011) Biodiesel synthesis and conformation of lipase from *Burkholderia cepacia* in room temperature ionic liquids and organic solvents. Bioresour Technol 102(22):10414–10418

Mahenthiralingam E et al (2005) The multifarious, multireplicon *Burkholderia cepacia* complex. Nature Rev Microbiol 3:144–156

Mahenthiralingam E et al (2008) *Burkholderia cepacia* complex bacteria: opportunistic pathogens with important natural biology. J Appl Microbiol 104(6):1539–1551

Martínez-Aguilar L, Díaz R, Peña-Cabriales JJ, Estrada-de los Santos P et al (2008) Multichromosomal genome structure and confirmation of diazotrophy in novel plant-associated *Burkholderia* species. Appl Environ Microbiol 74:4574–4579

Martínez-Aguilar L, Salazar-Salazar C, Méndez RD et al (2013) *Burkholderia caballeronis* sp.nov.,a nitrogen fixing species isolated from tomato (Lycopersiconesculen- tum) with the ability to effectively nodulate *Phaseolus vulgaris*. AntonieVan Leeuwenhoek 104:1063–1071

Moulin L, Munive A, Dreyfus B et al (2001) Nodulation of legumes by members of the beta-subclass of Proteobacteria. Nature 411:948–950

Nandakumar R, Shahjahan AK, Yuan XL et al (2009) *Burkholderia glumae* and *B. gladioli* cause bacterial panicle blight in rice in the southern United States. Plant Dis 93:896–905

Nierman WC, DeShazer D, Kim HS et al (2004) Structural flexibility in the *Burkholderia mallei* genome.Proc. Natl Acad Sci USA 101(39):14246–14251

Oren A, Garrity GM (2015a) List of new names and new combinations previously effectively, but not validly, published. Intl J Sys Evol Microbiol 65:2017–2025

Oren A, Garrity GM (2015b) List of new names and new combinations previously effectively, but not validly published. Intl J Sys Evol Microbiol 65:2777–2783

Otsuka Y, Muramatsu Y, Nakagawa Y et al (2011) *Burkholderia oxyphila* sp.nov. a bacterium isolated from acidic forest soil that catabolizes(+)-catechin and its putative aromatic derivatives. Intl J Sys Evol Microbiol 61:249–254

Palleroni NJ (2005) The genus Burkholderia. In: Brenner DJ, Krieg NR, Garrity GM, Staley JT (eds) Bergey's manual of systematic bacteriology: the proteobacteria; the alpha-, beta-,delta-, and epsilon proteobacteria, vol 2. Springer, East Lansing, pp 575–600

Parte AC (2013) LPSN–list of prokaryotic names with standing in nomenclature. Nucleic Acids Res:gkt1111

Perin L, Martinez-Aguilar L, Paredes-Valdez G et al (2006a) *Burkholderia silvatlantica* sp.nov. diazotrophic bacterium associated with sugarcane and maize. Intl J Sys Evol Microbiol 56:1931–1937

Perin L, Martinez-Aguilar L, Castro Gonzalez R et al (2006b) *Diazotrophic Burkholderia* species associated with filed grown maize and sugarcane. Appl Environ Microbiol 72:3103–3110

Reis VM, Estrada-de los Santos P, Tenorio-Salgado S et al (2004) *Burkholderia tropica* sp. nov. a novel nitrogen-fixing, plant-associated bacterium. Intl J Sys Evol Microbiol 54(6):2155–2162

Rokas A, Holland PW (2000) Rare genomic changes as a tool for phylogenetics. Trends Ecol Evol 15:454–459

Sawana A et al (2014) Molecular signatures and phylogenomic analysis of the genus *Burkholderia*: proposal for division of this genus into the emended genus *Burkholderia* containing pathogenic organisms and a new genus *Paraburkholderia* gen. nov. harboring environmental species. Front Genet 5:429

Sessitsch A, Coenye T, Sturz AV et al (2005) *Burkholderia phytofirmans* sp. nov., a novel plant-associated bacterium with plant-beneficial properties. Intl J Sys Evol Microbiol 55:1187–1192

Spilker T, Baldwin A, Bumford A et al (2009) Expanded multi locus sequence typing for *Burkholderia* species. J Clin Microbiol 47:2607–2610

Suárez-Moreno ZR, Caballero-Mellado J, Coutinho BG et al (2012) Common features of environmental and potentially beneficial plant-associated *Burkholderia*. Microb Ecol 63:49–266

Talbi C, Delgado MJ, Girard L et al (2010) *Burkholderia phymatum* strains capable of nodulating *Phaseolus vulgaris* are present in Moroccan soils. Appl Environ Microbiol 76:4587–4591

Urakami T, Ito-Yoshida C, Araki H et al (1994) Transfer of *Pseudomonas plantarii* and *Pseudomonas glumae* to *Burkholderia* as *Burkholderia* spp. and description of *Burkholderia vandii* sp. nov. Int J Syst Evol Microbiol 44:235–245

Ussery DW, Kiil K, Lagesen K et al (2009) The genus *Burkholderia*: analysis of 56 genomic sequences. Genome Dyn 6:140–457

Valverde A, Delvasto P, Peix A et al (2006) *Burkholderia ferrariae* sp.nov. isolated from an iron ore in Brazil. Intl J Sys Evol Microbiol 56:2421–2425

Van VT, Berge O, Ke SN (2000) Repeated beneficial effects of rice inoculation with a strain of Burkholderia vietnamiensis on early and late yield components in low fertility sulphate acid soils of Vietnam. Plant Soil 218(1–2):273–284

Vandamme P, Peeters C (2014) Time to revisit polyphasic taxonomy. Antonie Van Leeuwenhoek 106:57–65

Vandamme P, Holmes B, Vancanneyt M et al (1997) Occurrence of multiple genomovars of *Burkholderia cepacia* in cystic fibrosis patients and proposal of *Burkholderia* multivorans sp. nov. Intl J Sys Evol Microbiol 47:1188–1200

Vandamme P, Goris J, Chen WM et al (2002) *Burkholderia tuberum* sp. nov. and *Burkholderia phymatum* sp. nov. nodulate the roots of tropical legumes. Syst Appl Microbiol 25:507–512

Vandamme P, Opelt K, Knöchel N et al (2007) *Burkholderia bryophila* sp.nov.and *Burkholderia megapolitana* sp.nov. moss-associated species with anti fungal and plant-growth-promoting properties. Intl J Sys Evol Microbiol 57:2228–2235

Vanhaverbeke C et al (2003) Conformational analysis of the exopolysaccharide from *Burkholderia caribensis* strain MWAP71: impact on the interaction with soils. Biopolymers 69(4):480–497

Vanlaere E, LiPuma JJ, Baldwin A et al (2008) *Burkholderia latens* sp. nov. *Burkholderia diffusa* sp. nov. *Burkholderia arboris* sp. nov. *Burkholderia seminalis* sp. nov. and *Burkholderia metallica* sp. nov. novel species within the *Burkholderia cepacia* complex. Intl J Sys Evol Microbiol 58(7):1580–1590

Vanlaere E, Baldwin A, Gevers D et al (2009) Taxon K, a complex within the *Burkholderia cepacia* complex, comprises at least two novel species *Burkholderia contaminans* sp nov and *Burkholderia lata* sp nov. Intl J Sys Evol Microbiol 59(1):102–111

Viallard V, Poirier I, Cournoyer B et al (1998) *Burkholderia graminis* sp. nov. a rhizospheric *Burkholderia* species, and reassessment of *Pseudomonas phenazinium*, *Pseudomonas pyrrocinia* and *Pseudomonas glathei* as *Burkholderia*. Intl J Sys Evol Microbiol 48(2):549–563

White NJ(2003) Melioidosis. Lancet 361:1715

Whitlock GC, Estes DM, Torres AG (2007) Glanders: off to the races with *Burkholderia mallei*. FEMS Microbiol Lett 277:115–122. doi:10.1111/j.1574-6968.2007. 00949.x

Wong-Villarreal A, Caballero-Mellado J et al (2010) Rapid identification of nitrogen-fixing and legume-nodulating *Burkholderia* species based on PCR 16S rRNA species-specific oligonucleotides. Syst Appl Microbiol 33(1):35–43

Yabuuchi E, Kosako Y, Oyaizu H et al (1992) Proposal of *Burkholderia* gen. nov. and transfer of seven species of the genus *Pseudomonas* homology group II to the new genus, with the type species *Burkholderia cepacia* (Palleroni and Holmes 1981) comb. nov. Microbiol Immunol 36:1251–1275

Yabuuchi E, Kosako Y, Yano I et al (1995) Transfer of two *Burkholderia* and an *Alcaligenes* species to *Ralstonia* gen. nov. proposal of *Ralstonia pickettii* (Ralston, Palleroni and Doudoroff 1973) comb. nov. *Ralstonia solanacearum* (Smith 1896) comb. nov. and Ralstonia eutropha (Davis 1969) comb. nov. Microbiol Immunol 39:897–904

Yang HC, Im WT, Kim KK et al (2006) *Burkholderia terrae* sp. nov., isolated from a forest soil. Intl J Sys Evol Microbiol 56(2):453–457

Zhang H, Hanada S, Shigematsu T et al (2000) *Burkholderia kururiensis* sp.nov. a trichloroethylene (TCE)-degrading bacterium isolated from an aquifer polluted with TCE. Intl J Sys Evol Microbiol 50:743–749

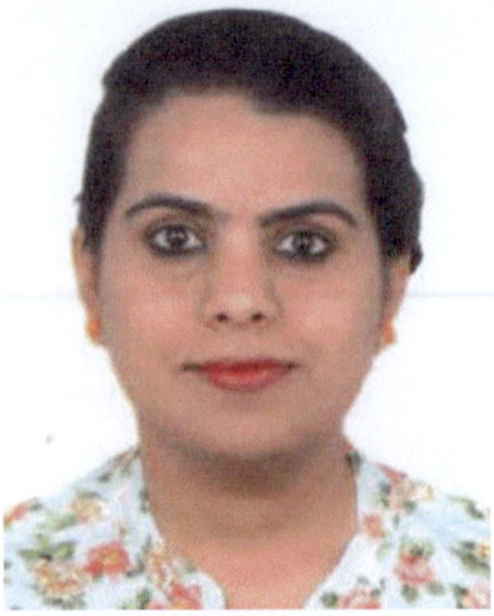

Chandandeep Kaur, M.Sc, is supported by the Department of Science and Technology, Government of India, under the DST-Women in Science programme at ICAR- Indian Institute of Horticultural Research, Bengaluru.

G. Selvakumar, Ph.D., works as Principal Scientist (Agrl. Microbiology) at the ICAR- Indian Institute of Horticultural Research, Bengaluru.

A.N. Ganeshamurthy, Ph.D., works as Principal Scienist (Soil Science) at ICAR- Indian Institute of Horticultural Research, Bengaluru.

WRKY Transcription Factors: Involvement in Plant–Pathogen Interactions

11

Lopamudra Satapathy, Dhananjay Kumar, and Kunal Mukhopadhyay

Abstract

Transcriptional reprogramming takes place as defense response to help plants overcome different stresses. The defense-related gene expression is elicited by interaction of a ligand–receptor complex that induces the expression of early response genes. A number of transcription factors (TFs) have been found to participate in defense responses in plants. Among these, the WRKY TFs are conspicuous, having a strictly conserved 60 amino acid regions comprising of the highly conserved WRKYGQK peptide sequence and a zinc fingerlike motif. These TFs have a recognition sequence of (C/T)TGAC(T/C), known as the W-box that is found in the promoter region of WRKY and other defense-related genes. WRKY TF family members are categorized mainly into three groups based on the number of WRKY domains and certain features associated with the zinc fingerlike motifs. These TFs are involved in priming diverse pathways such as defense against pathogens, trichome development, senescence, and biosynthesis of secondary metabolites. The present article focuses on the defense-related role of WRKY TFs during biotic stress in crop plants.

Keywords

WRKY transcription factors • Biotic stress • Abiotic stress • Wheat • Rice • *Arabidopsis*

L. Satapathy • D. Kumar • K. Mukhopadhyay (✉)
Department of Bio-Engineering, Birla Institute of Technology, Mesra, Ranchi 835215, Jharkhand, India
e-mail: lipi.lopa@gmail.com; dhananjaykumar4@gmail.com; kmukhopadhyay@bitmesra.ac.in

P. Shukla (ed.), *Recent Advances in Applied Microbiology*,
DOI 10.1007/978-981-10-5275-0_11

11.1 Introduction

Plants are sessile and, therefore, confront numerous kinds of biotic and abiotic stresses in their natural habitat which leads to up- or downregulation of several genes. Depending on the type of stress, appropriate defense-responsive genes get transcriptionally reprogrammed so that the plant can coordinate with the intricate network of stress-responsive signal transduction pathways to combat the stress conditions. The expressions of defense-responsive genes are elicited by ligand–receptor interaction that evokes the expression of primary response genes. The cellular reactions are controlled by the products of these regulatory genes.

The evolutionary trends put forward that plants have developed diverse defense mechanisms to fight infections and to combat diseases. The plant innate immunity is set off by any pathogen attack, and a response is generated in the form of either PTI [pathogen-associated molecular pattern (PAMP)-triggered immunity] or ETI (effector-triggered immunity), driven by plant disease resistance proteins (Chisholm et al. 2006; Pandey and Somssich 2009). Both PTI and ETI can generate local and systemic acquired resistance (SAR) (Durrant and Dong 2004). The defense response is a complex signaling network that may be triggered by defense signaling molecules like plant hormones, viz., jasmonic acid (JA), salicylic acid (SA), and ethylene (ET) (Chen and Chen 2000; Eulgem et al. 2000; Kim et al. 2000; Ulker and Somissch 2004). Also, an extensive and complex transcriptional reprogramming generates responses leading toward activation and deactivation of several defense-responsive genes and various transcription factors (TFs) (Eulgem and Somssich 2007). WRKY TFs are members of such regulatory TF family.

11.2 WRKY Transcription Factors

The WRKY TFs having zinc finger motifs belong to a large superfamily WRKY–GCM1 (Babu et al. 2006; Marquez and Pritham 2010; Wei et al. 2012). Members of this TF family contain at least one conserved DNA-binding region of 60 amino acids, designated as the WRKY domain that comprises the highly conserved WRKYGQK peptide sequence and a zinc fingerlike motif. The number of WRKY domains and the type of zinc fingerlike motif present in these TFs have been used in their classification (Eulgem et al. 2000; Rushton et al. 2012). WRKY proteins with two WRKY domains were classified as class I. Class II WRKY is marked by the presence of one WRKY domain with C2H2 zinc finger structure. Class II is additionally subgrouped into a–e based upon amino acid motifs present outside the WRKY domain. Class III WRKY proteins contain a single domain and differ from class I and II in its altered C2HC zinc finger motif ($C–X_7–C–X_{23}–H–X–C$) (Ulker and Somssich 2004; Ling et al. 2011). Most group I and subgroup IIc members, on the basis of WRKY zinc finger motif (C–Xm–C–Xn–H–X–H/C), show m value of 4 and n ranged from 21 to 23. Subgroups IIa, IIb, IId, and IIe have m value of 5 and n value of either 23 or 24. Subgroup IIIa members have the motif changed to C–X7–C–X23–H–X–C. The motif in subgroup IIIb is inconsistent with the value of m

Table 11.1 Classification of WRKY transcription factors

Type	Number of WRKY domains	Zinc finger motif	DNA binding
Class I	2	C2H2 ($C–X_{4–5}–C–X_{22–23}–H–X_1–H$)	C-terminal end
Class II	1	C2H2 ($C–X_{4–5}–C–X_{22–23}–H–X_1–H$)	C-terminal end
Class III	1	C2HC ($C–X_7–C–X_{23}–H–X–C$)	–

Fig. 11.1 The logo of WRKY motif

ranging from 6 to 9 and n ranging from 23 to 28. Based on the domain features of WRKY proteins described in previous research on *Arabidopsis* and rice, an evolutionary history of *WRKY* gene family in wheat was construed by Zhu et al. 2013. Subgroup Ia genes have been directly acquired from ancestral forms of the wheat *WRKY* genes, and the sub-group IIc and some members of subgroup Ib wheat WRKYs may have acquired from subgroup Ia genes by losing the C-terminus WRKY domain. Meanwhile, the subgroups IIa/IIb, IId/IIe, and other members of subgroup Ib TaWRKYs have been originated from subgroup Ia genes which lost the N-terminus WRKY domain. Subgroup IIIa and IIIb evolved from subgroup IId/IIe members. Detail on domain-based classification of WRKY TFs is shown in Table 11.1.

The WRKY proteins show tremendous binding affinity toward a DNA sequence known as the W-box, (C/T)TGAC(T/C), that is present on promoter regions of *WRKY* and other defense-related genes (Pandey and Somssich 2009, Kayum et al. 2015). A few alternative binding sites (TTTTCCAC) and (CGACTTTT) have been identified for NtWRKY12-BD of *Nicotiana tabacum* and AtWRKY70 of *Arabidopsis*, respectively. These binding sites are essential for activating gene expression (van Verk et al. 2008; Machens et al. 2013). Variants of the WRKYGQK signature motif also include different WRKY motifs such as WRKYGEK, WRKYGKK, WRKYGQE, WLKYGQK, LRKYGPK, WRNYGQN, WKKYGQK, WRKDGQK, WSKYGQK, WTKYGQK, GRKYGEK, WMKYGQK, WRKYGQR, WRKYGQN, WRKYGSK, WQKYGQK, WSKYGQM, WNKYGQK, WKRKGQK, WVKYGQK, WRRYGLK, WRKYEDK, WRKYGKR, WRKCGLK, WKKYGYK, WKKYGED, WLKYGQK, and WKKYEEK. The motif WRKDGQE has been identified in wheat WRKY TF proteins (Zhu et al. 2013; Okay et al. 2014; Satapathy et al. 2014). The logo of WRKY motif is shown in Fig. 11.1. The existence of several uncommon WRKY domains led to the proposition of the consensus sequence to be W(R/K)(K/R)Y (Xie et al. 2005). Fingerlike structures are formed

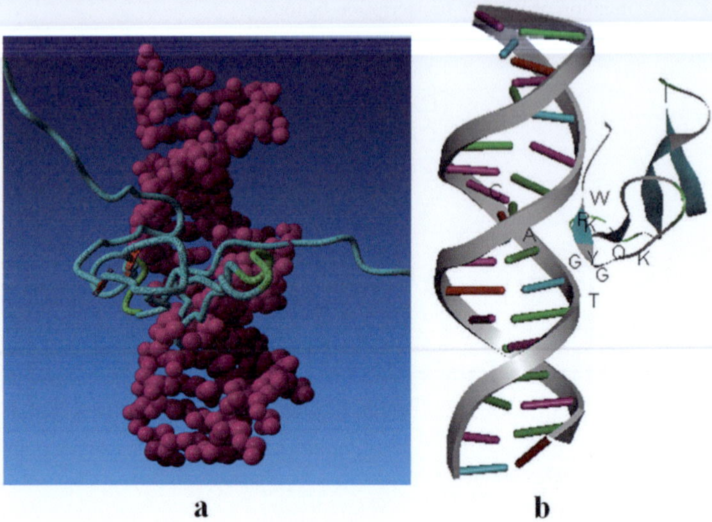

Fig. 11.2 (**a**) Pictorial representation of WRKY domain bound with the core sequence of W-box. (**b**) Docked model of WRKY protein–DNA complex

when cysteine and histidine residues bind to a zinc atom and for appropriate DNA binding; both WRKY and zinc finger motif are essential (Maeo et al. 2001). AtWRKY4 protein contains a novel zinc and DNA-binding region having conserved cys/his residues; its structure consists of four-stranded β-sheet with a zinc-binding pocket (Yamasaki et al. 2005). The crystal structure of another WRKY protein of *A. thaliana*, AtWRKY1-C, has five globular β-strands, with DNA-binding residues located at β2- and β3-strands (Duan et al. 2007). Some WRKY proteins of *Petroselinum crispum*, PcWRKY4 and PcWRKY5, have leucine zipper motifs, L–x6–V–x6–L–x6–M–x6–L and L–x6–L–x6–L–x6–I, respectively, instead of zinc fingerlike motif (Cormack et al. 2002). A different group, class IV, has also been reported in rice that contains a WRKY domain but no zinc fingerlike motif (Xie et al. 2005; Ross et al. 2007). Another class of WRKY protein has been found to bind with calmodulin through nonclassical calmodulin-binding domain (AcMBD) VSSFK(K/R) VISLL (Park et al. 2005). A pictorial representation showing binding affinity of WRKY domain toward W-box has been shown in Fig. 11.2. The WRKY proteins comprise of leucine zippers, putative basic nuclear localization signals, glutamine-rich region, serine–threonine-rich region, proline-rich region, kinase domains, and TIR-NBS-LRR structures that help WRKY proteins in playing vital roles in gene expression regulation (Chen et al. 2011). A phylogenetic tree depicting different classes of WRKY proteins in *Arabidopsis*, rice, and wheat and the evolutionary relationship among them are shown in Fig. 11.3.

Fig. 11.3 (continued) (NM_102668), OsWRKY45 (DQ298181), ATWRKY62 (AF224700), AtWRKY70 (AF421157), AtWRKY54 (AF426253), OsWRKY64 (BK005067), OsWRKY19 (BK005022), OsWRKY74 (BK005077), AtWRKY41 (NM117177), AtWRKY53 (NM118512), AtWRKY55 (NM129636), TaWRKY46 (EF368365), TaWRKY45 (EF397613), OsWRKY46 (BK005049), TaWRKY19-a (EF368362), TaWRKY19-b (EF397616), TaWRKY74-a (EF368359), TaWRKY74-b (EF397615), TaWRKY1-a (DQ334400), TaWRKY1-b (DQ334401), AtWRKY20 (NM_179119), OsWRKY78 (BK005212), TaWRKY78 (HM013818)

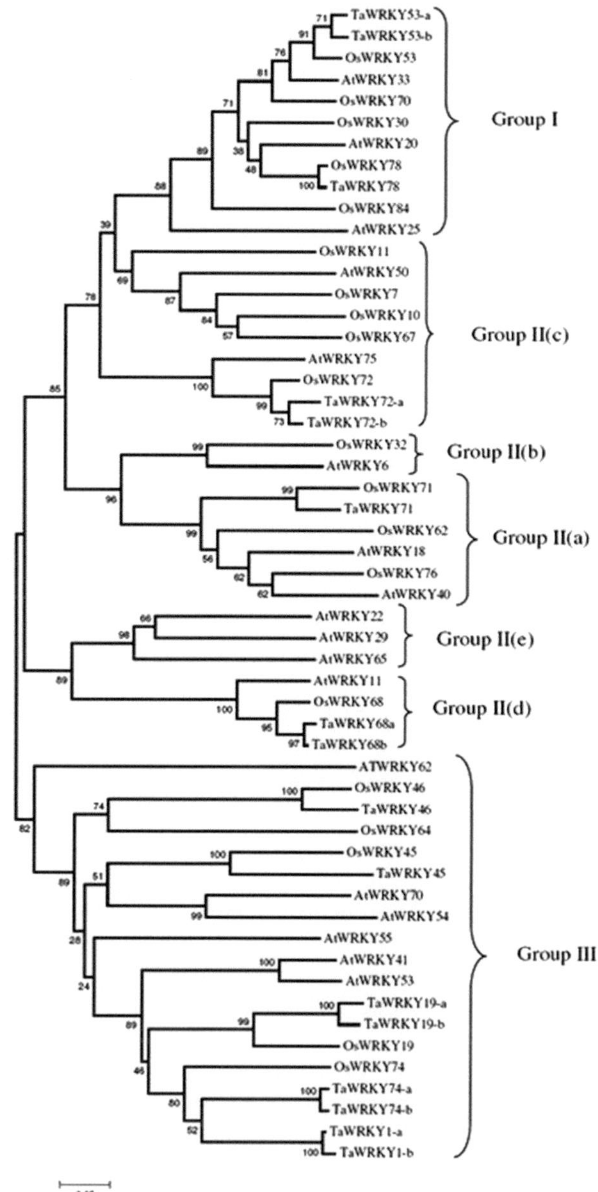

Fig. 11.3 Phylogenetic tree depicting different classes of WRKY proteins and evolutionary relationship among *Arabidopsis*, rice, and wheat WRKYs (Figures in parentheses are GenBank accession numbers)

AtWRKY33 (AF509499), OsWRKY53 (AY676929), OsWRKY84 (AF193802), AtWRKY25 (AF418309), OsWRKY30 (DQ298180), OsWRKY70 (DQ298184), TaWRKY53-a (EF368357), TaWRKY53-b (EF368364), OsWRKY62 (DQ298182), OsWRKY71 (AY676927), AtWRKY40 (NM106732), TaWRKY71 (EF368356), OsWRKY76 (DQ298185), OsWRKY11 (BK005014), AtWRKY18 (AF224698), AtWRKY6 (AF331713), OsWRKY32 (BK005035), OsWRKY10 (AY341854), OsWRKY67 (AK066252), AtWRKY75 (NM121311), OsWRKY72 (BK005075), TaWRKY72-a (EF368358), TaWRKY72-b (EF368363), AtWRKY50 (NM122518), OsWRKY7 (DQ298179), AtWRKY11 (NM179228), OsWRKY68 (BK005071), TaWRKY68-a (EF368360), TaWRKY68-b (EF397617), AtWRKY22 (AF442392), AtWRKY29 (AF442394), AtWRKY65

WRKY TFs have been studied in *Arabidopsis* (Dong et al. 2003; Kalde et al. 2003; Eulgem and Somssich 2007), rice (Ryu et al. 2006; Ramamoorthy et al. 2008), barley (Mangelsen et al. 2008), maize (Wei et al. 2012), *Brachypodium* (Tripathi et al. 2012), creosote bush (Zou et al. 2004), soybean (Zhou et al. 2008), *Brassica rapa* (Kayum et al. 2015), beans (Wang et al. 2016), banana (Shekhawat et al. 2011), pepper (Dang et al. 2013), and few lower plants like ferns, mosses, and green algae (Rushton et al. 2010).

11.3 Plant–Pathogen Interactions: Network of WRKY, MAPK, and Phytohormones

Plant pathogens use several strategies to successfully stay alive and infect plants. After entering inside plants through gas (stomata) or water pores (hydathodes) or wounds, pathogenic bacteria multiply in intercellular spaces (apoplast). Nematodes and aphids directly insert a style into the plant cell for sustenance. Fungi enter plant epidermal cells through stomata and spread out hyphae in between or through plant cells. Pathogenic as well as symbiotic fungi and oomycetes can produce feeding structures called haustoria that invaginate the host cell plasma membrane. Haustorial plasma membranes, extracellular matrix, and host plasma membranes form a close relation at which the interaction occurs. The pathogens secrete effector molecules (virulence factors) into the plant cell. This initiates and activates the plant immune system. Two types of reaction happen: the first is transmembrane pattern recognition receptors (PRRs) which upon recognition of an effector molecule respond as pathogen-associated molecular patterns (PAMPs), and the second response occurs inside the host cell through polymorphic nucleotide-binding (NB) and leucine-rich repeat (LRR) domains (NB–LRR proteins) encoded mostly by different resistance (*R*) genes. The effector molecules secreted by pathogens belonging to diverse kingdoms are recognized by NB–LRR proteins and activate host defense responses. NB–LRR-mediated disease resistance occurs only against obligate biotrophic and hemibiotrophic pathogens that can grow on living host tissue. But this mechanism is not in effect against necrotrophic pathogens that kill host tissue during colonization (Jones and Dangl 2006). Toll-interleukin-I receptor (TIR) coordinates with NB–LRR along with WRKY TF proteins to generate an alternate of the ETI-mediated defense activation pathway. PAMP activates mitogen-activated protein kinase (MAPK) cascade which comprises of MAP kinase, MAP kinase kinase, and MAP kinase kinase kinase. The MAP kinases are serine, threonine-specific protein kinases that respond to extracellular stimuli. The activation of MAP kinase cascades regulates *WRKY* and other TF genes which subsequently activates defense- and stress-responsive gene expression. A model depicting the role of *WRKY* genes in response to biotrophic fungal pathogenesis is shown in Fig. 11.4.

Phytohormones like SA, JA, ethylene, abscisic acid (ABA), gibberellic acid (GA), auxins, cytokinins, and brassinosteroids (BRs) are defense signaling molecules connected with plant immunity. The phytohormone signaling networks form a complex that coordinate multiple stress responses. SA is a positive regulator of

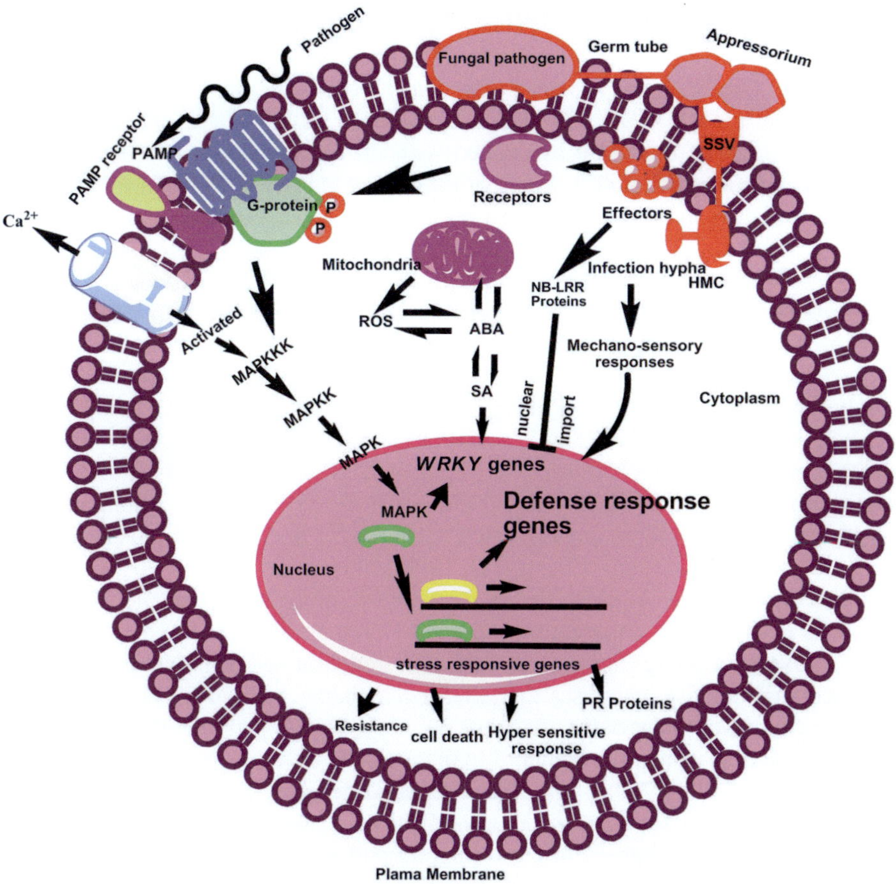

Fig. 11.4 A model depicting the role of *WRKY* genes in response to fungal pathogenesis. Cellular defense signaling is triggered by the recognition of pathogen-derived PAMPs via distinct plasma membrane-localized receptors. Mechanosensory responses transduce MAP kinase cascades that regulate *TaWRKY* genes that eventually activate defense gene expression. *SSV*, substomatal vesicles; *HMC*, haustorial mother cell; *G-protein*, guanine nucleotide-binding proteins; *PAMP*, pathogen-associated molecular pattern; *PR proteins*, pathogenesis-related proteins; *NB–LRR proteins*, nucleotide-binding leucine-rich repeat proteins

immunity against biotrophic and hemibiotrophic pathogens. JA and ethylene are positive regulators of immunity against necrotrophic pathogens. Other phytohormones participate in the central phytohormone signaling network and take part in various interactions. Phytohormone biosynthesis is modulated during pathogen attack and immunity, and the synthesized phytohormones are recognized by dedicated receptors which transduce signals to transcriptional complexes leading to hormonal cross talk at the level of TFs (Tsuda and Somssich 2015).

11.4 Defensive Roles of WRKY TFs in *Arabidopsis*

The WRKY family of TFs has been extensively studied for plant defense responses in the model plant *Arabidopsis thaliana*. As many as 74 *WRKY* genes had been identified, of them many are associated with different types of defense systems (Eulgem and Somssich 2007). Many of these *Arabidopsis WRKY* genes determine the balance between the defense pathways and signaling networks dependent on SA and JA pathways. Double and triple mutants of *AtWRKY18*, *AtWRKY40*, and *AtWRKY60* indicate their role in defense response against *Pseudomonas syringae* and *Botrytis cinerea* (Xu et al. 2006). *AtWRKY18 AtWRKY40 AtWRKY60* triple mutant and *AtWRKY18, AtWRKY40* and *AtWRKY18, AtWRKY60* double mutants were resistant to *P. syringae* but susceptible to *B. cinerea* than wild types illustrating partly repetitive functions in negative regulation toward *P. syringae* resistance (Xu et al. 2006). However, *AtWRKY18*, *AtWRKY53*, and *AtWRKY70* regulate positively to SAR, whereas *AtWRKY58* negatively controls SAR and acts downstream to natriuretic peptide receptor (Wang et al. 2006). Similarly, *AtWRKY33* is among positive regulators which confer resistance toward *Alternaria brassicicola* and *B. cinerea*; both are necrotrophic fungi (Zheng et al. 2006). Mutational analysis (loss or gain of function) for different WRKY TFs in *Arabidopsis* illustrated their role in defense response network as both negative and positive regulators (Eulgem and Somssich 2007). Overexpressing lines of *AtWRKY25* function as negative regulator of SA-mediated defense responses to *P. syringae* (Zheng et al. 2007). *AtWRKY3* has positive control on plant defense against the necrotrophic pathogen *B. cinerea*, but *AtWRKY4* acts as a negative regulator to plant resistance to the biotrophic pathogen *P. syringae* (Lai et al. 2008). This was illustrated by using T-DNA insertion single and double mutants for *AtWRKY3* and *AtWRKY4* (Chen et al. 2010). *AtWRKY38* and *AtWRKY62* overexpression cause suppression of defense and defense-related genes like SA-regulated PR1 protein, and thereby, *AtWRKY38* and *AtWRKY62* act as negative regulators of plant basal defense mechanism (Kim et al. 2008). Similarly, *AtWRKY48* also negatively regulates basal defense against *P. syringae* (Xing et al. 2008).

Studies on knockout mutants of *AtWRKY70* identified this TF as a convergence node for integrating SA- and JA-mediated defense signaling during pathogen attack (Li et al. 2006). *Arabidopsis* plants when infected with the pathogens, *Alternaria brassicicola* and *Erysiphe cichoracearum*, showed this converging role of *AtWRKY70* (Li et al. 2006). There was also evidence of the dual functionality of certain *AtWRKY* genes. *AtWRKY53* mutants exhibited increased susceptibility toward *P. syringae* and decreased resistance toward *Ralstonia solanacearum* (Murray et al. 2007; Hu et al. 2008). Another *AtWRKY* gene with dual functionality is *AtWRKY41* whose overexpression lines demonstrated decreased resistance to *Erwinia carotovora* but enhanced resistance to *P. syringae* (Higashi et al. 2008).

In *Arabidopsis*, WRKY25, WRKY26, and WRKY33 were found to positively regulate the cooperation between heat-shock protein and ethylene-activated signaling pathways that function in plant responses to heat stress (Li et al. 2011). Co-expression of downstream functional genes *AtbHLH17* and *AtWRKY28* in

Arabidopsis confers resistance during drought and oxidative stress (Babitha et al. 2013). From gene expression profiling studies, it has been confirmed that both *AtWRKY28* and *AtWRKY75* are crucial regulators of SA- and JA-/ET-dependent defense signaling pathways during transcription and are actively involved in JA/ET pathway to defend *Arabidopsis* against *S. sclerotiorum* and oxalic acid stress (Chen et al. 2013).

11.5 Defensive Roles of WRKY TFs in Rice

Significant research on WRKY TFs has also been conducted in rice (*Oryza sativa* L.) in the recent past to elucidate various roles of WRKY TFs under biotic stress. Till date, 107 *WRKY* genes are reported in the rice genome; most have cropped up through duplication and deletion events during the recent evolutionary history of rice (Ross et al. 2007; Ramamoorthy et al. 2008). Most of these genes respond to various phytohormones and biotic stresses (Ryu et al. 2006; Ramamoorthy et al. 2008). Enhanced resistance to *Xanthomonas oryzae* pv. *oryzae* (*Xoo*) was exhibited by overexpression of *OsWRKY13* and *OsWRKY71* genes (Liu et al. 2005, 2007). Overexpression of *OsWRKY13* showed enhanced resistance to *Xoo* and *Magnaporthe grisea*, the rice blast fungal pathogen, by suppressing JA signaling pathways and activating SA-dependent pathways (Qiu et al. 2007). The study showed upregulation of *OsWRKY10* and downregulation of *OsWRKY14*, *OsWRKY24*, *OsWRKY42*, *OsWRKY45*, *OsWRKY51*, and *OsWRKY68* (Qiu et al. 2007). Together with *OsWRKY13*, *OsWRKY31* also enhances resistance against *M. grisea.* Early defense response against infection was found by overexpression of *OsWRKY31.* The gene also showed constitutive expression of other defense-related genes like *PBZ1* (probenazole), *OsSci2*, pathogenesis-related (*PR*) gene, etc. (Zhang et al. 2008). However, overexpression of *OsWRKY45* gene showed enhanced resistance to *M. grisea* and *Xoo* (Shimono et al. 2007; Tao et al. 2009). Similarly, overexpressing lines of *OsWRKY53* are resistant to both *M. grisea* and *Xoo* showing that the gene is involved in basal defense response (Chujo et al. 2007). This was illustrated by constitutive expression of defense-related genes like *chitinase 1* and *PR5*. *OsWRKY89* not only confers resistance against fungal blast but also provides resistance against insects like white-backed plant hopper *Sogatella furcifera* (Wang et al. 2007). *Xa21*, a defense-related gene in rice which encodes for a receptor kinase and confers resistance against *Xoo* (Song et al. 1995), has been functionally linked with *OsWRKY62*, and the study suggests that overexpression of *OsWRKY62* suppresses defense gene expression and acts as negative regulator in basal defense (Peng et al. 2008). Thus substantial evidence shows that a few *WRKY* TF genes (viz., *OsWRKY13*, *OsWRKY31*, and *OsWRKY71*) exhibit enhanced resistance against bacterial rice blight pathogen *Xoo*, while a few others (*OsWRKY31*, *OsWRKY89*, *OsWRKY45*, *OsWRKY53*, and *OsWRKY13*) confer resistance to blast fungus, *M. grisea.*

The expression analysis of *OsWRKY28* showed enhanced susceptibility to blast fungus *M. oryzae*, strain Ina86-137, together with *OsWRKY71*, and exhibits an early-induced expression prior to the late-induced expressions of *OsWRKY62* and

OsWRKY76 (Chujo et al. 2013). Overexpression of *OsWRKY30* in rice upregulates an SA-responsive gene *OsWRKY45* and enhances disease resistance via an SA signaling pathway to *Xoo* (Han et al. 2013). These genes also function as positive regulators of various disease resistance in rice. It has also been reported that C-terminal region of OsWRKY30 enhanced resistance to pathogen by constitutively expressing the defense-related genes in transgenic *Arabidopsis* and rice and played a vital role in defense signaling (Lee et al. 2013).

11.6 Defensive Roles of WRKY TFs in Wheat

Presently research on WRKY TFs in wheat is very scanty. However, *Triticum aestivum* is presumed to have more *WRKY* genes than rice or *Arabidopsis* because of hexaploidy and huge genome size. *TaWRKY18* and *TaWRKY60* had overlapping negative effects on SA-mediated defense but had positive roles in JA-mediated defense (Xu et al. 2006). In wheat, 15 *WRKY* TF genes have been isolated and expression studied and categorized based on their corresponding orthologous sequence in rice (Wu et al. 2008). *TaWRKY53* has been identified and characterized to be differentially upregulated during wheat resistance response to the aphid, *Diuraphis noxia* (Botha et al. 2010). In vitro studies with *TaWRKY78* gene demonstrated its role in plant defense as it can bind to a region of the wPR4e promoter and thereby regulate *PR4* genes (Proietti et al. 2010). *TaWRKY45* has a positive role in defense response to *Fusarium* attack during *Fusarium* head blight, and its overexpression confers improved resistance toward *Fusarium graminearum* (Bahrini et al. 2011). This gene may also regulate expression of other disease resistance and stress response pathway genes in wheat.

A WRKY gene *TaWRKY71-1* which is specifically expressed in leaves was derived from a wheat introgression line SR3. The gene produced more mRNA in SR3 compared to parent wheat line JN177. Overexpression of *TaWRKY71-1* in *Arabidopsis* caused hyponastic rosette leaves with the hyponastic strength closely correlated with its transcription level (Qin et al. 2013). Full-length cDNA of TaWRKY1B was isolated from a wheat cultivar HD2329 having the leaf rust resistance gene *Lr28* (Kumar et al. 2014). A virulent race of leaf rust fungus infection stemmed 146-fold and 12-fold induction of the gene in resistant and susceptible plants, respectively, as equated to mock-inoculated controls. Using in silico transcriptomic approaches, Satapathy et al. 2014 have predicted 67 WRKY TF proteins involved in regulating gene expression in response to leaf rust disease in wheat. Many of these identified TFs were also associated with other developmental functions in wheat which has not been reported or studied before. Tag-based expression and microarray-based differential expression analysis revealed rust-specific *TaWRKY10*, *TaWRKY15*, *TaWRKY17*, and *TaWRKY56* genes. Therefore, the authors suggested that many of these TaWRKY TFs participate in cross talk in plant defense signaling pathways.

11.7 Defensive Roles of WRKY TFs in Tomato

WRKY72-type TFs in tomato and *Arabidopsis* have a partially conserved role in basal defense (Bhattarai et al. 2010). *Botrytis cinerea*-responsive WRKY gene *SlDRW1* (*Solanum lycopersicum* defense-related WRKY1) from tomato was identified by Liu et al. (2014) which is a nuclear-localized protein, and it was found to be a positive regulator of defense responses in tomato against *B. cinerea* and oxidative stress.

11.8 Role of WRKY TFs in Abiotic Stress and Nutrient Deficiency

Global climate change-induced abiotic stresses are the major limiting factor of plant growth and crop yields. Crop production is rarely free of environmental stresses. The major environmental stresses of coexisting economic importance worldwide that affect crops are heat, cold (chilling and freezing), drought, soil mineral deficiency, soil mineral toxicity, salinity, excess water (flooding), pollutants, radiations, oxidative stress (reactive oxygen species, ozone), and chemicals (Mahajan and Tuteja 2005). Plants constantly face a surge of abiotic stress and have evolved an intricate mechanism to sense and respond accordingly. WRKY TFs also play a central role in plant abiotic stress responses imparting stress tolerance besides its major role in biotic stress. Several findings and studies reveal the active role of WRKY in various plant species such as *Arabidopsis*, wheat, rice, and barley during abiotic stress.

Although more research on WRKY TFs in wheat in response to various abiotic stresses have been performed, their exact roles in abiotic stress tolerance are largely unknown in wheat (Niu et al. 2012). Overexpression of wheat WRKY genes *TaWRKY2* and *TaWRKY19* in transgenic *Arabidopsis* plants showed better tolerance to salt and drought (Niu et al. 2012). Multiple stress induced by PEG, NaCl, cold, and H_2O_2 in transgenic tobacco plants with *TaWRKY10* displayed upregulation of the gene and greater tolerance to drought and salt stress (Wang et al. 2013). The expression profiling of 18 *TaWRKY* genes subjected to salinity, PEG, or ABA revealed upregulation of the genes in all three stresses (Zhu et al. 2013).

Nutrient deficiency affects plant architecture and adaptability to adverse conditions. WRKY TFs are involved in nutrient deficiency-related response signaling pathways. In *Arabidopsis*, *AtWRKY75* was induced strongly during inorganic phosphate (Pi) deficiency. Plants were found to be susceptible to Pi stress upon suppressing expression of the gene (Devaiah et al. 2007). *AtWRKY6* also functions in plant response to low Pi stress by negatively regulating *Arabidopsis PHOSPHATE1* (*PHO1*) gene expression by binding to W-box located upstream of PHO1 promoter (Chen et al. 2009). During low Pi stress, both *AtWRKY75* and *AtWRKY6* regulate the expression of different downstream target genes involved in different regulatory pathways. *AtWRKY6* also acts as a positive regulator in low boron condition (Kasajima et al. 2010). *AtWRKY45* and *AtWRKY65* are involved in the regulation of carbon starvation (Wang et al. 2007). Studies have indicated that *OsWRKY72* is sensitive to sugar starvation. *HvWRKY46* mediates expression of *ISO1* and *SBEllb*, involved in sugar signaling in barley plants (Mangelsen et al. 2008).

11.9 WRKY and Reactive Oxygen Species (ROS)

WRKY TFs play an important role in ROS signaling web. In *Arabidopsis*, *AtWRKY30*, *AtWRKY75*, *AtWRKY38*, *AtWRKY49*, and *AtWRKY6* genes are significantly induced by H_2O_2 (Zhang et al. 2004; van der Auwera et al. 2007; Chen et al. 2010). *AtWRKY70* was constitutively expressed in ROS scavenging enzyme gene *Atapx1* mutant plants. In light stress expression of other WRKY TF genes, *AtWRKY6*, *AtWRKY18*, *AtWRKY25*, *AtWRKY33*, *AtWRKY40*, *AtWRKY46*, *AtWRKY54*, and *AtWRKY60* were elevated in *Atapx1* mutant plants (Rizhsky et al. 2004, Ciftci-Yilmaz et al. 2007).

11.10 Conclusion

On the basis of recent technological breakthroughs, enormous research has been done in the past several years on the role of WRKY TFs in model as well as several non-model crop plants:

- WRKY TFs play fundamental roles in regulating defense against pathogens in *Arabidopsis*, rice, wheat, barley, maize, and other crops.
- Some WRKY TFs act as negative regulators, whereas others act as positive regulators of plant defense.
- WRKY TFs are early-responsive genes that coordinately act with MAP kinase cascade.
- WRKY TFs also modulate expression of SA-, JA-, and ET-responsive genes during pathogen infection.
- Apart from pathogen defense, WRKY TFs are also involved in the regulation of several physiochemical processes controlling the expression of an array of phytohormones and other genes in abiotic stress conditions.

Acknowledgments This work was supported by the Department of Biotechnology, Government of India (Grant No. BT/PR6037/AGR/02/308/05), BTISNet SubDIC (BT/BI/04/065/04), and Center of Excellence – Technical Education Quality Improvement Programme (CoE-TEQIP-II) (Grant No. NPIU/TEQIP II/FIN/31/158). L. S. is grateful to the Department of Science and Technology-INSPIRE (Fellowship/2011/318) and D. K. to the Council of Scientific and Industrial Research [9/554 (0026) 2010-EMR-I] for Ph.D. fellowships.

Conflicts of Interest Authors declare no conflicts of interest.

References

Babitha KC, Ramu SV, Pruthvi V, Mahesh P, Nataraja KN, Udayakumar M (2013) Co-expression of AtbHLH17 and AtWRKY28 confers resistance to abiotic stress in *Arabidopsis*. Transgenic Res 22:327–341

Babu MM, Iyer LM, Balaji S, Aravind L (2006) The natural history of the WRKY-GCM1 zinc fingers and the relationship between transcription factors and transposons. Nucl Acids Res 34(22):6505–6520

Bahrini I, Sugisawa M, Kikuchi R, Ogawa T, Kawahigashi H, Ban T, Handa H (2011) Characterization of a wheat transcription factor, *TaWRKY45*, and its effect on *Fusarium* head blight resistance in transgenic wheat plants. Breed Sci 61(2):121–129

Bhattarai KK, Atamian HS, Kaloshian I, Eulgem T (2010) WRKY72-type transcription factors contribute to basal immunity in tomato and *Arabidopsis* as well as gene-for-gene resistance mediated by the tomato R gene *Mi-1*. Plant J 63:229–240

Botha AM, Swanevelder ZH, Lapitan NLV (2010) Transcript profiling of wheat genes expressed during feeding by two different biotypes of *Diuraphis noxia*. Environ Entomol 39(4):1206–1231

Chen C, Chen Z (2000) Isolation and characterization of two pathogen- and salicylic acid-induced genes encoding WRKY DNA-binding proteins from tobacco. Plant Mol Biol 42(2):387–396

Chen YF, Li LQ, Hu Q, Kong YH, Wang H, Wu WH (2009) The WRKY6 transcription factor modulates PHOSPHATE1 expression in response to low pi stress in *Arabidopsis*. Plant Cell 21:3554–3566

Chen LG, Zhang LP, Yu DQ (2010) Wounding-induced WRKY8 is involved in basal defense in *Arabidopsis*. Mol Plant-Microbe Interact 23:558–565

Chen H, Lai ZB, Shi ZW, Xiao Y, Chen ZX, Xu XP (2011) Roles of *Arabidopsis WRKY18, WRKY40* and *WRKY60* transcription factors in plant responses to Abscisic acid and abiotic stress. BMC Plant Biol 10:281–295

Chen X, Liu J, Lin G, Wang A, Wang Z, Lu G (2013) Overexpression of AtWRKY28 and AtWRKY75 in *Arabidopsis* enhances resistance to oxalic acid and *Sclerotinia sclerotiorum*. Plant Cell Rep 32:1589–1599

Chisholm ST, Coaker G, Day B, Staskawicz BJ (2006) Host-microbe interactions: shaping the evolution of the plant immune response. Cell 124(4):803–814

Chujo T, Takai R, Akimoto-Tomiyama C, Ando S, Minami E, Nagamura Y, Kaku H, Shibuya N, Yasuda M, Nakashita H (2007) Involvement of the elicitor-induced gene OsWRKY53 in the expression of defense-related genes in rice. Biochem Biophys Acta 1769(7–8):497–505

Chujo T, Miyamoto K, Shimogawa T, Shimizu T, Otake Y, Yokotani N, Nishizawa Y, Shibuya N, Nojiri H, Yamane H, Minami E, Okada K (2013) OsWRKY28, a PAMP-responsive transrepressor, negatively regulates innate immune responses in rice against rice blast fungus. Plant Mol Biol 82:23–37

Ciftci-Yilmaz S, Morsy MR, Song L, Coutu A, Krizek BA, Lewis MW, Warren D, Cushman J, Connolly EL, Mittler R (2007) The EAR-motif of the Cys2/His2-type zinc finger protein Zat7 plays a key role in the defense response of *Arabidopsis* to salinity stress. J Biol Chem 282(12):9260–9268

Cormack RS, Eulgem T, Rushton PJ, Kochner P, Hahlbrock K, Somssich IE (2002) Leucine zipper-containing WRKY proteins widen the spectrum of immediate early elicitor-induced WRKY transcription factors in parsley. Biochim Biophys Acta 1576(1–2):92–100

Dang FF, Wang YN, Yu L, Lai Y, Liu ZQ, Wang X, Qiu AL, Zhang TX, Lin J, Chen YS, Guan DY, Cai HY, Mou SL, He SL (2013) CaWRKY40, a WRKY protein of pepper, plays an important role in the regulation of tolerance to heat stress and resistance to *Ralstonia solanacearum* infection. Plant Cell Environ 36:757–774

Devaiah BN, Karthikeyan AS, Raghothama KG (2007) WRKY75 transcription factor is a modulator of phosphate acquisition and root development in *Arabidopsis*. Plant Physiol 143:1789–1801

Dong J, Chen CH, Chen ZX (2003) Expression profiles of the *Arabidopsis WRKY* gene superfamily during plant defense response. Plant Mol Biol 51(1):21–37

Duan MR, Nan J, Liang YH, Mao P, Lu L, Li L, Wei C, Lai L, Li Y, Su XD (2007) DNA binding mechanism revealed by high resolution crystal structure of *Arabidopsis thaliana* WRKY1 protein. Nucl Acids Res 35(4):1145–1154

Durrant WE, Dong X (2004) Systemic acquired resistance. Annu Rev Phytopathol 42:185–209

Eulgem T, Somssich IE (2007) Networks of WRKY transcription factors in defense signaling. Curr Opin Plant Biol 10(4):366–371

Eulgem T, Rushton PJ, Robatzek S, Somssich IE (2000) The WRKY superfamily of plant transcription factors. Trends Plant Sci 5(5):199–206

Han M, Ryu HS, Kim CY, Park DS, Ahn YK, Jeon JS (2013) *OsWRKY30* is a transcriptional activator that enhances rice resistance to the *Xanthomonas oryzae* pathovar *oryzae*. J Plant Biol 56:258–265

Higashi K, Ishiga Y, Inagaki Y, Toyoda K, Shiraishi T, Ichinose Y (2008) Modulation of defense signal transduction by flagellin-induced WRKY41 transcription factor in *Arabidopsis thaliana*. Mol Gen Genomics 279:303–312

Hu J, Barlet X, Deslandes L, Hirsch J, Feng DX, Somssich I, Marco Y (2008) Transcriptional responses of *Arabidopsis thaliana* during wilt disease caused by the soil-borne phytopathogenic bacterium, *Ralstonia solanacearum*. PLoS One 3:e2589

Jones JDG, Dangl JL (2006) The plant immune system. Nature 444:323–329

Kalde M, Barth M, Somssich IE, Lippok B (2003) Members of the *Arabidopsis* WRKY group III transcription factors are part of different plant defense signaling pathways. Mol Plant-Microbe Interact 16(4):295–305

Kasajima I, Ide Y, Yokota HM, Fujiwara T (2010) WRKY6 is involved in the response to boron deficiency in *Arabidopsis thaliana*. Plant Physiol 139:80–92

Kayum MA, Jung HJ, Park JI, Ahmed N, Saha G, Yang TJ, Nou IS (2015) Identification and expression analysis of *WRKY* family genes under biotic and abiotic stresses in *Brassica rapa*. Mol Gen Genomics 290:79–95

Kim CY, Lee SH, Park HC, Bae CG, Cheong YH, Choi YJ, Han C, Lee SY, Lim CO, Cho MJ (2000) Identification of rice blast fungal elicitor-responsive genes by differential display analysis. Mol Plant-Microbe Interact 13(4):470–474

Kim KC, Lai Z, Fan B, Chen Z (2008) *Arabidopsis* WRKY38 and WRKY62 transcription factors interact with histone deacetylase 19 in basal defense. Plant Cell 20:2357–2371

Kumar D, Kapoor A, Singh D, Satapathy L, Singh AK, Kumar M, Prabhu KV, Mukhopadhyay K (2014) Functional characterization of a WRKY transcription factor of wheat and its expression analysis during leaf rust pathogenesis. Functional Plant Biol 41:1295–1309

Lai Z, Vinod KM, Zheng Z, Fan B, Chen Z (2008) Roles of *Arabidopsis* WRKY3 and WRKY4 transcription factors in plant responses to pathogens. BMC Plant Biol 8:68

Lee H, Ko YJ, Cha JY, Park SR, Ahn I, Hwang DJ (2013) The C-terminal region of OsWRKY30 is sufficient to confer enhanced resistance to pathogen and activate the expression of defense-related genes. Plant Biotechnol Rep 7:221–230

Li J, Brader G, Kariola T, Tapio Palva E (2006) WRKY70 modulates the selection of signaling pathways in plant defense. Plant J 46:477–491

Li SJ, Fu QT, Cheng HG, Huang WD, Hu DQ (2011) *Arabidopsis thaliana* WRKY25, WRKY26, and WRKY33 coordinate induction of plant thermotolerance. Planta 233:1237–1252

Ling J, Jiang WJ, Zhang Y, Yu HJ, Mao ZC, Gu XF, Huang SW, Xie BY (2011) Genome-wide analysis of WRKY gene family in *Cucumis sativus*. BMC Genomics 12:471

Liu XQ, Bai XQ, Qian Q, Wang XJ, Chen MS, Chu CC (2005) *OsWRKY03*, a rice transcriptional activator that functions in defense signaling pathway upstream of OsNPR1. Cell Res 15(8):593–603

Liu X, Bai X, Wang X, Chu C (2007) *OsWRKY71*, a rice transcription factor, is involved in rice defense response. J Plant Physiol 164(8):969–979

Liu B, Honga Y, Zhang Y, Li X, Huanga L, Zhanga HJ, Li DY, Song FM (2014) Tomato WRKY transcriptional factor SlDRW1 is required for disease resistance against *Botrytis cinerea* and tolerance to oxidative stress. Plant Sci 227:145–156

Machens F, Becker M, Umrath F, Hehl R (2013) Identification of a novel type of WRKY transcription factor binding site in elicitor-responsive cis-sequences from *Arabidopsis thaliana*. Plant Mol Biol 84(4–5):371–385

Maeo K, Hayashi S, Hisae K-S, Morikami A, Nakamura K (2001) Role of conserved residues of the WRKY domain in the DNA-binding of tobacco WRKY family proteins. Biosci Biotechnol Biochem 65(11):2428–2436

Mahajan S, Tuteja N (2005) Cold, salinity and drought stresses: an overview. Arch Biochem Biophys 444:139–158

Mangelsen E, Kilian J, Berendzen KW, Kolukisaoglu UH, Harter K, Jansson C, Wanke D (2008) Phylogenetic and comparative gene expression analysis of barley (*Hordeum vulgare*) WRKY transcription factor family reveals putatively retained functions between monocots and dicots. BMC Genomics 9:194

Marquez CP, Pritham EJ (2010) Phantom, a new subclass of Mutator DNA transposons found in insect viruses and widely distributed in animals. Genetics 185(4):1507–1517

Murray SL, Ingle RA, Petersen LN, Denby KJ (2007) Basal resistance against *Pseudomonas syringae* in *Arabidopsis* involves WRKY53 and a protein with homology to a nematode resistance protein. Mol Plant-Microbe Interact 20:1431–1438

Niu C, Wei W, Zhou Q, Tian A, Hao Y, Zhang W, Ma B, Lin Q, Zhang Z, Zhang J, Chen S (2012) Wheat *WRKY* genes *TaWRKY2 and TaWRKY19* regulate abiotic stress tolerance in transgenic *Arabidopsis* plants. Plant Cell Environ 35(6):1156–1170

Okay S, Derelli E, Unver T (2014) Transcriptome-wide identification of bread wheat WRKY transcription factors in response to drought stress. Mol Gen Genomics 289(5):765–781

Pandey SP, Somssich IE (2009) The role of WRKY transcription factors in plant immunity. Plant Physiol 150(4):1648–1655

Park CY, Leea JH, Yooa JH, Moona BC, Choia MS, Kanga YH, Leea SM, Kima HS, Kanga KY, Chunga WS, Lima CO, Cho MJ (2005) WRKY group IId transcription factors interact with calmodulin. FEBS Lett 579(6):1545–1550

Peng Y, Bartley LE, Chen X, Dardick C, Chern M, Ruan R, Canlas PE, Ronald PC (2008) *OsWRKY62* is a negative regulator of basal and Xa21-mediated defense against *Xanthomonas oryzae* pv. *oryzae* in rice. Mol Plant 1(3):446–458

Proietti S, Bertini L, Vander ES, Reyes LA, Pieterse CMJ, Tucci M, Caporale C, Caruso C (2010) Cross activity of orthologous WRKY transcription factors in wheat and *Arabidopsis*. J Exp Bot 62(6):1975–1990

Qin Z, Lv H, Zhu X, Meng C, Quan T (2013) Ectopic expression of a wheat WRKY transcription factor gene TaWRKY71-1 results in hyponastic leaves in *Arabidopsis thaliana*. PLoS One 8(5):e63033

Qiu D, Xiao J, Ding X, Xiong M, Cai M, Cao Y, Li X, Xu C, Wang S (2007) *OsWRKY13* mediates rice disease resistance by regulating defense-related genes in salicylate- and jasmonate-dependent signaling. MPMI 20(5):492–499

Ramamoorthy R, Jiang SY, Kumar N, Venkatesh PN, Ramachandran S (2008) A comprehensive transcriptional profiling of the WRKY gene family in rice under various abiotic and phytohormone treatments. Plant Cell Physiol 49(6):865–879

Rizhsky L, Davletova S, Liang HJ, Mittler R (2004) The zinc finger protein Zat12 is required for cytosolic ascorbate peroxidase 1 expression during oxidative stress. J Biol Chem 279(12):11736–11843

Ross CA, Liu Y, Shen QJ (2007) The WRKY gene family in rice (*Oryza sativa*). J Integr Plant Biol 49(6):827–842

Rushton PJ, Somssich IE, Ringler P, Qingxi JS (2010) WRKY transcription factors. Trends Plant Sci 15:247–258

Rushton DL, Tripathi P, Rabara RC, Lin J, Ringler P, Boken AK, Langum TJ, Smidt L, Boomsma DD, Emme NJ, Chen X, Finer JJ, Shen QJ, Rushton PJ (2012) WRKY transcription factors: key components in abscisic acid signaling. Plant Biotechnol J 10:2–11

Ryu HS, Han M, Lee SK, Cho JI, Ryoo N, Heu S, Lee YH, Bhoo SH, Wang GL, Haln TR, Jeon JS (2006) A comprehensive expression analysis of the WRKY gene superfamily in rice plants during defense response. Plant Cell Rep 25(8):836–847

Satapathy L, Singh D, Ranjan P, Kumar D, Kumar M, Prabhu KV, Mukhopadhyay K (2014) Transcriptome-wide analysis of WRKY transcription factors in wheat and their leaf rust responsive expression profiling. Mol Gen Genomics 289:1289–1306

Shekhawat UKS, Ganapathi TR, Srinivas L (2011) Cloning and characterization of a novel stress-responsive WRKY transcription factor gene (*Musa*WRKY71) from *Musa* spp. Cv. Karibale Monthan (ABB group) using transformed banana cells. Mol Biol Rep 38:4023–4035

Shimono M, Sugano S, Nakayama A, Jiang CJ, Ono K, Toki S, Takatsuji H (2007) Rice *WRKY45* plays a crucial role in benzothiadiazole-inducible blast resistance. Plant Cell 19(6):2064–2076

Song WY, Wang GL, Chen LL, Kim HS, Pi LY, Holsten T, Gardner J, Wang B, Zhai WX, Zhu LH, Fauquet C, Ronald P (1995) A receptor kinase-like protein encoded by the rice disease resistance gene, Xa21. Science 270(5243):1804–1806

Tao Z, Liu H, Qiu D, Zhou Y, Li X, Xu C, Wang S (2009) A pair of allelic WRKY genes play opposite roles in rice-bacteria interactions. Plant Physiol 151(2):936–948

Tripathi P, Rabara RC, Langum TJ, Boken AK, Rushton DL, Boomsma DD, Rinerson CI, Rabara J, Reese RN, Chen X, Rohila JS, Rushton PJ (2012) The WRKY transcription factor family in *Brachypodium distachyon*. BMC Genomics 13:270

Tsuda K, Somssich IE (2015) Transcriptional networks in plant immunity. New Phytol 206:932–947

Ulker B, Somssich IE (2004) WRKY transcription factors: from DNA binding towards biological function. Curr Opinion Plant Biol 7(5):491–498

van der Auwera S, Zimmermann P, Rombauts S, Vandenabeele S, Langebartels C, Wang HJ, Wan AR, Hsu CM, Lee KW, Yu SM, Jauh GY (2007) Transcriptomic adaptations in rice suspension cells under sucrose starvation. Plant Mol Bio 63:441–443

van Verk MC, Pappaioannou D, Neeleman L, Bol JF, Linthorst HJM (2008) A novel WRKY transcription factor is required for induction of PR-1a gene expression by salicylic acid and bacterial elicitors. Plant Physiol 146:1983–1995

Wang D, Amornsiripanitch N, Dong X (2006) A genomic approach to identify regulatory nodes in the transcriptional network of systemic acquired resistance in plants. PLoS Pathol 2:e123

Wang H, Hao J, Chen X, Hao Z, Wang X, Lou Y, Peng Y, Guo Z (2007) Overexpression of rice WRKY89 enhances ultraviolet B tolerance and disease resistance in rice plants. Plant Mol Biol 65:799–815

Wang C, Deng P, Chen L, Wang X, Ma H (2013) A wheat WRKY transcription factor TaWRKY10 confers tolerance to multiple abiotic stresses in transgenic tobacco. PLoS One 8(6):e65120

Wang N, Xia EH, Gao LZ (2016) Genome-wide analysis of WRKY family of transcription factors in common bean, *Phaseolus vulgaris*: chromosomal localization, structure, evolution and expression divergence. Plant Gene 5:22–30

Wei KF, Chen J, Chen YF, Wu LJ, Xie DX (2012) Molecular phylogenetic and expression analysis of the complete WRKY transcription factor family in maize. DNA Res 19:153–164

Wu H, Ni Z, Yao Y, Guo SQ (2008) Cloning and expression profiles of 15 genes encoding WRKY transcription factor in wheat (*Triticum aestivum* L.) Prog Nat Sci 18:697–705

Xie Z, Zhang Z, Zou XL, Huang J, Ruas P, Thompson D, Shen QJ (2005) Annotations and functional analyses of the rice *WRKY* gene superfamily reveal positive and negative regulators of abscisic acid signaling in aleurone cells. Plant Physiol 137(1):176–189

Xing DH, Lai ZB, Zheng ZY, Vinod KM, Fan BF, Chen ZX (2008) Stress- and pathogen-induced Arabidopsis WRKY48 is a transcriptional activator that represses plant basal defense. Mol Plant 1:459–470

Xu X, Chen C, Fan B, Chen Z (2006) Physical and functional interactions between pathogen-induced *Arabidopsis* WRKY18, WRKY40, and WRKY60 transcription factors. Plant Cell 18:1310–1326

Yamasaki K, Kigawa T, Inoue M, Tateno M, Yamasaki T, Yabuki T, Aoki M, Seki E, Matsuda T, Tomo Y, Hayami N, Terada T, Shirouzu M, Tanaka A, Seki M, Shinozaki K, Yokoyamab

S (2005) Solution structure of an *Arabidopsis* WRKY DNA binding domain. Plant Cell 17(3):944–956
Zhang ZL, Xie Z, Zou XL (2004) A rice WRKY gene encodes a transcriptional repressor of the gibberellin signaling pathway in aleurone cells. Plant Physiol 134:1500–1513
Zhang J, Peng Y, Guo Z (2008) Constitutive expression of pathogen-inducible *OsWRKY31* enhances disease resistance and affects root growth and auxin response in transgenic rice plants. Cell Res 18:508–521
Zheng Z, Qamar SA, Chen Z, Mengiste T (2006) Arabidopsis WRKY33 transcription factor is required for resistance to necrotrophic fungal pathogens. Plant J 48:592–605
Zheng Z, Mosher S, Fan B, Klessig D, Chen Z (2007) Functional analysis of *Arabidopsis* WRKY25 transcription factor in plant defense against *Pseudomonas syringae*. BMC Plant Biol 7:2
Zhou QY, Tian AG, Zou HF, Xie ZM, Lei G, Huang J, Wang CM, Wang HW, Zhang JS, Chen SY (2008) Soybean WRKY-type transcription factor genes, GmWRKY13, GmWRKY21, and GmWRKY54, confer differential tolerance to abiotic stresses in transgenic *Arabidopsis* plants. Plant Biotechnol J 6:486–503
Zhu X, Liu S, Meng C, Qin L, Kong L, Guangmin X (2013) WRKY transcription factors in wheat and their induction by biotic and abiotic stress. Plant Mol Biol Rep 31:1053–1067
Zou X, Seemann JR, Neuman D, Shen QJ (2004) A *WRKY* gene from creosote bush encodes an activator of the abscisic acid signaling pathway. J Biol Chem 279:55770–55779

Lopamudra Satapathy is M.Sc. (Biotechnology) from Annamalai University, Tamil Nadu, India and has completed Ph.D. from Department of Bio-Engineering, Birla Institute of Technology, Mesra, India with Prof. Kunal Mukhopadhyay. Her current research interests are wheat transcription factors and their involvement in biotic and abiotic stress.

Dhananjay Kumar completed his Ph.D. from Department of Bio-Engineering, Birla Institute of Technology, Mesra, Ranchi-835215 with Prof. Kunal Mukhopadhyay in the area of WRKY transcription factors, miRNAs and their role in leaf rust pathogenesis.

Kunal Mukhopadhyay completed his Ph.D from University of Calcutta and did his Post doctoral research on Plant Biotechnology from University of Georgia, Athens USA. Presently his research focus is on elucidation of molecular aspects of leaf rust fungus interaction with wheat.

Part IV

Computational Approaches in Microbiology

Humoral Responses of In Silico Designed Immunodominant Antigenic Peptide Cocktails from Anthrax Lethal Toxin Components

12

Nagendra Suryanarayana, Vanlalhmuaka, Kulanthaivel Thavachelvam, and Urmil Tuteja

Abstract

To date, efforts of developing successful vaccine that targets specific epitopes of anthrax lethal toxin components, viz., protective antigen (PA) and lethal factor (LF), have been limited as evident by a handful of research publication on efficacious peptide vaccines against anthrax. Present study aims in this direction, and as a preliminary step in the development of vaccine, humoral response of peptide cocktails consisting of antigenic epitope sequences of anthrax lethal toxin components has been evaluated. Four peptide cocktail combinations were made from five PA peptides and two LF peptide sequences which were known to be antigenic and software predicted B-cell epitopes. Enzyme-linked immunosorbent assay result from all four peptide cocktail combinations revealed that two combinations PC 4 + 1 and PC 4 + 4 showed good anti-PA and anti-LF IgG antibody response with predominant IgG1 and IgG2b antibodies. Both anti-PA and anti-LF avidity of peptide cocktails were determined and were found to be higher than their respective native proteins. Overall, the results obtained from this study give a promising hope of developing a successful peptide vaccine candidate against anthrax.

Keywords

Avidity • Immunodominant • Lethal factor • Peptide cocktail • Protective antigen

N. Suryanarayana (✉) • K. Thavachelvam • U. Tuteja
Microbiology Division, Defence Research and Development Establishment, Jhansi Road, Gwalior 474002, Madhya Pradesh, India
e-mail: nagendranavada06@gmail.com

Vanlalhmuaka
Entomology Division, Defence Research Laboratory, Tezpur 784001, Assam, India
e-mail: vanlalh@gmail.com

P. Shukla (ed.), *Recent Advances in Applied Microbiology*,
DOI 10.1007/978-981-10-5275-0_12

12.1 Introduction

The progression and impact of the disease anthrax can be effectively managed with vaccine and antibiotics at pre- and postexposure stages, respectively (Chitlaru et al. 2010). Antibiotics work by killing the circulating bacterium but fail to protect at later stage of infection when toxin molecules are already produced in huge amounts (Altboum et al. 2002). Antitoxin antibodies and small molecule inhibitors have made their way in recent times to combat anthrax toxin molecule but still under clinical trials (Hu et al. 2008; Malkevich et al. 2013). As far as prophylaxis is concerned, vaccination is the only approach which is known to provide protection against the disease. Developed during the early 1970s, anthrax vaccine adsorbed (AVA), the only FDA-approved anthrax vaccine for humans, is still in use due to its robust protective efficacy in spite of its adverse immunological side effects (Weiss et al. 2007; Vietri et al. 2006). Progress in this regard with the aim to obtain a better vaccine molecule that reduces the immunization schedule with no immunological side effects is being carried out from past few years. Recombinant subunit vaccine consisting of only PA protein of *Bacillus anthracis* has been under clinical trial and is likely to replace the existing vaccine (Bellanti et al. 2012). Since a better vaccine with a long-lasting memory and increased shelf life is a major area of concern, current research efforts are focused toward all possible safe alternatives including DNA- and peptide-based vaccine approaches. The availability of protein sequence information and large number of computational methodologies to predict B-cell and T-cell epitopes has made it an easy approach for developing therapeutics and vaccine molecules against infectious diseases (Verma et al. 2015; Backert and Kohlbacher 2015; Tambunan et al. 2016). In silico design with wet lab validation of epitope vaccine has shown to be a promising approach for combating diseases like malaria, cancer, multiple sclerosis, etc. (Lopez et al. 2001; Knutson et al. 2001; Bourdette et al. 2005).

Considerable progress has also been made toward peptide vaccine against anthrax. Peptide vaccine consisting of B-cell epitopes were found to be effective as these epitopes enhance humoral antibody responses that is sufficient enough to provide protection against the disease anthrax.

We had previously described the antigenic epitope sequences from PA and LF (Suryanarayana et al. 2015) which is known to be the software-predicted B-cell epitopes. Seven peptides proved to be immunodominant by *in vitro* analysis, and these peptides (Table 12.1), five from PA and two from LF, were further immunized

Table 12.1 Immunodominant PA and LF peptide sequences reactive to their native proteins

Immunodominant PA peptide sequences	Immunodominant LF peptide sequences
PA 4 – VDVKNKRTFLSPWIS	LF 1 – PVLVIQSSEDYVENT
PA 11 – QDGKTFIDFKKYNDK	LF 4 – RNDSEGFIHEFGHAV
PA 12 – GKTFIDFKKYNDKLP	
PA 13 – KTFIDFKKYNDKLPL	
PA 14 – FKKYNDKLPLYISNP	

to Balb/c mice as domain-specific four peptide cocktail combinations. The combinations were made by mixing PA-D1 spanning peptides with LF-D1 and LF-D4 spanning peptides, and PA-D4 peptides with LF-D1 and LF-D4 spanning peptides. They were designated as PC 1 + 1, PC 1 + 4, PC 4 + 1, and PC 4 + 4, respectively. Serum IgG, IgG isotyping, and antibody avidity of immunized serum were determined by ELISA. Present work is one of the stepping stones in developing a successful peptide vaccine candidate against anthrax.

12.2 Determination of Serum IgG

Indirect ELISA was performed as per the standard protocol of Engvall and Perlman (1971) for the titration of IgG. ELISA plates (Nunc, Denmark) were coated with PA and LF (500 ng/well) in triplicates using 0.1 M carbonate-bicarbonate buffer (pH 9.6). The plates were sealed and incubated overnight at 4 °C. The coated wells were blocked with 5% skim milk for 1 h at 37 °C. Following washing thrice with PBS, wells were incubated with 100 µl of serially twofold diluted immunized sera (1:100 to 1:3200) at room temperature for 1 h. The wells were then washed thrice with PBS-T (PBS with 0.05% Tween 20) and incubated for 1 h at 37 °C with 1:10,000 dilutions of anti-mouse IgG-HRP conjugate and 1:2000 dilution of IgG1/ IgG2a/ IgG3-HRP conjugates . Following three washings as mentioned above, the reactions were developed with 100 µl of o-phenylenediamine (OPD). The enzyme reaction was stopped by adding 50 µl of 1N H_2SO_4. The absorbance was recorded at 490 nm by UV-Vis microplate reader. End point titer was calculated as the highest serum dilution showing the absorbance twice than that of the control ±2 SD. All the four peptide cocktail combinations produced IgG antibody response against both PA and LF antigens. The anti-PA IgG response obtained from PC 4 + 1 and PC 4 + 4 was robust, and the reciprocal end point titer was found to be 2400 and 4000, respectively (Fig. 12.1A). In comparison, the anti-PA IgG response against PC 1 + 1 and PC 1 + 4 was minimal with reciprocal end point titer of 200 for each (Fig. 12.1A). Similarly, anti-LF IgG response to PC 4 + 1 and PC 4 + 4 was high in comparison to PC 1 + 1 and PC 1 + 4 (Fig. 12.1B). Overall, the anti-LF IgG response was weak in comparison to anti-PA IgG response. All the combinations produced anti-PA IgG1 antibodies with predominant levels observed in groups of animals immunized with PC 4 + 1 and PC 4 + 4. Further both these peptide cocktail groups showed significant levels of IgG2b response (Fig. 12.2). All the experiments were performed in triplicate, and the figure represents the mean of triplicate values, and all the values were subtracted from the values of unimmunized naïve mice sera.

12.3 Determination of Antibody Avidity

To determine the antibody avidity of peptide-immunized serum, ammonium thiocyanate (NH_4SCN) was used as a chaotrope to disrupt the antigen-antibody complex (Pullen et al. 1986). Briefly, two 96 well plates were coated with recombinant PA and

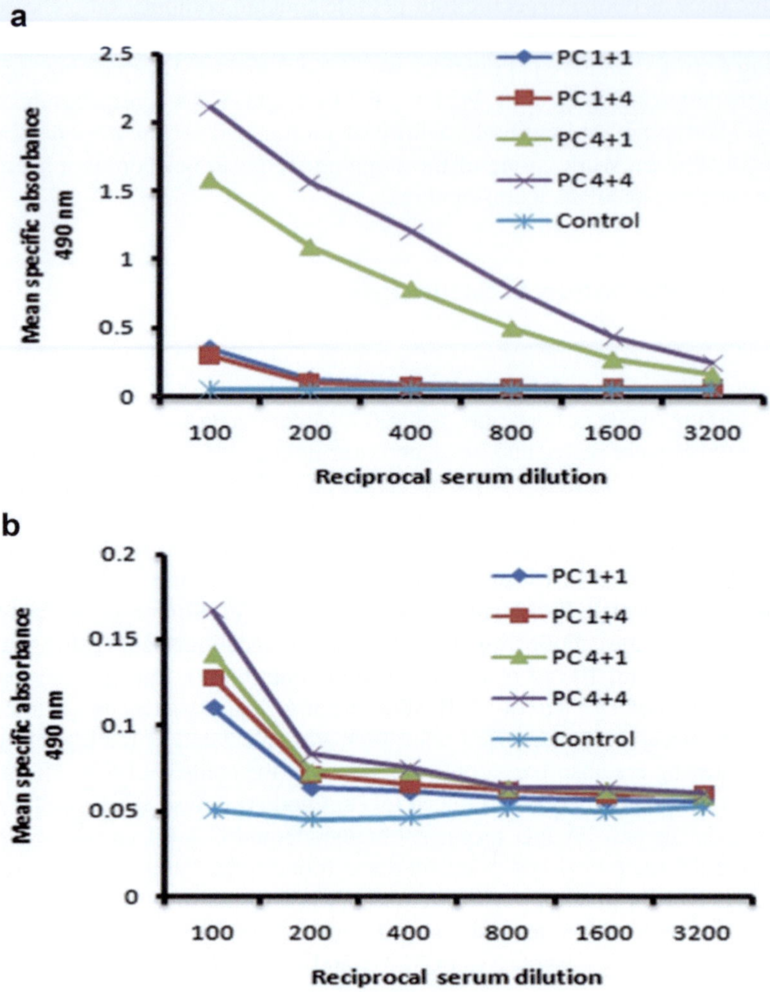

Fig. 12.1 Anti-PA and Anti-LF IgG response of a peptide cocktail combination at 35th day of immunization: (**a**) anti-PA IgG response and (**b**) anti-LF IgG response

LF and incubated with sera obtained from peptide-immunized mice. After incubation, plates were washed TBS-T and treated with different concentrations of NH_4SCN starting from 0 M to 5 M for 15 min at room temperature. The plates were again washed thrice with TBS-T and incubated with anti-mouse HRP conjugate for 1 h at room temperature. Finally the reactions were developed using OPD as substrate as mentioned above. One hundred percent binding was considered as the antigen-antibody interaction in the absence of NH_4SCN. Avidity index was represented by NH_4SCN molarity required for 50% reduction in absorbance corresponding to total binding. The avidity of anti-PA and anti-LF antibodies to peptide cocktail combinations was found to be higher than that of PA and LF. Among peptide cocktail

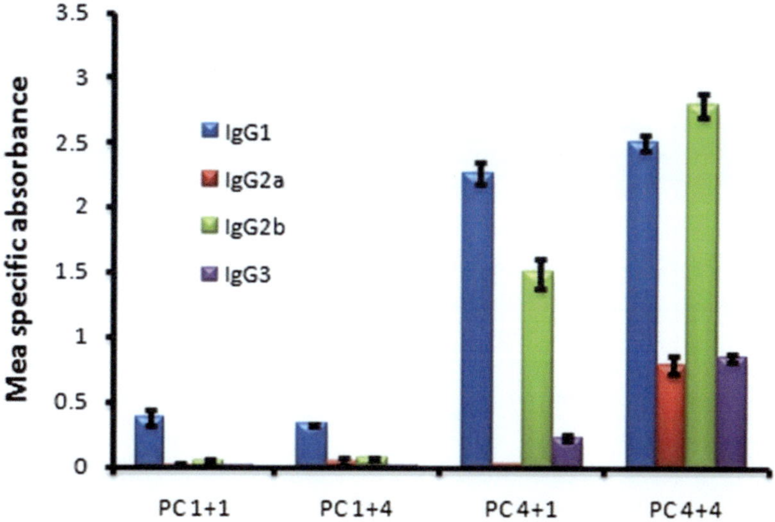

Fig. 12.2 IgG isotyping of the peptide cocktail immunized mice sera

combinations, PC 4 + 4 and PC 1 + 4 showed higher avidity than that of PC 1 + 1 and 4 + 1. The concentration of NH_4SCN required to release 50% of immune complexes of PC 1 + 1, PC 1 + 4, PC 4 + 1, and PC 4 + 4 from anti-PA IgG was 0.667, 0.7, 0.867, and 1.47 M, respectively, while for PA-immunized sera, it was 0.62 M (Fig. 12.3 A). Similarly anti-LF avidity of PC 1 + 1, PC 1 + 4, PC 4 + 1, and PC 4 + 4 was 1.52, 1.62, 1.8, and 2.2 M, respectively, and for LF-immunized sera, it was 0.55 M (Fig. 12.3 B).

12.4 Discussion

Recent years have witnessed considerable advance in the field of anthrax research largely due to the aftermath of 2001 US mail attack. With the availability of protein sequence information and computer-aided vaccine designing platform, development of epitope-based peptide vaccine has become easy in terms of time and cost. Although present vaccine AVA is known to confer protection against anthrax, its long immunization schedule, undefined composition, and immunological side effects have made researchers to look for safer alternatives. Peptide vaccines are one such approach which is being investigated in this regard.

Peptide vaccines have several advantages. It provides an opportunity to include only protective epitopes while excluding suppressive epitopes. It also avoids usage of any infectious agents that can possibly revert to its virulent state or integrate into host genome as debated in case of DNA vaccines. Peptide compositions during manufacturing will be well defined chemically and thereby undergoing quality control. Peptides are relatively economical, stable, easy to manufacture and can be lyophilized to overcome transportation and cold-chain storage facility (Purcell et al. 2007).

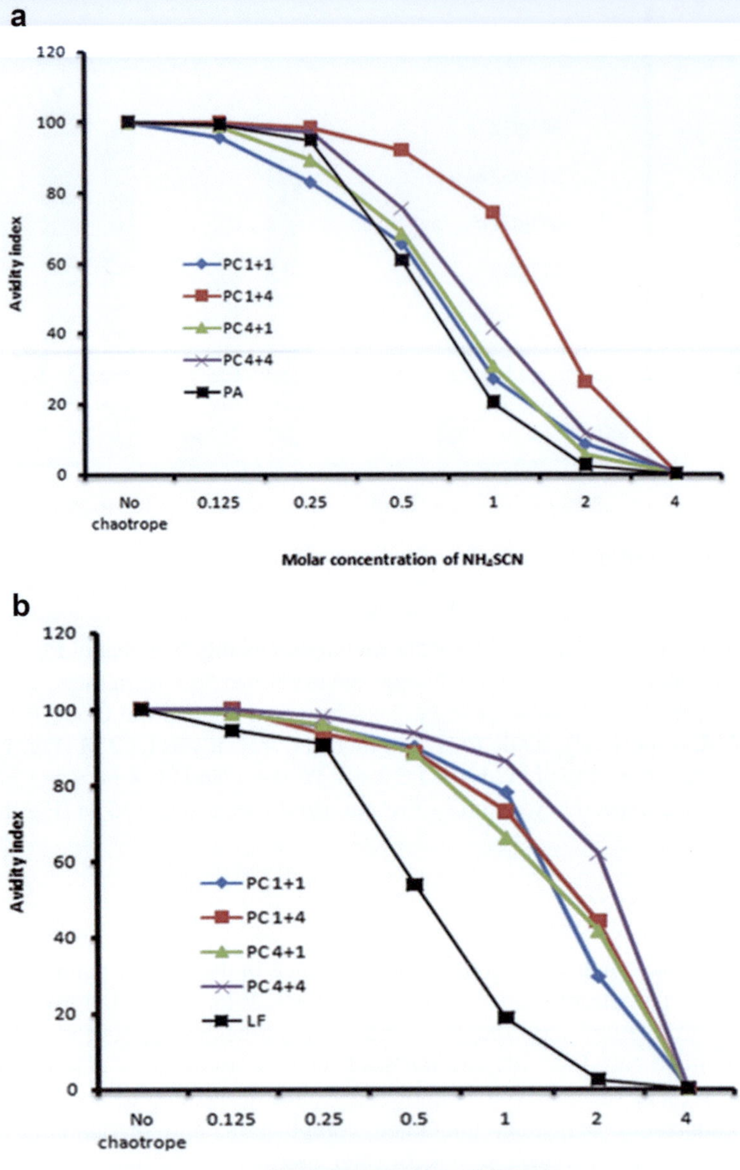

Fig. 12.3 Determination of antibody avidity. (**a**) Anti-PA avidity and (**b**) anti-LF avidity

The present study was aimed to find out the promising immunodominant B-cell epitopes contained in the protein sequences of PA and LF antigens that could serve as peptide vaccine against anthrax. For this purpose, the entire lengths of both PA and LF proteins were screened in silico to design a series of 15-mer peptide sequences having characteristics with specificity, antigenicity, surface accessibility, and B-cell epitopes using web servers and protein database. A total of 26 peptide sequences were then shortlisted and custom synthesized. Of these, seven peptides could prove the prediction of having B-cell epitopes as they were able to elicit significant level of antibody response in Balb/c mice against their respective native proteins in ELISA.

Since all the immunodominant peptide sequences were found to be present in the region of domain one and domain four of both the proteins PA and LF, four combinations were made to assess their immune responses. The peptide cocktails PC 4 + 1 and PC 4 + 4 comprising of four PA peptides (PA-11 to PA-14) and one LF peptide (LF-1/LF-4) showed a robust IgG1 and IgG2b antibody titers against PA in comparison to PC 1 + 1 and PC 1 + 4 comprising of PA peptide PA-1 and LF peptide LF-1/LF-4. There are reports in the literature which describes that the increased antibody response can be obtained by immunizing multiple peptides compared to individual peptides (Slingluff 2011). The results in the present study also indicate the same. The PC 4 + 1 and PC 4 + 4 peptide cocktails comprised of multiple peptides (5 numbers) in comparison to two peptides in PC 1 + 1 and 1 + 4. The anti-LF response in all peptide cocktail-immunized sera was less compared to anti-PA response. This again can be attributed due to the presence of four PA peptide sequences in comparison to one LF peptide sequence in peptide cocktails. The IgG isotyping revealed the presence of dominant IgG1 response in all the four peptide cocktails, thus showing the presence of dominant Th2 response. The avidity of antibodies in all the peptide cocktail-immunized sera was measured with an aim to assess the strength of immune complex formed between antibody and antigen. It was noted that all the peptide cocktail antibodies had high avidity in comparison to PA-/LF-immunized sera. It has been reported that peptides generating high affinity and avidity antibodies will mimic the continuous epitopes on their respective native protein. In addition higher avidity and affinity of an antibody are directly indicative of an effective and long-lasting immunological memory response (Sundaram et al. 2004).

In the present study, peptide cocktail combinations were successful in triggering an effective IgG immune response against both PA and LF. In addition, the anti-PA and anti-LF avidity of all the immunized peptide cocktail sera were high than that of their native protein. To conclude, the immunodominant peptide cocktail combinations used in this study can be further evaluated to determine its vaccine potential against anthrax.

12.5 Conclusion

- Humoral response of immunodominant peptide cocktails against anthrax was determined in terms of total IgG, IgG isotyping, and IgG avidity.
- Among the four cocktail combinations, PC 4 + 1 and PC 4 + 4 showed high anti-PA and anti-LF IgG response. In particular, IgG1 and IgG2b antibody response was predominant.

- Antibody avidity of peptide cocktail-immunized sera was high in comparison to the native proteins PA and LF.
- The peptide cocktail combinations can be further evaluated to determine its vaccine potential against anthrax.

References

Altboum Z, Gozes Y, Barnea A, Pass A, White M, Kobiler D (2002) Postexposure prophylaxis against anthrax: evaluation of various treatment regimens in intranasally infected guinea pigs. Infect Immun 70:6231–6241

Backert L, Kohlbacher O (2015) Immunoformatics and epitope prediction in the age of genomic medicine. Genome Med 7:119

Bellanti JA, Lin FYC, Chu C, Shiloach J, Leppla SH, Benavides GA, Karpas A, Moayeri M, Guo C, Robbins JB, Schneerson R (2012) Phase I study of recombinant mutant protective antigen of Bacillus anthracis. Clin Vaccine Immunol 19:140–145

Bourdette DN, Edmonds E, Smith C, Bowen JD, Guttmann CR, Nagy ZP, Simon J, Whitham R, Lovera J, Yadav V, Mass M, Spencer L, Culberstone N, Bartholomew RM, Theofan G, Milano J, Offner H, Vandenbark AA (2005) A highly immunogenic trivalent T cell receptor peptide vaccine for multiple sclerosis. Mult Scler 11:552–561

Chitlaru T, Altboum Z, Reuveny S, Shafferman A (2010) Progress and novel strategies in vaccine development and treatment of anthrax. Immunol Rev 239:221–236

Engvall E, Perlman P (1971) Enzyme-linked immunosorbent assay (ELISA). Quantitative assay of immunoglobulin G. Immunochemistry 8:871–874

Hu M, Li J, Yao SQ (2008) *In-situ* click assembly of small molecule matrix metalloprotease inhibitor containing zinc chelating groups. Org Lett 10:5529–5531

Knutson KL, Schiffman K, Disis ML (2001) Immunization with a HER-2/neu helper peptide vaccine generates HER-2/neu CD8 T-cell immunity in cancer patients. J Clin Invest 107:477–484

Lopez JA, Weilenman C, Audran R, Roggero MA, Bonelo A, Tiercy JM, Spertini F, Corradin G (2001) A synthetic malaria vaccine elicits a potent CD8(+) and CD4(+) T lymphocyte immune response in humans. Implications for vaccine strategies. Eur J Immunol 31:1989–1998

Malkevich NV, Basu S, Rudge TL, Clement KH, Chakrabarti AC, Aimes RT, Nabros GS, Skiadopoulos MH, Ionin B (2013) Effect of anthrax immune globulin on response to biothrax (Anthrax vaccine adsorbed) in New Zealand white rabbits. Antimicrob Agents Chemother 57:5693–5696

Pullen GR, Fitzgerald MG, Hosking CS (1986) Antibody avidity determination by ELISA using thiocyanate elution. J Immunol Methods 86:83–87

Purcell AW, McClusley J, Rossjohn J (2007) More than one reason to rethink the use of peptides in vaccine design. Nat Rev Drug Discov 6:404–414

Slingluff CL Jr (2011) The present and future of peptide vaccines for cancer: single or multiple, long or short, alone or in combination? Cancer J 17:343–350

Sundaram R, Lynch MP, Rawale SV, Sun Y, Kazanji M, Kaumaya PT (2004) *De novo* design of peptide immunogens that mimic the coiled coil region of human T-cell leukemia virus type-1 glycoprotein 21 transmembrane subunit for induction of native protein reactive neutralizing antibodies. J Biol Chem 279:24141–24151

Suryanarayana N, Vanlalhmuaka, Thavachelvam K, Tuteja U (2015) *In-silico* design and antibody response to the peptide sequences from protective antigen and lethal factor toxins of *Bacillus anthracis*. Ind J Curr Microbiol Adv Sci 4:859–867

Tambunan USM, Sipahutar FRP, Parikesit AA, Kerami D (2016) Vaccine design for H5N1 based on B- and T-cell epitope prediction. Bioinform Biol Insights 10:27–35

Verma SK, Yadav S, Kumar A (2015) *In-silico* prediction of B- and T- cell epitope on Lassa virus proteins for peptide based subunit vaccine design. Adv Biomed Res 4:201

Vietri NJ, Purcell BK, Lawler JV, Leffel EK, Rico P, Gamble CS, Twenhafel NA, Ivins BE, Heine HS, Sheeler R, Wright ME, Friedlander AM (2006) Short-course postexposure antibiotic prophylaxis combined with vaccination protects against experimental inhalational anthrax. Proc Natl Acad Sci 103:7813–7816

Weiss MM, Weiss PD, Weiss JB (2007) Anthrax vaccine and public health policy. Am J Public Health 97:1945–1951

Dr. Nagendra Suryanarayana is an M.Sc. (Biochemistry) from Bangalore University, India, and has completed his Ph.D. (Biological Sciences) at Defence Research and Development Establishment, Gwalior, MP, India. His Ph.D. work included vaccine studies on anthrax. Presently he is engaged in developing a diagnostic kit for Candida speciation.

Dr. Vanlalhmuaka is a young scientist who obtained his M.V.Sc. and doctoral degree from IVRI, Izatnagar, Uttar Pradesh. His expertise includes veterinary virology with immense knowledge of molecular biology and immunology. Currently he is heading the entomology division at Defence Research Laboratory, Tezpur.

Dr. Kulanthaivel Thavachelvam finished his doctoral degree from ICAR, New Delhi. He is specialized in plant genetics and biochemistry. Currently he is working on DNA based portable detection kits for *Leptospira*, *Vibrio cholera*, *Yersinia pestis* and *Bacillus anthracis* at the microbiology division of Defence Research and Development Establishment, Gwalior, Madhya Pradesh.

Dr. Urmil Tuteja obtained her doctoral degree from Dr. H.S. Gaur University, Sagar, India, while being recruited as scientist at Defence Research and Development Establishment, Gwalior, Madhya Pradesh. Her expertise includes microbiology and hybridoma technology. Currently she is one among the senior women scientists in DRDO and holds a respectable position and responsibilities.

Microbial Enzyme Engineering: Applications and Perspectives

13

Vishal Kumar, Mehak Baweja, Hao Liu, and Pratyoosh Shukla

Abstract

Enzymes are fascinating the researchers because of their enormous power of catalysis and eco-friendly nature. In biotechnological processes, diversity of microbes is studied, and different metabolic reactions entitle a potential repository that direct valuable production of desirable products. Since community demands are getting more intensified, there is a continuous need to evolve the enzymes. There has been an immense development in techniques and computational tools that has developed the industries to meet the growing demands. The techniques such as protein engineering help in development of quality products by mutating the amino acids to make more stable and efficient product. Further, the techniques like enzyme immobilization give the opportunity to reuse the used enzyme with the same efficiency, thus a cost-effective measure for the industrial enzyme. Nanotechnology and CLEA formation are also incorporated in enzyme engineering to increase enzyme efficiency and their characteristics.

Keywords

Protein engineering • Immobilization • Nanotechnology • Microbial enzyme

V. Kumar • M. Baweja • P. Shukla (✉)
Enzyme Technology and Protein Bioinformatics Laboratory, Department of Microbiology, Maharshi Dayanand University, Rohtak 124001, Haryana, India
e-mail: vkaggarwal180@gmail.com; mehakbaweja1988@gmail.com; pratyoosh.shukla@gmail.com

H. Liu
State Key Laboratory of Pulp and Paper Engineering, South China University of Technology, Guangzhou, China
e-mail: feliuh@scut.edu.cn

P. Shukla (ed.), *Recent Advances in Applied Microbiology*,
DOI 10.1007/978-981-10-5275-0_13

13.1 Introduction

The microbial enzymes are natural biocatalysts which are produced by different forms of microorganisms such as bacteria, fungi, algae, yeast, and actinomycetes to accelerate the rate biochemical reactions taking place in vivo system (Baweja et al. 2016; Kumar et al. 2014; Singh et al. 2016). These enzymes have been involved in all the processes such as DNA replication, transcription, translation, protein synthesis, metabolic reactions, and cycles which are essential for life, and their unique ability to carry out substrate-specific transformation has made them suitable for industrial processes. Enzymes are omnipresent ranging from living organisms to industries which catalyze the biochemical reactions to maintain the living system or synthesis of a product or degradation of a substance (Shrivastava et al. 2012, 2013; Kumar et al. 2014, 2016). The enzymes have replaced the chemicals from the industries, which are more eco-friendly and help to step up toward green environment. The industries are now totally enzyme dependent, and there is continuous demand for a number of microbial enzymes with novel characteristics and stability under industrial extreme conditions (Kumar and Shukla 2015). There is a continuous progress in microbiology, biotechnology, and bioinformatics to provide microbial enzymes with novel characteristics for the development of better industrial processes (Singh and Shukla 2012, 2015). The techniques of recombinant DNA technology, protein engineering, immobilization, nanotechnology, and metabolic engineering have progressed enough to support the industrial enzyme load. The industry demands operation stable enzymes with novel biochemical properties which should be economic and environment friendly. The continuous efforts are being done to meet all objectives of industrial demand of enzymes by utilizing a combination of various modern techniques of biotechnology, bioinformatics, microbiology, and nanotechnology. The enzyme engineering is proving as one of the finest techniques to improve the enzyme properties like operational stability, physical stability of enzyme, substrate specificity, and enhanced activity. There are several enzyme engineering techniques such as protein engineering, metabolic engineering, immobilization, and nanotechnology, and in silico methods; in this chapter, all these aspects will be described briefly. It will be helpful to understand the revolution in the microbial enzyme by changing at certain level or complete redesigning of the enzymes.

13.2 Improvement of Enzymes by Tailoring Their Protein Sequences

The industries continuously demand enzymes with high stabilities and substrate specificities. Since the native enzyme lacks the efficient system that can cope with hostile industrial conditions, there is a need to improve the enzyme that could withstand the harsh industrial conditions. With progress in recombinant technology and enzyme engineering, it has become possible to obtain customized enzymes. The improved novel enzymes that can fit into industry can be achieved by genetic manipulated microorganisms such as recombinant insulin production by using

Escherichia coli as host. The recombinant DNA technology makes possible to 100-fold increase in production of enzymes than the native expression, making them available at low cost and in large quantities. Thus, various food-processing enzymes and laundry enzymes can be tailored as per demand of the industrial process.

The protein engineering allows modification in protein itself to improve the properties for suitability to industrial process. In protein engineering, mutation is the key to improve the enzyme properties and to explore protein function. It is a method to alter a protein sequence to obtain a desired effect, such as change in the substrate specificity with increased stability toward extremes of pH and the temperature and in organic solvents. The protein engineering is divided into two types: (1) site-directed mutagenesis or rational design and (2) random mutagenesis.

13.2.1 Improvement of Enzyme Properties by Site-Directed Mutagenesis

The protein engineering is proving as stupendous technique to modify the enzyme to achieve customized biocatalyst. The major drawback with wild-type enzyme is that they cannot bear harsh experimental conditions. To overcome the snag of native enzyme, researchers are continuously adopting various methods to obtain refined industrial enzymes. Protein engineering is one such technique that works at the level of nucleotide to evolve the functional aspect of the protein. A comparison of native and engineered enzyme production has been described in Fig. 13.1. Site-directed mutagenesis, as the name suggests, is site-specific technique to improve enzyme. Thus, it is the technique for the proteins with full knowledge of structure and mechanism of action. The mutation type may vary as per requisite like point mutation, insertion, deletion, and substitution. Depending upon the number of mutation in a gene, it can be single site-directed mutagenesis and multiple site-directed mutageneses. The major application of site-directed mutagenesis is to introduce novel properties like enhanced specificity, stability, activity, solubility, expression, etc. to the biocatalyst. There are number of reports on improvement of industrial enzyme. A study was conducted to improve α-galactosidase features; the protein was mutated at specific site that improved the enzyme properties and also added the structural and functional information (Xu et al. 2014). In a similar study, the thermostability of the immobilized protease was improved by introducing Cys residues on surface of a cysteine-free mutant of a thermolysin-like protease from *B. stearothermophilus* and thus facilitated the site-directed immobilization of protease via single thiol group onto thiol Sepharose (Eijsink et al. 1995). It was reported by Rahimi et al. 2016 that mutation at nearby active site region is more promising in improving the protein function. A study was conducted to improve the keratinase enzyme to enhance its application at industrial level. Although the native enzyme itself had immense activity and pH stability, a truncation of PPC domain improved the tolerance to alkalinity, salt, chaotropic agents, and detergents (Fang et al. 2012). A study deduced that substitution of conserved residue Asn by arginine of γ-glutamyltranspeptidase (*Bl*GGT) by site-directed mutagenesis resulted in

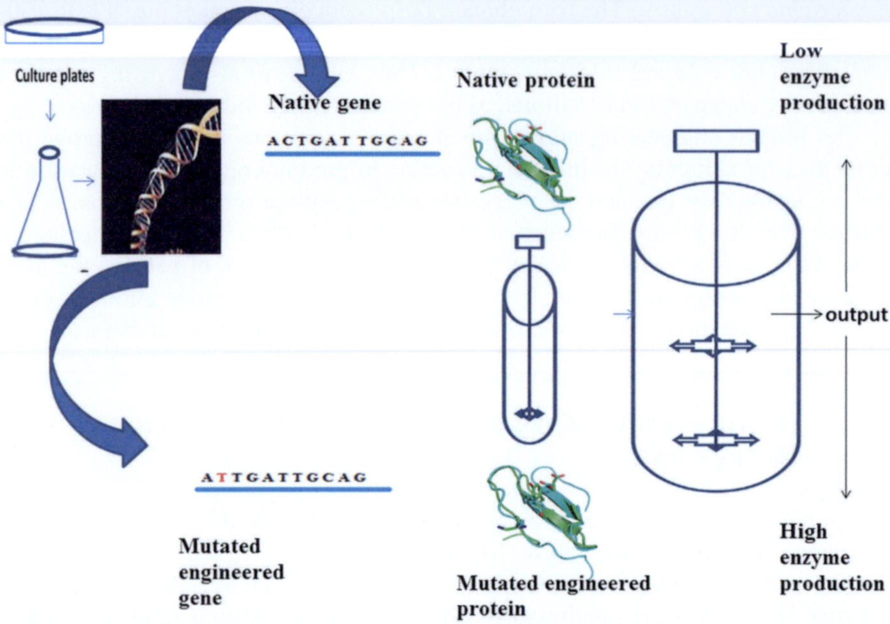

Fig. 13.1 A snapshot of industrial process describing high enzyme production using protein engineering

reduction in the catalytic activity (Lin et al. 2016). Wang and coworkers also elucidated the role of conserved amino acid residues by generating mutants by site-directed mutagenesis (Wang et al. 2015). The thermophilic archeal protein ST0452 was studied to comprehend the molecular machinery; after analyses, the researchers identified certain amino residues important for the glucosamine-1-phosphate and galactosamine-1-phosphate activities, viz., His308 is necessary for both GalN-1-P and GlcN-1-P AcTase activities, whereas Asn331 and Tyr311 are important only for the GalN-1-P AcTase activity (Zhang et al. 2015).

13.2.2 Improvement of Enzyme Properties by Random Mutagenesis

Random mutagenesis mimics the nature's process of variant generation following the unbiased approach. Since it involves mutation in randomized manner, thus it is a method of choice for those proteins whose structure has not been deduced (Baweja et al. 2016). Thus, it is quite an easy technique to employ but has cumbersome screening process since a number of variants produced are often large. There are various physical and chemical methods to create random mutation, such as chemical agents like ethyl methane sulfonate (EMS), methylnitronitrosoguanidine (MNNG), and ethylnitrosourea (ENU), using error-prone PCR by using less

FideliTaq polymerase instead of using pfu polymerase; using base analogs, altering the concentration of nucleotides; using heavy water during PCR, mutator strain, and many more; or using genetic recombination techniques like those based on gene recombination which are DNA shuffling, random chimeragenesis on transient templates (RACHITT), staggered extension process (StEP), recombined extension on truncated templates (RETT), and iterative truncation for the creation of hybrid enzymes (ITCHY) (Sen et al. 2007; Rasila et al. 2009; Baweja et al. 2015). The error-prone PCR is the most common technique in random mutagenesis with high success rates. The primary motto behind random mutagenesis is to characterize the open reading frame (ORF) and to modify the gene to obtain the desired product (Ramli et al. 2011).

There are various computational tools available that guide the library diversity and design, viz., ConSurf-HSSP GLUE, PEDEL, DRIVeR, and SCHEMA (Labrou 2010). There are various bioinformatics techniques available that reduce the cumbersome process of screening out libraries. Techniques like modeling and docking of enzymes prescreen the variants by giving the docking score that evaluates the enzyme-substrate relationship and effectiveness of their binding. The modeling and docking studies have been done in various enzymes like inulinase and xylanase (Karthik and Shukla 2012; Singh et al. 2016). The molecular dynamics simulation helps evaluate the stability of particular protein in particular milieu and thus filters out the variants during library screening (Singh et al. 2016).

13.3 Changing Pathways: Synthetic Metabolic Engineering

Industrial biotechnology promises to revolutionize conventional chemical manufacturing in the years ahead, largely owing to the excellent progress to reengineer cellular metabolism. It was evidenced that production of food stuffs and biofuels was enhanced after the era of metabolic engineering (Yadav et al. 2012). Metabolic engineering involves modification of metabolic pathways to screen the effect on the production of desired metabolite. To owe successful metabolic engineering, the first step is to crack metabolic pathway involved in the production of particular metabolite and to preclude the rate-limiting step in the reaction. The alteration in the rate-limiting step can be done by either overexpression of heterologous gene contributing the rate limitation or inhibiting the pathway in the network that halts the formation of desired product as shown in Fig. 13.2. To renew the production of metabolite, various experimental and computational tools are used to add beneficial traits to the system (Stafford and Gregory 2001). Metabolic flux plays an important role in valuation of the cellular phenotype; thus flux determination methods are inevitable components of metabolic engineering. Nowadays isotopic tracers are used to evaluate the balance of intracellular and extracellular metabolites. The 13C and 14C compounds were used to track the changes in and out of the cell (Klapa et al. 1999; Schmidt et al. 1997; Stafford et al. 2001). The introduction of multigene pathways into a host for heterologous production often faces flux imbalance because the host usually does not possess complex regulatory machinery to maintain such pathways.

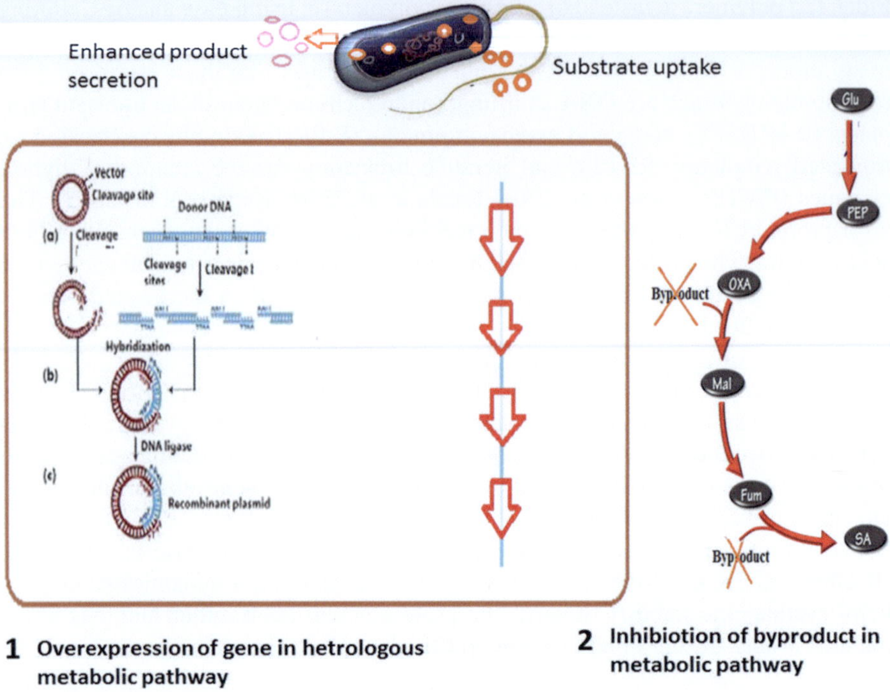

Fig. 13.2 Approaches of metabolic engineering for enhanced product secretion

However efforts have been done to resolve such problems by combining metabolic engineering tools with combinatorial genetics (Ajikumar et al. 2010; Lee et al. 2011). Although metabolic flux helps to deduce the pathway feature, it is not enough to decipher the system. Various high-throughput probes are used for complete revelation of metabolic networks. Among various molecular biological tools, efficient transformation system, viz., plasmids, hosts, and efficient promoters, is crucial for efficient product development and its modification using gene-editing approaches (Gupta and Shukla 2016; Keasling 1999). Gupta and Shukla (2015) described *E.coli* as suitable host during transformation.

13.4 Immobilization of Microbial Enzyme

Microbial enzymes catalyze a number of biochemical reactions efficiently and selectively; that's why they possess the ability to synthesize or to convert one compound to another. Immobilized forms of microbial enzymes have several industrial applications that are clear-cut as they provide a recycling method for the production of various compounds through biocatalyzed reactions. An immobilized microbial enzyme is the stable form of enzyme which is bound to an inert, insoluble material such as silica, chitosan, calcium alginate, copper alginate, agarose, polyacrylamide, etc. which provides increased stability in changing conditions of pH or temperature during industrial processes. It allows enzymes to stay held on supported material throughout the reaction following which they are separated for recycling and reuse. Along with synthesis of desired compounds, immobilized microbial enzymes also have the ability to decompose harmful compounds making them suitable for additional field of industrial application in bioremediation and purification. Besides these applications, the enzyme immobilization techniques are basis to synthesize a number of biotechnological products that have various applications in biosensors and bioaffinity chromatography and diagnostics (Guibault et al. 1991). A therapeutic application of immobilized enzymes is in extracorporeal shunts (Chang 1991). In the history of three or four decades, immobilization techniques have been developed swiftly, but still there is a need for further development. Immobilization technique is systematically studied with the probability of modification and improvement of enzyme stability and characteristics for economic purposes. There are a number of microbial enzymes such as xylanase, phytase, laccase, inulinase, cellulase, and amylase which have been immobilized on various materials. Among this series, microbial xylanase, which catalyzes the hydrolysis of xylan, is considered one of the most significant hydrolases. It has numerous applications, but most extensively it is utilized in paper and pulp industry as a biobleaching and biodeinking agent. Kapoor and Kuhad in 2007 used a number of matrices to immobilize the xylanase enzyme using various methods from *Bacillus pumilus* such as entrapment using gelatin, physical adsorption on chitin, ionic binding with Q-sepharose, and covalent binding with HP-20 beads with maximum xylanase immobilization efficiency. Similarly, Nagar et al. (2012) used the immobilized xylanase enzymes to improve the digestibility of poultry feed. Aluminum oxide pellets charged with glutaraldehyde were used for the immobilization which results in increase of enzyme temperature optima from 50 to 60 °C and V_{max} from 3333.33 to 5000 IU/mL. Immobilized xylanase was biochemically active up to ten consecutive cycles with 60% of its initial activity. In the same series, xylanase enzyme has also been covalently immobilized on the beads of glutaraldehyde-alginate exteriorly which retains their efficiency more than 91% with an increase in kinetic parameters V_{max} (7092–8000 IU/ml) and K_m (0.9–1.49%) and an increase in pH optima 5–5.5 and temperature optima from 40 to 45 ° (Pal and Khanum 2011). The enzyme has been reused five times while retaining >85% of its starting activity. Recently, matrix entrapment method was carried out by Bibi et al. (2015) to immobilize microbial endo-β-1,4-xylanase produced by *Geobacillus stearothermophilus* KIBGE-IB29 within agar-agar gel beads.

Among the industrial enzymes, protease has taken a pivotal position in detergent industry and leather industry. The alkaline protease from *Bacillus mycoides* was immobilized on different carriers using various immobilization methods including physical adsorption, covalent binding, entrapment, and ionic binding. An alkaline protease preparation was physically adsorbed on chitosan, entrapped in 2% cross-linked polyacrylamide, covalently bonded on chitin and ionically bonded on Amberlite IR-120 that were observed with highest activities by Abdel-Naby et al. (1998). In previous year, chitosan-immobilized protease from *Bacillus licheniformis* was applied in therapeutic use by Elchinger et al. (2015). They synthesized protease gains anti-biofilm activities after immobilization and was explored against biofilms formed by *Listeria monocytogenes*, *Pseudomonas aeruginosa*, *Staphylococcus aureus*, etc. Similarly, immobilization techniques also employed for laccase enzyme to make them suitable for various bioremediation application and wastewater treatment. The laccase beads were synthesized by immobilizing laccase enzyme on copper-alginate beads, and additionally Fe_2O_3 was incorporated in the bead through magnetic force. These lac beads have been used for bioremediation of triclosan and Remazol Brilliant Blue R and subsequently for wastewater treatment (Thanh le et al. 2016). Laccase from *Trametes versicolor* was also covalently immobilized on the composite polymer particles of poly(2-chloroethyl acrylate), p(CEA), which were grafted on zeolite particles via surface-initiated atom transfer radical polymerization (SI-ATRP). The immobilized laccase on the zeolite-g-p (CEA) particles was applied in biodegradation of dye Reactive Red 120. Besides these enzymes, immobilization was carried out with several other microbial enzymes which have been summarized in Table 13.1 with their application and support material being used.

13.4.1 Microbial Enzyme Immobilization Using Nanotechnology

Nanoparticles exhibit some attractive properties like elevated surface reactivity, high catalytic efficiency, tough adsorption ability, and great surface-to-volume ratio which make them attractive agent for immobilization. Adsorption of microbial enzymes on nanoparticles leads to enhanced performance of microbial enzymes in terms of its catalytic activity (Lynch and Dawson 2008). The application of enzyme immobilized nanoparticles was started during the 1980s (Pereira et al. 2002; Soriano et al. 2005). A number of microbial enzymes such as xylanase, protease, amylases, and phytase have been immobilized on various nanoparticles such as Fe_3O_4-coated chitosan, 1,3,5-triazine-functionalized $Fe_3O_4@SiO_2$ nanoparticles, gold nanoparticles, carbon nanoparticle, etc. In order to characterize the structure, size, and magnetic properties of the immobilized xylanase, Fourier transform infrared spectra (FTIR), thermo-gravimetric analysis (TGA), transmission electron microscopy (TEM), vibrating sample magnetometer (VSM), and X-ray photoelectron spectroscopy (XPS) were used for analysis. The enzyme activity, thermostability, storage stability, pH stability, and reusability of the nanoparticles of microbial enzymes have exhibited significant superiority to the free microbial enzymes. The xylanase MNPs showed quite impressive stability after nine reaction cycles with about 65% of its initial activity (Soozanipour et al. 2015). Experimental results by Soozanipour et al. (2015) suggested that the 1,3,5-triazine-functionalized $Fe_3O_4@SiO_2$ nanoparticles could be the novel convenient magnetic carrier for xylanase immobilization. Similarly, recently, Shahrestani et al. (2016) synthesized 1,3,5-triazine-functionalized silica encapsulated magnetic nanoparticles to immobilize xylanase enzyme to apply in clarification of bear and juices with impressive stability even after ten reaction cycle. In the same sequence, xylanase from *Aspergillus niger* that

Table 13.1 Microbial enzyme immobilization on various supporting matrix and their application

Sr. No.	Enzyme	Supporting matrix	Type of immobilization	Application	References
1.	Xylanase	Aluminum oxide pellets charged with glutaraldehyde	Covalent	Digestibility of poultry feed	Nagar et al. (2012)
2.	Xylanase	Glutaraldehyde-alginate beads	Covalent	–	Pal and Khanum (2011)
3.	Xylanase	Agar-agar	Matrix entrapment method	Biodegradation of xylan	Bibi et al. (2015)
4.	Protease from *Bacillus licheniformis*	Chitosan	Surface adsorption	Anti-biofilm activities	Elchinger et al. (2015)
5.	Laccase	Copper-alginate beads	Entrapment	Bioremediation or waste water treatment	Thanh le et al. (2016)
6.	Laccase from *Trametes versicolor*	Poly(2-chloroethyl acrylate) zeolite-g-p(CEA) particles	Covalent adsorption	Biodegradation of Reactive Red	Celikbicak et al. (2014)
7.	Inulinase from *Aspergillus niger*	Chitosan beads	Covalent immobilization	Continuous inulin hydrolysis	Yewale et al. (2013)
8.	Inulinase from *Aspergillus niger*	Polyurethane foam	–	–	Silva et al. (2013)
9.	α-Amylase	Silica nanoparticles	Covalent adsorption	Formulation of detergent	Soleimani et al. (2012)
10.	α-Amylase from *Bacillus stearothermophilus*	Poly (urethane urea) (PUU) microparticles	Covalent attachment	–	Strakšys et al. (2016)

has been immobilized covalently on the surface Fe_3O_4-coated chitosan magnetic nanoparticles showed a high binding capacity (Liu et al. 2015). Xylanase MNPs can be used in a number of industrial applications under broader pH and temperature ranges, having long-term storage capability and permitting magnetically recycling of the enzyme for purification or reuse of the product. Similarly, cellulase enzyme has also been physically adsorbed through ionic bond on superparamagnetic nanoparticles with binding efficiency of 95% and used for long-term storage (Khoshnevisan et al. 2011). Amylase enzymes from *Streptomyces sp.* MBRC-82 also have been immobilized on gold nanoparticles which have various medicinal applications by Manivasagan et al. (2015).

13.4.2 Immobilization by Forming CLEAs

CLEAs are insoluble enzyme aggregates which are formed by cross-linking of protein precipitates using cross-linking reagents such as glutaraldehyde. CLEAs exhibited high stability and high activity in aqueous medium as well as in nonaqueous medium. These enzyme aggregates have also showed a high stability at high temperature (Sheldon 2007). CLEAs may have a combination of several enzyme activities; such CLEAs are called multipurpose CLEAs or combi-CLEAs. Extent of cross-linking often influences their activity morphology, stability, and enantioselectivity. Nadar et al. (2016) evaluated the effects of various cross-linkers and precipitating agent on amylase activity recovery of macromolecular cross-linked enzyme aggregates (M-CLEAs) of α-amylase. Precipitates of amylase enzyme cross-linked by dextran showed 91% activity, ammonium sulfate used as precipitating agent, but glutaraldehyde CLEAs (G-CLEAs) exhibited only 42% activity. Recently, Mahmod et al. (2015) manufactured multipurpose cross-linked enzyme aggregate (multi-CLEA) with lipase and protease activity.

13.5 Homologue Augmentation and Substitution

The rate-limiting step is the major issue in enhanced production of desired metabolite which can be altered by importing the homologous enzyme from different hosts, the process known as homologous augmentation. Furthermore, turnover of the heterologous pathway can also be increased by using homologues of nonnative enzymes, known as homologue substitution. Both of these techniques were employed to deduce the carotenoid production in *E. coli* (Yoon et al. 2009). Yadav et al. proposed chimeric pathways involving each enzyme from different host to construct MVA pathway for carotenoid production.

13.6 Conclusion and Future Perspectives

In this chapter a number of enzyme engineering techniques have been briefed to improve the enzymatic characteristics together with its recycling methods. Microbial enzymes with enhanced physiological properties have greater commercial application in the industries. Protein engineering technology along with other molecular techniques and nanotechnology approaches has occupied the major position in the proteomic studies. Here it is shown that microbial enzymes, isolated from a number of sources such as fungi, bacteria, actinobacteria, yeast, and metagenomic sample, were subjected to protein engineering to modify their characteristics such as activity, temperature, and pH stability to make them economic for industrial applications. A number of modern genomic techniques such as genome-walking PCR, TAIL-PCR, error-prone PCR, StEP recombination, and metagenomic approaches have been proved as successful tools to create structural modifications at protein translational level. A few reports on site-directed mutagenesis and directed evolution have been explained as successful techniques to enhance thermostability and pH stability of microbial enzyme. The main focus of the chapter is laid upon modern enzyme engineering techniques to obtain a microbial protein with greater operational stability together with reusability. A combinatorial approach of metagenomic, proteomics, genomics, nanotechnology, and bioinformatics is required to obtain our goal of enzyme engineering.

References

Abdel-Naby M, Ismail AMS, Ahmed SA, Abdel Fattah AF (1998) Production and Immobilization of alkaline protease from *Bacillus Mycoides*. Bioresour Technol 64(3):205–210

Ajikumar PK, Xiao W-H, Keith EJ, Tyo YW, Fritz S, Effendi L, Oliver M, Too HP, Blaine P, Gregory S (2010) Isoprenoid pathway optimization for Taxol precursor overproduction in *Escherichia coli*. Science 330:70–74

Baweja M, Singh PK, Shukla P (2015) Enzyme technology, functional proteomics and systems biology towards unravelling molecular basis for functionality and interactions in biotechnological processes. In: Shukla P (ed) Frontier discoveries and innovations in interdisciplinary microbiology. Springer/Verlag, Berlin/Heidelberg, pp 207–212

Baweja M, Nain L, Kawarabayasi Y, Shukla P (2016) Current technological improvements in enzymes toward their biotechnological applications. Front Microbiol 7:965. doi:10.3389/fmicb.2016.00965

Bibi Z, Shahid F, Ul Qader SA, Aman A (2015) Agar–agar entrapment increases the stability of endo-β-1,4-xylanase for repeated biodegradation of xylan. Int J Biol Macromol 57:121–127

Celikbicak O, Bayramoglu G, Yılmaz M, Ersoy G, Bicak N, Salih B, Arica MY (2014) Immobilization of laccase on hairy polymer grafted zeolite particles: Degradation of a model dye and product analysis with MALDI–ToF-MS. Microporous Mesoporous Mater 199:57–65

Chang MS (1991) Therapeutic applications of immobilized proteins and cells. Bioprocess Technol 14:305–318

Eijsink VGH, Veltman OR, Aukema W, Vriend G, Venema G (1995) Structural determinants of the stability of thermolysin-like proteinases. Nat Struct Biol 2:374–379

Elchinger PH, Delattre C, Faure S, Roy O, Badel S, Bernardi T, Taillefumier C, Michaud P (2015) Immobilization of proteases on chitosan for the development of films with anti-biofilm properties. Int J Biol Macromol 72:1063–1068

Fang ZM, Li TL, Chang F, Zhou P, Fang W, Hong YZ, Zhang XC, Peng H, Xiao YZ (2012) A new marine bacterial laccase with chloride-enhancing, alkaline-dependent activity and dye decolorization ability. Bioresour Technol 111:36–41

Guibault GG, Kauffmann JM, Patriarche GJ (1991) Immobilized enzyme electrodes as biosensors. Bioprocess Technol 14:209–262

Gupta SK, Shukla P (2015) Advanced technologies for improved expression of recombinant proteins in bacteria: perspectives and applications. Crit Rev Biotechnol 18:1–10

Gupta SK, Shukla P (2016) Microbial platform technology for recombinant antibody fragment production. A review. Crit Rev Microbiol 43:1–12. doi:10.3109/1040841X.2016.1150959

Kapoor M, Kuhad RC (2007) Immobilization of xylanase from *Bacillus pumilus* strain MK001 and its application in production of xylo-oligosaccharides. Appl Biochem Biotechnol 142(2):125–138

Karthik MVK, Shukla P (2012) Computational strategies towards improved protein function prophecy of xylanases from *Thermomyces lanuginosus*. Springer, New York. doi:10.1007/978-1-4614-4723-8

Keasling JD (1999) Gene expression tools for the metabolic engineering of bacteria. Trends Biotechnol 17:452–460

Khoshnevisan K, Bordbar AK, Zare D, Davoodi D, Noruzi M, Barkhi M, Tabatabaei M (2011) Immobilization of cellulase enzyme on superparamagnetic nanoparticles and determination of its activity and stability. Int J Biol Macromol 171(2):669–673

Klapa MI, Park SM, Sinskey AJ (1999) Stephanopoulos G: metabolite and isotopomer balancing in the analysis of metabolic cycles: I. Theory. Biotechnol Bioeng 62:375–391

Kumar V, Shukla P (2015) Functional aspects of xylanases toward industrial applications. In: Shukla P (ed) Frontier discoveries and innovations in interdisciplinary microbiology. Springer/Verlag, Berlin/Heidelberg, pp 157–165

Kumar V, Pandey P, Gupta S, Shukla P (2014) A reviving preliminary evoke on few xylanase producing fungal isolates from different ecological niche. Int J Curr Microbiol App Sci 3(4):501–506

Kumar V, Marín-Navarro J, Shukla P (2016) Thermostable microbial xylanases for pulp and paper industries: trends, applications and further perspectives. World J Microbiol Biotechnol 32(2):34. doi:10.1007/s11274-015-2005-0

Labrou NE (2010) Random Mutagenesis Methods for *In Vitro* Directed Enzyme Evolution. Curr Protein Pept Sci 11:91–100

Lee JY, Yang KS, Jang SA, Sung BH, Kim SC (2011) Engineering butanol-tolerance in *Escherichia coli* with artificial transcription factor libraries. Biotechnol Bioeng 108:742–749

Lin MG, Chi MC, Chen YY, Wang TF, Lo HF, Lin LL (2016) Site-directed mutagenesis of a conserved Asn450 residue of *Bacillus licheniformis* γ-glutamyl transpeptidase. Int J Biol Macromol 91:416–425

Liu MQ, Huo WK, Xu X, Jin DF (2015) An immobilized bifunctional xylanase on carbon-coated chitosan nanoparticles with a potential application in xylan-rich biomass bioconversion. J Mol Catal B Enzym 120:119–126

Lynch I, Dawson KA (2008) Protein-nanoparticle interactions. NanoToday 3:40–47

Mahmod SS, Yusof F, Jami MS, Khanahmadi S, Shah H (2015) Development of an immobilized biocatalyst with lipase and protease activities as a multipurpose cross-linked enzyme aggregate (multi-CLEA). Process Biochem 50(12):2144–2157

Manivasagan P, Venkatesan J, Kang KH, Sivakumar K, Park SJ, Kim SK (2015) Production of α-amylase for the biosynthesis of gold nanoparticles using *Streptomyces sp.* MBRC-82. Int J Biol Macromol 72:71–78

Nadar SS, Muley AB, Ladole MR, Joshi PU (2016) Macromolecular cross-linked enzyme aggregates (M-CLEAs) of α-amylase. Int J Biol Macromol 84:69–78

Nagar S, Mittal A, Kumar D, Gupta VK (2012) Immobilization of xylanase on glutaraldehyde activated aluminum oxide pellets for increasing digestibility of poultry feed. Process Biochem 47(9):1402–1410

Pal A, Khanum F (2011) Covalent immobilization of xylanase on glutaraldehyde activated alginate beads using response surface methodology: characterization of immobilized enzyme. Process Biochem 46(6):1315–1322

Pereira JF, de Queiroz MV, Gomes EA, Muro-Abad JI, de Ara'ujo EF (2002) Molecular characterization and evaluation of pectinase and cellulose production of *Penicillium* spp. Biotechnol Lett 24(10):831–838

Rahimi M, van der Meer JY, Geertsema EM, Poddar H, Baas BJ, Poelarends GJ (2016) Corrigendum: mutations closer to the active site improve the promiscuous aldolase activity of 4-oxalocrotonate tautomerase more effectively than distant mutations. Chem Bio Chem 17(13):1290. doi:10.1002/cbic.201600321

Ramli AM, Muhammad NM, Rabu A, Munir A, Murad A, Diba FAB, Illias RM (2011) Molecular cloning, expression, and biochemical characterisation of a cold-adapted novel recombinant chitinase from *Glaciozyma antarctica* PI12. Microb Cell Factories 10:94. doi:10.1186/1475-2859-10-94

Rasila TS, Pajunen MI, Savilahti H (2009) Critical evaluation of random mutagenesis by error-prone polymerase chain reaction protocols, *Escherichia coli* mutator strain, and hydroxylamine treatment. Anal Biochem 388:71–80

Schmidt K, Carlsen M, Nielsen J, Villadsen J (1997) Modeling isotopomer distributions in biochemical networks using isotopomer mapping matrices. Biotechnol Bioeng 55:831–840

Sen S, Dasu VV, Mandal B (2007) Developments in directed evolution for improving enzyme functions. Appl Biochem Biotechnol 143:212–223

Shahrestani H, Taheri-Kafrani A, Soozanipour A, Tavakoli O (2016) Enzymatic clarification of fruit juices using xylanase immobilized on 1,3,5-triazine-functionalized silica-encapsulated magnetic nanoparticles. Biochem Eng J 109:51–58

Sheldon RA (2007) Cross-linked enzyme aggregates (CLEAs): stable and recyclable biocatalysts. Biochem Soc Trans 35(6):1583–1587

Shrivastava S, Lata S, Shukla P (2012) An insight on recent advances on immobilization methods for Industrial enzymes and its relevance to xylanases. Dyn Biochem Process Biotechnol Mol Biol 6(1):57–61

Shrivastava S, Shukla P, Deepalakshmi PD, Mukhopadhyay K (2013) Characterization, cloning and functional expression of novel xylanase from *Thermomyces lanuginosus* SS-8 isolated from self-heating plant wreckage material. World J Microbiol Biotechnol 12:2407–2415

Silva MF, Rigo D, Mossi V, Treichel H (2013) Evaluation of enzymatic activity of commercial inulinase from *Aspergillus niger* immobilized in polyurethane foam. Food Bioprod Process 91(1):54–59

Singh PK, Shukla P (2012) Molecular modeling and docking of microbial inulinases towards perceptive enzyme-substrate interactions. Indian J Microbiol 52:373–380

Singh P, Shukla P (2015) Systems biology as an approach for deciphering microbial interactions. Brief Funct Genomics 14:166–168

Singh PK, Joseph J, Goyal S, Grover A, Shukla P (2016) Functional analysis of the binding model of microbial inulinases using docking and molecular dynamics simulation. J Mol Model 22(4):1–7

Soleimani M, Khani A, Najafzadeh K (2012) α-Amylase immobilization on the silica nanoparticles for cleaning performance towards starch soils in laundry detergents. J Mol Catal B Enzym 74(1–2):1–5

Soozanipour A, Taheri-Kafrani A, Isfahani AL (2015) Covalent attachment of xylanase on functionalized magnetic nanoparticles and determination of its activity and stability. Chem Eng J 270:235–243

Soriano M, Diaz P, Pastor FI (2005) Pectinolytic systems of two aerobic sporogenous bacterial strains with high activity on pectin. Curr Microbiol 50(2):114–118

Stafford DE, Gregory S (2001) Metabolic engineering as an integrating platform for strain development. Curr Opin Microbiol 4:336–340

Stafford DE, Yanagimachi KS, Stephanopoulos G (2001) Metabolic engineering of indene bioconversion in *Rhodococcus sp.* Adv Biochem Eng Biotechnol 73:85–101

Strakšys A, Kochane T, Budriene S (2016) Catalytic properties of maltogenic α-amylase from *Bacillus stearothermophilus* immobilized onto poly (urethane urea) microparticles. Food Chem 211:294–299

Thanh le T, Murugesan K, Lee CS, Vu CH, Chang YS, Jeon JR (2016) Degradation of synthetic pollutants in real wastewater using laccase encapsulated in core-shell magnetic copper alginate beads. Bioresour Technol 216:203–210

Wang S, Fu X, Liu Y, Liu X, Wang L, Fang J, Wang PG (2015) Probing the roles of conserved residues in uridyltransferase domain of *Escherichia coli* K12 GlmU by site-directed mutagenesis. Carbohydr Res 413(2):70–74

Xu H, Qin Y, Huang Z, Liu Z (2014) Characterization and site-directed mutagenesis of an α-galactosidase from the deep-sea bacterium *Bacillus megaterium*. Enzym Microb Technol 56:46–52

Yadav VG, Mey MD, Lim CG, Ajikumar PK, Stephanopoulos G (2012) The future of metabolic engineering and synthetic biology: towards a systematic practice. Metab Eng 14(3):233–241

Yewale T, Singhal RS, Vaidya A (2013) Immobilization of inulinase from *Aspergillus niger* NCIM 945 on chitosan and its application in continuous inulin hydrolysis. Biocatal Agric Biotechnol 2:96–101

Yoon SH, Lee SH, Das A, Ryu HK, Jang HJ, Kim JY, Oh DK, Keasling JD, Kim SW (2009) Combinatorial expression of bacterial whole mevalonate pathway for the production of beta-carotene in *E. coli*. J Biotechnol 140:218–226

Zhang Z, Shimizu Y, Kawarabayasi Y (2015) Characterization of the amino acid residues mediating the unique amino-sugar-1-phosphate acetyltransferase activity of the archaeal ST0452 protein. Extremophiles 19:417–427

Vishal Kumar completed his M.Sc. (Microbiology) from the C.C.S. Haryana Agricultural University, Hisar Haryana, India, and pursued Ph.D. (Microbiology) from Maharshi Dayanand University, Rohtak Haryana, India. His current research interest is thermostable xylanase production from *Thermomyces lanuginosus* and its application in paper and pulp industry.

Mehak Baweja completed her M.Sc. (Applied Microbiology and Biotechnology) from the Banasthali Vidhyapeeth, Vanasthali, Rajasthan, India, and pursued Ph.D. (Microbiology) from Maharshi Dayanand University, Rohtak, Haryana, India. Her current research interest is protease production and its application in detergent industry.

Dr. Hao Liu is working as an Associate Professor in the State Key Laboratory of Pulp and Paper Engineering, South China University of Technology. His research covers a range of enzymatic cellulose hydrolysis, enzyme-assisted pulp bleaching and refining, direct fermentation of lignocellulose to ethanol, focusing on lignocellulytic enzymes including cellulases, xylanases and laccases. He obtained his PhD in 2011 at South China University of Technology, focusing on thermochemical pre-treatment and enzymatic hydrolysis of woody biomass. He visited the Department of Biological Systems Engineering, University of Wisconsin at Madison, USA as a visiting scholar, and also served as a DAAD short-term visiting scholar at RWTH Aachen University. He did his postdoctoral studies in Department of Civil and Environmental Engineering, the Hong Kong Polytechnic University. Currently, Dr. Liu's major research projects is about the enzymes from cultivation of edible mushroom. Dr. Liu has 15-year research experience in the field of pulp and paper engineering and enzyme biotechnology. He has published over 30 research papers and most of which were related with industrial enzymes.

Pratyoosh Shukla completed his M.Sc. (Applied Microbiology Biotechnology) from Dr. H.S. Gour University, Sagar, India, and Ph.D. in Microbiology and is presently working as Professor and Head at Department of Microbiology, Maharshi Dayanand University, Rohtak, India. He was awarded with Indo-US Visiting Research Professorship by American Society of Microbiology (ASM) in 2014. His current interests are bioenergy, enzyme technology, and protein bioinformatics.

Fungal Chondroitinase: Production and Prospects for Therapeutic Application

14

Kasinathan Narayanan, Nagarathenam Sivagurunathan, Volety Mallikarjuna Subrahmanyam, and Josyula Venkata Rao

Abstract

Chondroitinase is useful in treating glial scars and controlling tumour progression. Chondroitinase being a glycosylated protein, fungi could be an ideal host for production of chondroitinase. However, the work reported on bioprospecting of fungi capable of utilizing chondroitin sulphate and improvement of the cultural conditions is sparse. In this study, soil samples collected around various places of Udupi, India, were screened for isolating fungi capable of utilizing chondroitin sulphate and improving the enzyme yield. Seventy-six isolates obtained through soil enrichment were checked for chondroitinase activity by rapid plate technique. Among the 15 positive isolates, MSSS-1 was selected for further studies based on the isolate stability, consistency, yield and growth pattern. The enzyme yield was improved systematically through optimization of the cultural conditions, viz. one-factor-at-a-time method, factorial design and

K. Narayanan
Biocon Research Limited, Biocon Ltd, Bengaluru 560099, India
e-mail: narayanan158@gmail.com

N. Sivagurunathan
Department of Biomedical Sciences, College of Pharmacy, Shaqra University, Al Dawadmi, Kingdom of Saudi Arabia
e-mail: sivagurunathann@gmail.com

V.M. Subrahmanyam (✉) • J. Venkata Rao
Department of Pharmaceutical Biotechnology, Manipal College of Pharmaceutical Sciences, Manipal University, Manipal, 576104, Karnataka, India
e-mail: vm.subra@manipal.edu; rao.josyula@gmail.com

P. Shukla (ed.), *Recent Advances in Applied Microbiology*,
DOI 10.1007/978-981-10-5275-0_14

Box-Behnken design. The crude enzyme obtained by salt precipitation at 80% saturation controlled proliferation of HepG2, MCT-116 and MCF-7 cells. Maximum activity was observed against MCF-7 cells lines with an IC50 of 0.68 U/ml.

Keywords
Box-Behnken design • Chondroitinase • Factorial design • *Mucor irregularis* • Plackett-Burman design

14.1 Introduction

Chondroitinase catalyses the cleavage of β-glycosidic bond of chondroitin sulphate into their constituent units of N-acetyl galactosamine and glucuronic acid (Sugahara et al. 2003; Vázquez et al. 2013). Invasiveness and pathogenicity of the microbes including fungal species such as *Aspergillus niger*, *Candida* sp., *Paracoccidioides brasiliensis* and *Malassezia pachydermatis* depend on their ability to utilize chondroitin sulphate (Kasinathan et al. 2015; Hershon 1971; Shimizu et al. 1995; de Assis et al. 2003; Gu et al. 1995; Coutinho and Paula 2000). Over the last decade, chondroitinase has received increased attention due to its application as a therapeutic agent in treating cancer (Denholm et al. 2001) and spinal cord injury (Bradbury et al. 2002), improving recovery of autonomic functions (Caggiano et al. 2005) and vitreo-retinal pathology (Sebag 2002; Kasinathan et al. 2016) (Fig. 14.1). These applications were established using commercially available chondroitinase produced by bacterial sources such as *Flavobacterium heparinum* and *Proteus vulgaris* (Blain et al. 2002; Denholm et al. 2001; Bradbury et al. 2002). These bacteria were

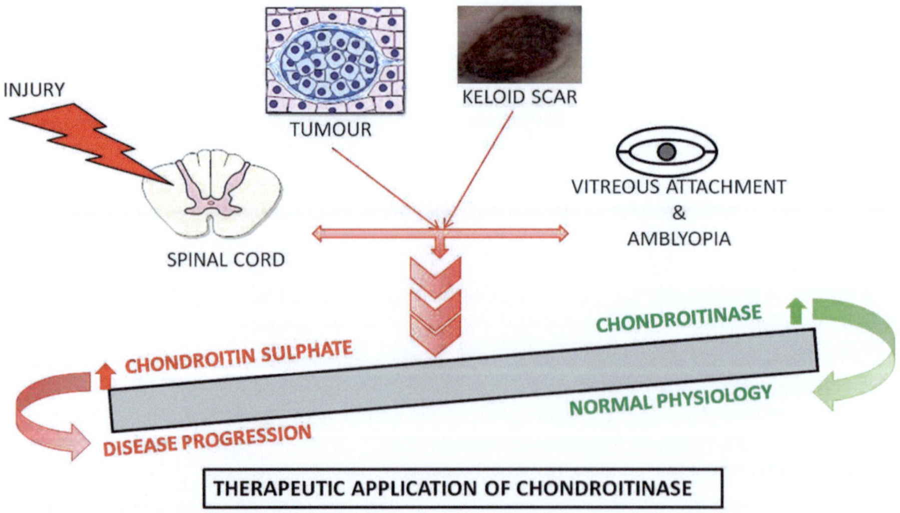

Fig. 14.1 Therapeutic applications of chondroitinase

first obtained either through extensive screening of natural habitats such as soil and estuaries or by allowing bacteria to adapt to a medium containing chondroitin sulphate (Kitamikado and Lee 1975; Ke et al. 2005). The productivity of these organisms is very low, and therefore the production cost has been high. In addition, the post-translational modification makes it difficult to clone these enzymes in organisms such as *E. coli* (Blain et al. 2002). These limitations have made chondroitinase less attractive for large-scale commercialization and clinical applications. As such, there is always a need to isolate newer organism expecting that these microbes will have an inherent capacity to produce high amount of enzyme. This yield is further improved using various approaches such as cultural condition optimization, strain improvement through mutation and recombinant technology.

Chondroitin sulphate is distributed throughout the animal kingdom and is an important component of extracellular matrix (Vázquez et al. 2013). Therefore, fungi, opportunistic pathogens, would be able to utilize chondroitin sulphate (de Assis et al. 2003; Coutinho and Paula 2000). There are no extensive studies on screening and production of chondroitinase by fungi species. So far, chondroitin sulphate-utilizing fungi, particularly yeast, were isolated either from infected patients or from infected animals (Coutinho and Paula 2000; Shimizu et al. 1995; de Assis et al. 2003). Chondroitinase is a glycosylated enzyme. Prokaryote and eukaryote differ in their post-translational modification. Therefore, isolation of chondroitinase from a eukaryote, fungi in this case, would be useful in understanding the difference in properties of chondroitinases produced by prokaryotes.

In a study made at Manipal, Udupi, India, fungal isolate (MSS-1) later on identified as *Mucor irregularis* obtained through bioprospecting was utilized for production of chondroitinase. After isolating a microbe from its natural environment, it is necessary that conditions be optimized so that a consistent yield is achieved. Being a wild strain, the conditions in submerged fermentation would be different from natural habitat conditions. Therefore, it is important to develop cultural conditions that are most suitable condition to get the maximum yield from the isolated microbe. Normally, the optimization involves a sequence of optimization steps involving various statistical-based designs that would assist in reducing the time required in identification of the most suitable condition. The general scheme of things followed during this study is given in Fig.14.2.

The effect of variables (nutrients) for production of chondroitinase by the *Mucor irregularis* species was sequentially studied using one factor at a time (OFAT), Plackett-Burman design (PBD) and response surface methodologies (RSM). There are few reports on chondroitinase production by bacteria. However, there were no reports on optimization of chondroitinase production by isolated *Mucor irregularis*.

The following sections will discuss:

- Screening and isolation of fungi capable of utilizing chondroitin sulphate
- Optimization of cultural conditions for production of chondroitinase
- Studies on enzyme purification and cytotoxicity profile

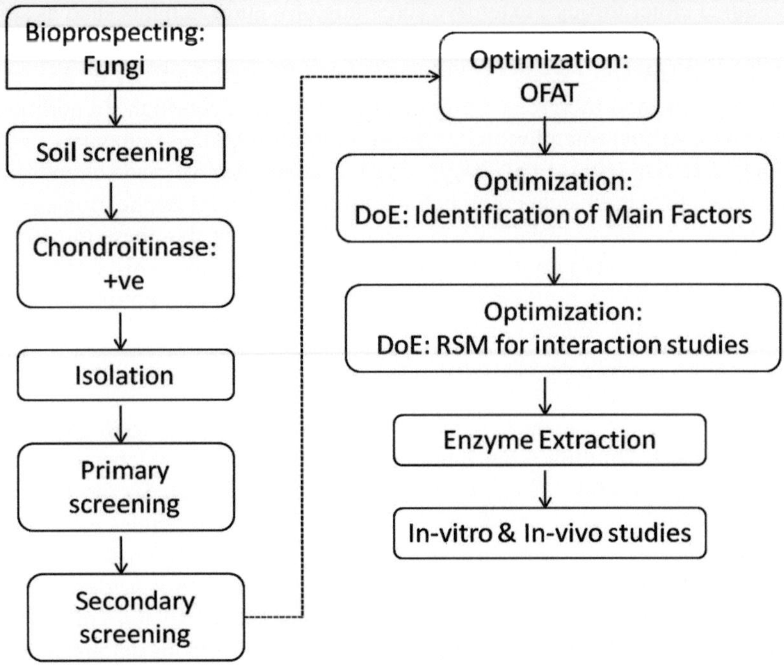

Fig. 14.2 General scheme for identification of inoculation and production medium composition for achieving higher enzyme yield

14.1.1 Screening and Isolation of Fungi Capable of Utilizing Chondroitin Sulphate

The soil is still a good source for isolating microbes capable of producing metabolite of interest (Smrithi et al. 2013). In the study, soil samples (marine, garden and soil) were collected from 15 different locations in and around Udupi, India. The soil samples were collected in sterile tubes from a depth of 3–5 cm of the surface (Mueller et al. 2004). The collected samples were screened using rapid plate method for fungi capable of breaking down chondroitin sulphate (Smith and Willett 1968). Soil samples were collected from places that differed in their characteristics. The locations included seashore, plantations near sea shore, locations used for drying of fish, poultry (specifically chicken carcass decomposing area) and portion of the soil where the kitchen waste seeps.

Enrichment technique was used to stimulate the growth of microorganism, which could utilize chondroitin sulphate. To enrich the medium, aliquots of 3 ml of chondroitin sulphate (0.04% w/v) was added to 1 g of soil sample. The tubes were incubated for 3 days at 27 °C at 150 rpm. On the third day, 10 ml of sterile saline (0.85%) was added to each of the test tubes and mixed thoroughly to obtain clear suspension. Serially diluted suspension was added to chondroitin sulphate-enriched Sabouraud dextrose agar (CSDA). The plates were incubated at 27 °C.

From the third day of incubation onwards, individual fungal colonies were picked and screened for chondroitinase activity. The individual colonies were subcultured on CSDA slants. A small quantity of each of the fungal specimen isolated was streaked on the surface of CSDA-containing bovine serum albumin fraction V (1% w/v) (BSA) and 0.04% w/v of chondroitin sulphate and incubated for 2–4 days at 27 °C. Glacial acetic acid (2 N) was added to each of the plates and incubated for 10 min. Formation of clear zone around the colony indicated breakdown of chondroitin sulphate. The enzyme activity was determined using modified Assis et al. method (de Assis et al. 2003).

The fungal isolates capable of utilizing chondroitin sulphate were maintained on CSDA at 4–8 °C. It was ensured that the selected isolate differed morphologically from each other with respect to margin, soluble pigments, aerial mycelium, reverse mycelium colour and shape.

The primary screening resulted in isolation of 76 isolates. Among these isolates, only 15 were found to produce chondroitinase. Out of the 15, five isolates were unstable. Therefore, the remaining ten isolates were selected for secondary screening by submerged fermentation.

14.1.1.1 Submerged Fermentation

The isolates showing positive results during primary screening were further investigated for their ability to produce chondroitinase by submerged fermentation.

The isolates were cultivated in an inoculation medium (w/v) containing dextrose (1%), yeast extract (0.5%), peptone (0.5%), K_2HPO_4 (0.1%), $MgSO_4$ (0.05%), KCl (0.05%), NaCl (0.05%) and $FeSO_4$ (0.001%). The production medium was prepared using 0.5% dextrose and 0.04% chondroitin sulphate as inducer and keeping the remaining composition as same as the cultivation medium. The production medium was inoculated with 5% inoculum. After 3 days, aliquots of samples were withdrawn in the interval of 24 h till tenth day and centrifuged at 5000 g for 10 min to check the enzyme activity of the supernatant.

14.1.1.2 Estimation of Chondroitinase Activity

Utilization of chondroitin sulphate was assessed based upon Muir et al.'s method (Reissig et al. 1955; Muir et al. 2010).

Enzyme activity was determined as described by Seikagaku Corporation, Japan, with some modifications. Release of N-acetyl galactosamine from chondroitin sulphate upon incubation with chondroitinase was used for estimating chondroitinase activity (Muir et al. 2010; Kasinathan et al. 2015). Based on the enzyme yield, at the end of secondary screening, MSSS-1 (Fig. 14.3) was found to be most suitable for further studies. It was identified as *Mucor irregularis* by using 18S rDNA technique (carried out by Gujarat State Biotechnology Mission, Gujarat) and was not previously reported as capable of producing chondroitinase. It was, therefore, selected for further optimization studies.

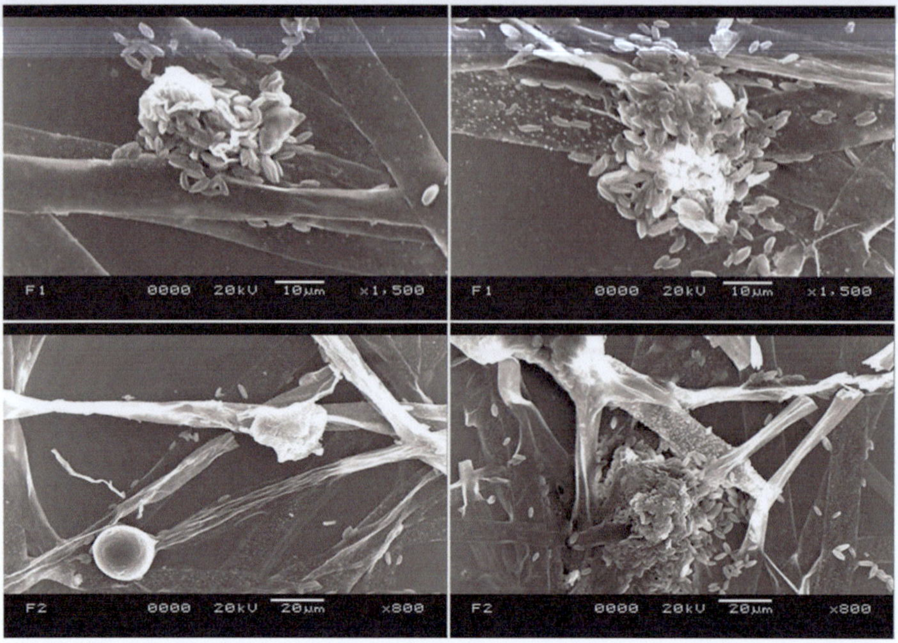

Fig. 14.3 Scanning electron microscopy image of the MSSS-1

14.1.2 Optimization of Cultural Conditions for Chondroitinase Production

MSSS-1 was grown using in the medium used for secondary screening. The production medium was prepared using 0.5% dextrose and 0.04% chondroitin sulphate as inducer and keeping the remaining components as same as the growth medium (pH maintained at 6.2 ± 0.2). The production medium was inoculated at 5% level. Aliquots of samples were withdrawn on each day until 6 day. The samples were centrifuged at 5000 g for 10 min, and the supernatant was used for further activity assays.

The cultivation conditions were optimized in order to improve the extracellular production of chondroitinase. Previous reports on chondroitinase production by bacteria show that chondroitinase production in bacteria could be either intracellular (Salyers and Kotarski 1980) or extracellular (Takegawa et al. 1991). There are no extensive studies on chondroitinase production in fungi to understand if chondroitinase production in fungi is extracellular or intracellular. However, it is believed that lytic enzymes expressed in fungi are extracellular (Linhardt et al. 1987). Although the chondroitinase activity with this isolate was observed to be present both in intracellular fraction and extracellular fraction, conditions were optimized for improving the extracellular enzyme yield as optimization of conditions for intracellular production of enzyme is costly, tedious and time-consuming.

The effect of cultivation medium, viz. inoculum and production conditions on enzyme yield, was initially analysed using OFAT design (Table 14.1). Inoculum conditions included effect of various organic nitrogen sources (1% w/v) and carbon sources (1% w/v). Enzyme yield was consistent when yeast extract and dextrose (each at 1% w/v) were used as nitrogen and carbon sources, respectively (Table 14.1).

The variables used under production conditions included:

Various organic nitrogen sources (1% w/v)
Inorganic nitrogen sources (1% w/v)
Yeast extract-peptone combinations
pH (3, 4, 5, 6, 6.5 7, 7.5, 8, 9)
Surfactants (low 0.01% w/v and high 0.1% w/v)
Inoculum level (1, 5, 10, 15 and 20%)
Antibiotics-ciprofloxacin (2.5 μg/ml)
Streptomycin (5 μg/ml) and tetracycline (2 μg/ml)
Carbon sources (0.5% w/v with 0.04% w/v of chondroitin sulphate and glucose (0.5% w/v) in absence of chondroitin sulphate)
Concentrations of chondroitin sulphate as the only carbon source
Concentrations of phosphate (0.05, 0.1, 0.25, 0.5 and 1.0%w/v)
Levels of free headspace
Inoculum age (12, 24, 36, 48, 60 and 72 h)
Different concentrations of chondroitin sulphate as both carbon and nitrogen source (0.5, 1, 2 and 3% w/v) (Table 14.1)

During optimization studies for identification of an ideal production medium, production medium was seed at 5% level. The production medium contained 0.5% w/v of chondroitin sulphate instead of dextrose, and other conditions remained the same. Aliquots of samples were withdrawn aseptically every 24 h and centrifuged, and the supernatants were analysed for enzyme activity.

The effect of variables on enzyme yield studied using OFAT studies is given in Table 14.1. OFAT studies showed that the combination of yeast extract and peptone at 1:0.5% w/v induced maximum amount of enzyme production, while the studies on pH showed that the yield was highest at pH 8. It was observed that the enzyme yield decreased as the acidity of the production medium was increased. MSSS-1 was isolated from marine seashore where the pH of soil is alkaline. Therefore the isolate, MSSS-1, could have adapted itself to alkaline conditions. Inorganic nitrogen source as the only nitrogen source reduced the enzyme yield. This shows that the presence of growth factors present in the organic nitrogen source is important for enzyme yield (Narayanan et al. 2013). Inoculum size affects the duration of lag phase that has direct effect on production of metabolites (Sen and Swaminathan 2004). In this study, the difference in enzyme yield with 20% and 10% was not high. Therefore, further studies were carried out using inoculum at 10% level.

MSSS-1 produced chondroitinase even in the presence of simple sugars. Various studies indicate that chondroitinase production in bacteria could be constitutive or inducible. Chondroitinase expression is constitutive in bacteria such as *B.*

Table 14.1 Effect of cultural conditions on enzyme yield studied using one-factor-at-a-time approach

S.No	Parameter	Observation (order of effect on enzyme yield)
Inoculation conditions		
1	Organic nitrogen	Peptone ≥ yeast extract >casein ≥ malt extract ≥ meat extract > control (ye + peptone)
2	Carbon	Lactose > glucose > fructose > sucrose > soluble starch
Production conditions		
1	Organic nitrogen	Control (YE + peptone) > meat extract > peptone > yeast extract > casein >tryptone > soya bean > albumin (BSA fraction V) > malt extract > gelatin
2	Inorganic nitrogen	Control (YE + peptone) > ammonium ferrous sulphate > ammonium dihydrogen phosphate > ferric nitrate > ammonium acetate > sodium nitrate ≥ potassium nitrate ≥ ammonium sulphate ≥ ammonium chloride > ammonium nitrate
3	Yeast extract-peptone combination (% w/v)	1.0% + 0.5% > 1.0% + 0.25% > 1% + 1% > 0.25% + 1% > 0.5% + 1% > 0.5% + 0.5% > 0.25% + 0.5% > 0.5% + 0.25% > 0.25% + 0.25%
4	pH	8 > 7.5 > 7 > 6.5 > 6 > 5 > 4 > 3
5	Surfactant (0.01 and 0.1%w/v)	Control > Tween-20 > Tween-80 > SDS > Triton X-100
6	Inoculum level (%)	20% ≥10% ≥ 15% > 5% > 1%
7	Antibiotics	Tetracycline ≥ control > streptomycin > ciprofloxacin
8	Carbon source with inducer chondroitin sulphate (% w/v) as the only carbon source	lactose > fructose ≥ CS (1.0%) > CS (2.0%) > CS (0.5%) ≥ only glucose > glucose > sucrose >raffinose > starch soluble > glycerol
9	Phosphate	0.5% > 0.25% > 1% > 0.1% >1% (ammonium dihydrogen orthophosphate) > 0.05%
10	Medium volume (250 ml EM flask)	100 ml (60% headspace) > 50 ml (80% headspace) > 75 ml (70% headspace) > 50 ml in 500 ml EM flask (90% headspace)
11	Inoculum age (h)	36 h > 48 h ≥ 60 h > 72 h > 12 h > 24 h
12	Chondroitin sulphate (as carbon and nitrogen source)	Control (0.5% glucose +0.04% chondroitin sulphate) > > 3% > 2% > 1% ≥ 0.5%

thetaiotaomicron and *F. heparinum* (Salyers and Kotarski 1980; Linhardt et al. 1987). These bacteria produce high level of chondroitinase when they come in contact with chondroitin sulphate. In the present study, the enzyme yield was highest in a medium containing lactose and chondroitin sulphate (inducer). Enzyme yield was also comparatively high in the presence of chondroitin sulphate as the only carbon source. The isolate produced chondroitinase in the medium containing dextrose as the only carbon source even though the yield was low when compared to the yield obtained in presence of chondroitin sulphate either alone or in combination. This shows that enzyme production by the present isolate is constitutive and addition of inducer would enhance the yield.

Surfactants are known to improve the enzyme yield by increasing the cell membrane permeability, causing the release of membrane-bound proteins, and/or by dispersing the mycelium within the cultivation medium (Reese and Maguire 1969; Pardo 1996). In the present study, addition of surfactant did not improve the enzyme yield. Inclusion of Tween-20 (0.1 and 0.01% v/v) in the cultivation medium did not affect the enzyme yield. Addition of Tween-80, Triton X-100 and SDS reduced the enzyme yield. Further studies using lower concentration of surfactant would help in understanding whether reduction in enzyme yield is due to the concentration of the surfactant or due to the inhibitory action of the surfactant itself.

Enzyme yield improved with increasing concentration of phosphate. Maximum yield observed when the concentration was 0.5% w/v. Addition of antibiotic into the bacterial cultivation medium reduces chondroitinase production (Salyers and Kotarski 1980). The presence of antibiotic in the fungal cultivation medium would help in controlling bacterial contamination. Therefore, this effect was studied. Addition of tetracycline to cultivation medium did not affect the enzyme yield. However, addition of other two antibiotics, viz. ciprofloxacin and streptomycin, in the cultivation medium reduced the enzyme yield. Although ciprofloxacin and streptomycin are ineffective against fungi, they can influence protein production mechanism resulting in decreased enzyme yield. Growth of the culture has a major dependence on the age of the inoculum (Jaapar et al. 2011). Enzyme yield was better with 36-h-old inoculum. There was no significant change in the enzyme yield with 48-h-and 60-h-old inocula. The use of 12-, 24- and 72-h-old inocula reduced the enzyme yield.

In a shake flask culture, the liquid-to-flask-volume ratio determines the headspace. The headspace has direct effect on oxygen transfer. The role of headspace becomes significant during the growth phase (Liu et al. 2006; Nikakhtari and Hill 2006). In the present study, under shake flask conditions with 150 rpm, enzyme yield was better with 60% free headspace.

Chondroitin sulphate contains both carbon and nitrogen elements. Therefore, it could serve as both as a carbon and nitrogen source. OFAT study showed that organism requires organic nitrogen source for production of high amount of enzyme (Table 14.1).

Further studies on identification of main and interaction effect of the selected nutrients were carried out using statistical model (factorial and RSM) using Minitab 16 and Design-Expert (Stat-Ease) software.

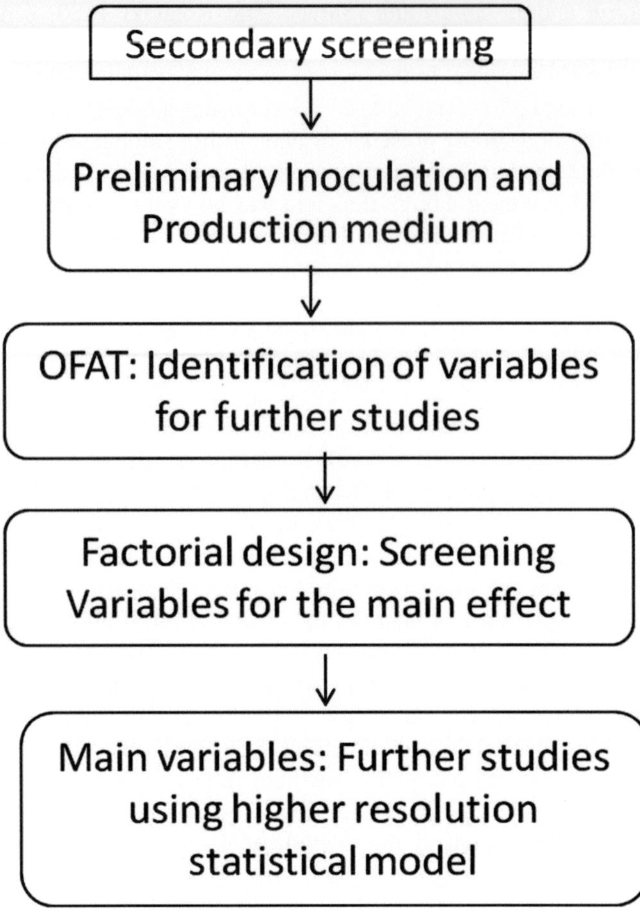

Fig. 14.4 General scheme for identification of important variables

14.1.2.1 Factorial Design Studies

OFAT studies showed that a combination of yeast extract and peptone, chondroitin sulphate, lactose and di-potassium hydrogen phosphate could play an important role in improving the enzyme yield. The level of significance and percentage contribution of each of the factors selected from OFAT studies on enzyme yield was studied using fractional factorial design consisting of eight trials with zero centre points (Fig. 14.4). The low and high level of selected variables were calculated after the central point for yeast extract-peptone, chondroitin sulphate, lactose (with chondroitin sulphate as inducer) and di-potassium hydrogen phosphate were fixed at 1.5, 1, 0.5 and 0.5% w/v, respectively.

All the factors included in the factorial design had a positive effect on the enzyme yield indicating that enzyme yield would be augmented by increasing the level of these variables.

Among these factors, combination of yeast extract-peptone had the most significant effect (p-value <0.05) on enzyme yield with more than 78% of the enzyme yield dependent on the level of yeast extract-peptone combination. While 11% of the enzyme yield was dependent on chondroitin sulphate, lactose (with inducer) and di-potassium hydrogen phosphate contributed 4.7% and 0.04%, respectively. Therefore, yeast extract-peptone and chondroitin sulphate were selected for further studies by response surface design.

Although lactose was able to induce enzyme production, chondroitin sulphate was selected for further studies as percentage contribution towards increasing the enzyme yield was higher. In addition, inclusion of chondroitin sulphate as the only carbon source could reduce production of non-specific enzyme. Separate set of studies on analysis of various types of enzymes produced in the presence of chondroitin sulphate and lactose would help in understanding this hypothesis. Although K_2HPO_4 improved the enzyme yield, there was no significant increase in the enzyme yield (p-value >0.05) at the level studied. Therefore, the level of K_2HPO_4 was fixed at 0.5% w/v as determined during OFAT studies.

14.1.2.2 Plackett-Burman Design

The significance of three major minerals, viz. NaCl, Ca $(OAc)_2$ and $MgSO_4$ included at 0.1% w/v, and eight minor minerals, viz. $ZnSO_4$, $FeSO_4$, $CuSO_4$, $BaCl_2$, $CoSO_4$, $Mn_3(PO_4)_2$, $AlCl_3$ and $SrCl_2$ at 0.01% w/v, on chondroitinase production was studied under this design consisting 12 trials. The absence of the selected mineral was considered as low level (Fig. 14.5).

A study on the effect of minerals by PBD showed that among all the minerals studied, only copper had significant effect (p-value <0.05) on enzyme yield with

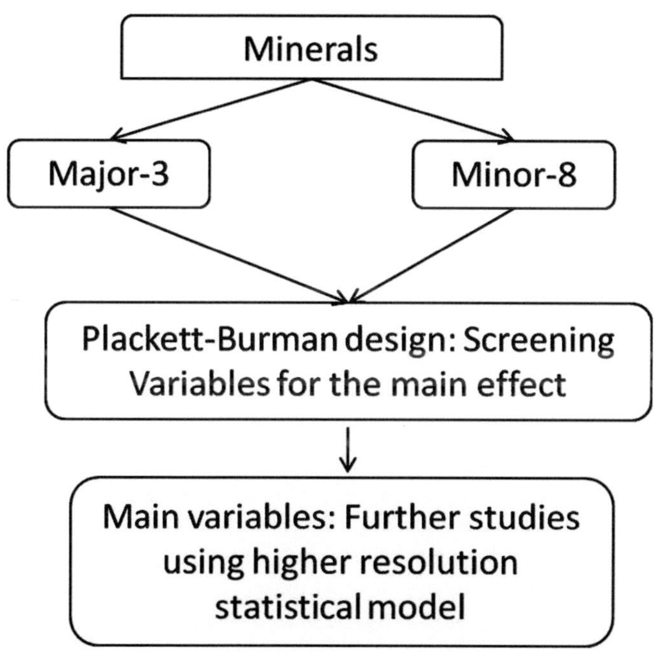

Fig. 14.5 General scheme for identification of important minerals affecting chondroitinase yield

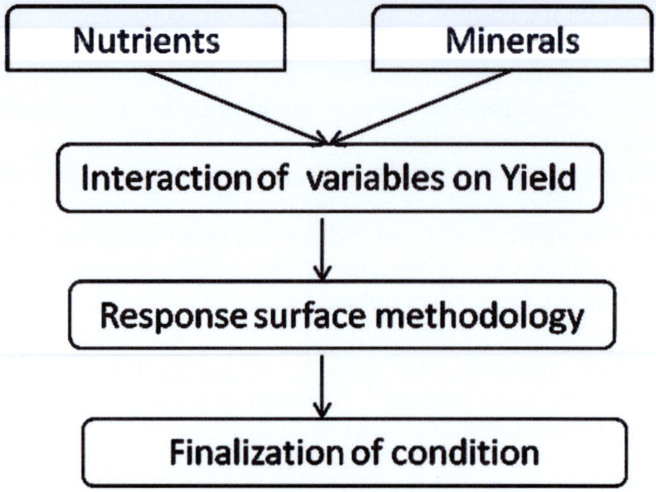

Fig. 14.6 General scheme for finalizing medium composition

78% of the enzyme yield being dependent on the amount of copper present in the cultural medium. The effect of the remaining minerals was insignificant (p-value >0.05). Except for copper and cobalt, only 5% of the outcome was dependent on the level of the remaining minerals. Na^{+}, Ca^{2+}, Mg^{2+} and Fe^{2+} had negative effect on the enzyme yield, i.e. addition of any of these minerals in the cultural medium reduced the enzyme yield. Therefore, these minerals were removed from the cultural medium during RSM studies. Because of the higher percentage contribution of Cu^{2+} and Co^{2+} in improving enzyme yield, Cu^{2+} and Co^{2+} were selected for further studies by RSM.

14.1.2.3 Box-Behnken Design

The effect of interaction among the factors, viz. yeast extract-peptone, chondroitin sulphate, $CuSO_4$ and $CoSO_4$, was investigated using BBD (Fig. 14.6). The design matrix is comprised of 27 trials for the selected four factors with three centre points, one base block and one replicate. The middle level for these variables was fixed based on the effect of their levels on enzyme yield.

This study showed that although the interaction of variables selected using factorial study and PBD did not produce significant effect on enzyme yield. The interaction study further confirmed that between the individual factors yeast-peptone combination and Cu^{2+} were the most significant factors affecting the enzyme yield. Further from the study, it was clear that enzyme yield could be enhanced if the level of yeast extract-peptone combination, Co^{2+} and Cu^{2+} is maintained high, middle and low, respectively. In such conditions, the level of chondroitin sulphate did not have any significant effect on enzyme yield. As such, chondroitin sulphate was employed at 0.5% w/v level when the levels of yeast extract-peptone combination, Co^{2+} and Cu^{2+} were maintained at 3.0, 0.05 and 0.01% w/v, respectively. A consistent enzyme yield was achieved using the conditions identified at the end of optimization studies.

Further improvement in the yield could be achieved through rDNA technology if the genes coding for the chondroitinase (in this strain) are identified.

14.1.3 Studies on Enzyme Separation and Cytotoxicity Profile

Chondroitinase has been previously obtained from the culture supernatant of *Curto bacterium*/*Aureobacterium* species using ammonium sulphate at 90% saturation (Takegawa et al. 1991). In this, ammonium sulphate at 80% saturation was used for precipitation of chondroitinase. The protein-ammonium sulphate mixtures were centrifuged at 9000 g at 4 °C for 20 min to collect the precipitates. The precipitates were suspended in 50 mM of TrisHCl (pH 7.5) and dialysed. The dialysed fractions were checked for enzyme activity.

The dialysed protein was centrifuged to remove debris. The supernatant was added successively to Amicon® centrifugal membrane filters with 100 kDa, 50 kDa and 30 kDa and centrifuged. The retentates were collected, dissolved separately in tri-buffer (pH-7.5) and stored at 2–8 °C until further use. Each of the solubilized retentate was checked for chondroitinase activity using the assay procedure described earlier.

Enzyme activity of the retentates obtained after centrifugation with Amicon® ultrafilters with 100, 50 and 30 kDa cut-off was 7.8, 15.75 and 2.13 U/ml, respectively. The protein content 100, 50 and 30 kDa cut-off (Amicon®) retentate was 375.12, 0.62 and 5.78 mg/ml. The amount of protein less than 30 kDa was 46.1 mg/ml. As 50 kDa retentate was showing higher enzyme activity than the other fractions, 50 kDa retentate was used for further studies including evaluation of therapeutic activity. SDS PAGE of 50 kDa retentate showed less number of protein bands compared to the number of protein bands observed in case of ammonium sulphate precipitated dialysed fraction.

Cytotoxicity of partially purified enzyme was studied using MTT assay against HepG2, HCT-116, MCF-7 and Vero cells (Koob 1989). The test enzyme controlled proliferation of tested cancerous cell lines. Among the cell lines studied (HepG2, HCT-116 and MCF-7), MCF-7 was most susceptible to the test enzyme (IC50 value was 0.68 U/ml).

14.2 Conclusion

The study showed that:

- Fungi are an alternate source for production of chondroitinase and could be commercially exploited.
- *Mucor irregularis* is capable of producing at least 15 U/ml of chondroitinase.
- Among the nutrients studied, yeast extract, chondroitin sulphate, phosphate, copper sulphate and cobalt sulphate are important in production of chondroitinase.
- The therapeutic potential and efficacy of fungal chondroitinase need to be realized completely.

14.3 Conflicts of Interest

The authors declare that there is no conflict of interest with any organization regarding the material discussed in this manuscript.

Acknowledgement The authors thank authorities of Manipal University for providing facilities to carry out the work outlined above.

References

Blain F, Tkalec AL, Shao Z, Poulin C, Pedneault M, Gu K, Eggimann B, Zimmermann J, Su H (2002) Expression system for high levels of GAG lyase gene expression and study of the hepA upstream region in Flavobacterium heparinum. J Bacteriol 184(12):3242–3252

Bradbury EJ, Moon LD, Popat RJ, King VR, Bennett GS, Patel PN, Fawcett JW, McMahon SB (2002) Chondroitinase ABC promotes functional recovery after spinal cord injury. Nature 416(6881):636–640. doi:10.1038/416636a416636a. [pii]

Caggiano AO, Zimber MP, Ganguly A, Blight AR, Gruskin EA (2005) Chondroitinase ABCI improves locomotion and bladder function following contusion injury of the rat spinal cord. J Neurotrauma 22(2):226–239. doi:10.1089/neu.2005.22.226

Coutinho SD, Paula CR (2000) Proteinase, phospholipase, hyaluronidase and chondroitin-sulphatase production by *Malassezia pachydermatis*. Med Mycol 38(1):73–76

de Assis CM, Gandra RF, Gambale W, Shimizu MT, Paula CR (2003) Biosynthesis of chondroitinase and hyaluronidase by different strains of *Paracoccidioides brasiliensis*. J Med Microbiol 52(Pt 6):479–481

Denholm EM, Lin YQ, Silver PJ (2001) Anti-tumor activities of chondroitinase AC and chondroitinase B: inhibition of angiogenesis, proliferation and invasion. Eur J Pharmacol 416(3):213–221

Gu K, Linhardt RJ, Laliberte M, Zimmermann J (1995) Purification, characterization and specificity of chondroitin lyases and glycuronidase from *Flavobacterium heparinum*. Biochem J 312(2):569–577

Hershon LE (1971) Elaboration of hyaluronidase and chondroitin sulfatase by microorganisms inhabiting the gingival sulcus: evaluation of a screening method for periodontal disease. J Periodontol 42(1):34–36

Jaapar SZS, Kalil MS, Ali E, Anuar N (2011) Effects of age of inoculum, size of inoculum and headspace on hydrogen production using *Rhodobacter sphaeroides*. Bacteriol J 1(1):16–23

Kasinathan N, Subrahmanyam VM, Rao VJ (2015) Design of cultural conditions for production of chondroitinase by *Aspergillus niger* using design of experiments. Indian Drugs 52(6):24–32

Kasinathan N, Volety SM, Josyula VR (2016) Chondroitinase: a promising therapeutic enzyme. Crit Rev Microbiol 42(3):474–484. doi:10.3109/1040841X.2014.959893

Ke T, Zhangfu L, Qing G, Yong T, Hong J, Hongyan R, Kun L, Shigui L (2005) Isolation of *Serratia marcescens* as a chondroitinase-producing bacterium and purification of a novel chondroitinase AC. Biotechnol Lett 27(7):489–493. doi:10.1007/s10529-005-2538-7

Kitamikado M, Lee YZ (1975) Chondroitinase-producing bacteria in natural habitats. Appl Microbiol 29(3):414–421

Koob TJ (1989) Effects of chondroitinase-ABC on proteoglycans and swelling properties of fibrocartilage in bovine flexor tendon. J Orthop Res 7(2):219–227. doi:10.1002/jor.1100070209

Linhardt RJ, Galliher PM, Cooney CL (1987) Polysaccharide lyases. Appl Biochem Biotechnol 12(2):135–176

Liu Y-S, Wu J-Y, K-p H (2006) Characterization of oxygen transfer conditions and their effects on *Phaffia rhodozyma* growth and carotenoid production in shake-flask cultures. Biochem Eng J 27(3):331–335

Mueller GM, Bills GF, Foster MS (2004) Biodiverstiy of fungi: inventory and monitoring methods. Elsevier Academic Press, Boston. Access online via Elsevier

Muir EM, Fyfe I, Gardiner S, Li L, Warren P, Fawcett JW, Keynes RJ, Rogers JH (2010) Modification of N-glycosylation sites allows secretion of bacterial chondroitinase ABC from mammalian cells. J Biotechnol 145(2):103–110. doi:10.1016/j.jbiotec.2009.11.002

Narayanan K, Chopade N, Raj PV, Subrahmanyam VM, Rao JV (2013) Fungal chitinase production and its application in biowaste management. J Sci Ind Res 72:393–399

Nikakhtari H, Hill GA (2006) Closure effects on oxygen transfer and aerobic growth in shake flasks. Biotechnol Bioeng 95(1):15–21

Pardo AG (1996) Effect of surfactants on cellulase production by *Nectria catalinensis*. Curr Microbiol 33(4):275–278

Reese ET, Maguire A (1969) Surfactants as stimulants of enzyme production by microorganisms. Appl Microbiol 17(2):242–245

Reissig JL, Strominger JL, Leloir LF (1955) A modified colorimetric method for the estimation of N-acetylamino sugars. J Biol Chem 217(2):959–966

Salyers AA, Kotarski SF (1980) Induction of chondroitin sulfate lyase activity in *Bacteroides thetaiotaomicron*. J Bacteriol 143(2):781–788

Sebag J (2002) Is pharmacologic vitreolysis brewing? Retina 22(1):1–3

Sen R, Swaminathan T (2004) Response surface modeling and optimization to elucidate and analyze the effects of inoculum age and size on surfactin production. Biochem Eng J 21(2):141–148

Shimizu MT, Jorge AO, Unterkircher CS, Fantinato V, Paula CR (1995) Hyaluronidase and chondroitin sulphatase production by different species of *Candida*. J Med Vet Mycol 33(1):27–31

Smith RF, Willett NP (1968) Rapid plate method for screening hyaluronidase and chondroitin sulfatase-producing microorganisms. Appl Microbiol 16(9):1434–1436

Smrithi S, Narayanan K, Kamath VB (2013) Design of cultural conditions for enhancement of alkaline protease production. Pharm Int J Adv Pharm Sci 4(6):1438–1446

Sugahara K, Mikami T, Uyama T, Mizuguchi S, Nomura K, Kitagawa H (2003) Recent advances in the structural biology of chondroitin sulfate and dermatan sulfate. Curr Opin Struct Biol 13(5):612–620

Takegawa K, Iwahara K, Iwahara S (1991) Purification and properties of chondroitinase produced by a bacterium isolated from soil. J Ferment Bioeng 72(2):128–131

Vázquez JA, Rodríguez-Amado I, Montemayor MI, Fraguas J, González MP, Murado MA (2013) Chondroitin sulfate, hyaluronic acid and chitin/chitosan production using marine waste sources: characteristics, applications and eco-friendly processes: a review. Mar Drugs 11(3):747–774

Kasinathan Narayanan is M. Pharm (Biotechnology) and Ph.D. from Manipal College of Pharmaceutical Sciences, Manipal University, Manipal, Udupi, Karnataka, India. His current research interests Fermentation technology, Formulation and development of Biosimilars and Scale Up. Presently, he is working for a private biopharmaceutical company.

Nagarathenam Sivagurunathan is M.Pharm. (Pharmaceutical Biotechnology) from The Tamilnadu Dr. MGR Medical University, Chennai, India and Ph.D. (Pharmaceutical Sciences) from Manipal University, Manipal, Karnataka, India. His current research interests are screening natural sources for microorganisms and their products with medical/industrial applications, Novel drug delivery approaches.

Volety Mallikarjuna Subrahmanyam is M.Tech (Biotechnology) and Ph.D from Andhra University, Visakhapatnam, India. His current research interests are in niche areas of Bioprocess and Enzyme technology.

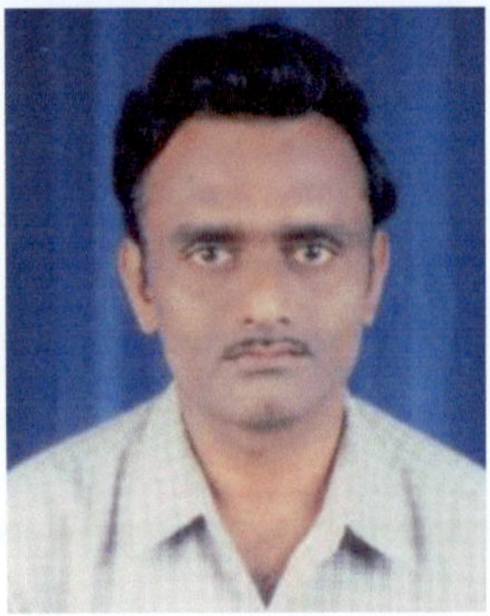

Josyula Venkata Rao is M.Pharm. (Fermentation Technology) and Ph.D. (Pharmaceutical Sciences) from Andhra University, Visakhapatnam, India. His current research interests are in the areas of Fermentation technology and Nano-drug delivery.

Erratum to: Recent Advances in Applied Microbiology

Pratyoosh Shukla

Erratum to:
P. Shukla (ed.), *Recent Advances in Applied Microbiology,* https://doi.org/10.1007/978-981-10-5275-0

The affiliation of the below authors was incorrectly published in the List of Contributors in the original version of the book and they have been corrected now.

D.K. Jha
Department of Botany, Gauhati University, Guwahati, Assam, India

Mohamed Hijri
Department of Biological Sciences, Institut de recherche en biologie végétale (IRBV), University De Montreal, Montreal, Canada

The updated original online version for this book can be found at https://doi.org/10.1007/978-981-10-5275-0

P. Shukla (ed.), *Recent advances in Applied Microbiology*,
DOI 10.1007/978-981-10-5275-0_15

MIX
Papier aus verantwortungsvollen Quellen
Paper from responsible sources
FSC® C105338

If you have any concerns about our products,
you can contact us on
ProductSafety@springernature.com

In case Publisher is established outside the EU,
the EU authorized representative is:
Springer Nature Customer Service Center GmbH
Europaplatz 3, 69115 Heidelberg, Germany

Printed by Libri Plureos GmbH
in Hamburg, Germany